Technical Editing in the 21st Century

Technical Editing in the 21st Century

Nicole Amare

Department of English, University of South Alabama

Barry Nowlin

*Department of Interdisciplinary Studies,
University of South Alabama*

Jean Hollis Weber

Documentation Project, OpenOffice.org

Prentice Hall

Boston   Columbus   Indianapolis   New York   San Francisco   Upper Saddle River
Amsterdam   Cape Town   Dubai   London   Madrid   Milan   Munich   Paris   Montreal   Toronto
Delhi   Mexico City   Sao Paulo   Sydney   Hong Kong   Seoul   Singapore   Taipei   Tokyo

Editor in Chief: Vernon Anthony
Senior Acquisitions Editor: Gary Bauer
Editorial Assistant: Megan Heintz
Director of Marketing: David Gesell
Senior Marketing Manager: Leigh Ann Sims
Senior Marketing Coordinator: Alicia Wozniak
Marketing Assistant: Les Roberts
Senior Managing Editor: JoEllen Gohr
Project Manager: Christina Taylor
Operations Specialist: Deidra Skahill

Senior Art Director: Jayne Conte
Manager, Rights and Permissions: Zina Arabia
Image Permission Coordinator: Jan Marc Quisumbing
Cover Designer: Suzanne Duda
Full-Service Project Management: Pavithra Jayapaul,
TexTech International
Composition: TexTech International
Printer/Binder: Edwards Brothers
Cover Printer: Lehigh-Phoenix Color/Hagerstown
Text Font: Times 10.5/12

Photo Credits: Addison Wesley Longman, Inc., Glenview: p. 197; Bernard Boutrit/Woodfin Camp, Jupiter Images-FoodPix–Creatas: p. 88; Luciano Corbella/© Dorling Kindersley: p. 196; © Dorling Kindersley: p. 285, p. 395; Harris Group, Inc. p. 199; Carlyn Iverson/Prentice Hall School Division: p. 197; Nick Lipscombe and Gary Biggin/© Dorling Kindersley: p. 197; Jules Selmes and Debi Treloar/© Dorling Kindersley: p. 395.

Library of Congress Cataloging-in-Publication Data

Amare, Nicole.
   Technical editing in the 21st century / Nicole Amare, Barry Nowlin, Jean Hollis Weber.
      p. cm.
   Includes bibliographical references and index.
   ISBN 0-13-119677-4 (alk. paper)
   1. Technical editing.   I. Nowlin, Barry.   II. Weber, Jean Hollis.   III. Title.
T11.4.A53 2011
601'.4—dc22

                                               2009044401

10 9 8 7 6 5 4 3 2 1

Prentice Hall
is an imprint of

www.pearsonhighered.com

ISBN 10:   0-13-119677-4
ISBN 13: 978-0-13-119677-3

BRIEF CONTENTS

CONTENTS

PREFACE

The field of technical communication has grown rapidly in the past thirty years, largely due to the growth in technology. Along with an increase in industry and governmental jobs, specialization, and prestige, technical communication has also experienced tremendous development in academe. Scholarship, coursework, and degree programs in technical communication have increased dramatically. With this educational advance comes the need for more specialized courses and training for students and textbooks to guide these students toward career fulfillment. We have designed *Technical Editing in the 21st Century* to fill this need.

Not only has the field of technical communication grown quickly in size and in specialization, but the editing industry has also developed rapidly with the influx of technology. *Technical Editing in the 21st Century* speaks to two needs: (1) an industry need for more skilled editors and writers in technical fields, particularly computing, engineering, government policies, and medicine, and (2) a need for subject matter specialists to understand more fully what technical editors actually *do*. Our textbook is aimed primarily at students currently enrolled in technical communication courses and at students taking editing courses for editing minors or majors. However, the information in *Technical Editing in the 21st Century* is current and valuable enough for a novice technical writer or editor who needs a reference guide, and this textbook also contains examples appropriate for more experienced technical communicators.

What makes this book unique is that in addition to incorporating successful approaches to teaching technical editing—the levels of edit, production management issues, ethical and legal issues of editing, and style—it also includes coverage of editing using computer technology. The most successful technical editors of today and tomorrow must know much more than style, grammar, and copymarking; they must be willing and able to quickly learn and use electronic editing tools. Authors and writing teams will work better with technical editors who are comfortable with the technology of editing and technological terminology, and this knowledge makes for better documents and more satisfied users.

Technical Editing in the 21st Century focuses not only on the process of editing and the product that is being edited but also on the process of becoming an editor. Other textbooks have addressed this issue by earmarking the levels of edit and sometimes designating these different levels to distinct editorial jobs (each editor on the editorial team performs one level of edit), but realistically technical editing students need to be aware of all of these roles, because they may be the only editors who are expected to accomplish all the levels and/or roles of editing or they may be subject matter experts (SMEs) who will need to understand what the entire editing team does.

Our approach allows the technical editing student to

- Apply a team approach to editing when necessary.
- Emphasize the editor's role throughout the document production process.
- Learn the process of editing a text while at the same time learn what it means to become an editor in every sense of the word.
- Discover where the field of technical editing originated and what the future may hold for technical editors.
- Learn relationship-building strategies with the author and the design team.

- Edit a technical document while employing the levels of edit.
- Address the numerous rhetorical issues involved in each stage of serving as an editor on a project, whether as a one-time freelancer or as an in-house editor.

Technical Editing in the 21st Century guides the technical editing student through each level of editing, each stage of becoming an editor, and each aspect of production. However, the technical editing teacher and student are not restricted to using the book chapters sequentially. Teachers may begin their course instruction from any chapter, and students (as well as novice technical editors) may refer to the chapters and units separately.

In short, *Technical Editing in the 21st Century* offers a distinctive yet invaluable approach to technical editing by providing the reader with

1. Learning objectives for each chapter.
2. Flowcharts that illustrate the stages of editing documents and graphics using the levels of edit.
3. Paradigms for becoming a better editor by applying the levels of edit.
4. Task analyses to guide the reader on how to conduct a needs assessment of the client as well as the users.
5. Discussions of electronic editing software, instruction manuals, and process-oriented documents.
6. Standard guidelines for usability studies of technical material.
7. Executive summaries of materials that the editor should supply to the client.

Technical Editing in the 21st Century also offers both hypothetical and real-world scenarios, sample assignments, exercises, and documents. These features, along with the organization, style, and layout of the text, are more student- and teacher-friendly. Students learn, chapter by chapter, what they need to know, what skills they should have, and what they need to consult in order to be a good technical editor. Students also discover the steps they need to take *as students about to become editors* in order to be successful in their class and, more importantly, as editors and most likely writers in the field of technical communication.

Technical Editing in the 21st Century provides information to technical editing students and novice technical editors in a logical, reader-friendly format. By using the fundamentals of instructional design—instructional theory, learner analysis, educational objectives, concrete examples, exercises, and self-diagnostics—the book provides a step-by-step guide for those who want to work as editors and writers in technical communication fields.

ORGANIZATION OF THE TEXTBOOK

We created this textbook from our own experiences that directly address the needs of students. We recognize that many students may not desire to pursue a career in technical editing, whereas others may consider it after reading this book. Therefore, we have designed this text to serve both ends—to provide a basis for good editing and, by extension, good writing for all students who desire to improve their communication and rhetorical skills regardless of their ultimate career choices. Furthermore, *Technical Editing in the 21st Century* covers not only the fundamentals of editing but also uniquely integrates emerging editing issues, such as electronic editing and editing materials to be delivered online. Thus, depending on the individual needs of their students, instructors can pick and choose from this book the specific areas of editing they feel are the most important for their students.

Unit One, "The Essentials of Editing," introduces students to editing by defining what it means *to edit*, differentiating between editing and technical editing, discussing theories of technical editing (Chapter 1), introducing the dynamics of the editing team, and providing some basic terms (Chapter 2). Chapter 3 covers hard-copy and soft-copy markup as well as a much-needed introduction to using styles in word processing documents. Advanced students will find these chapters a welcome overview of the editor's role and function, and novice students will benefit from reading the chapters, studying the terms, and completing all the exercises in the unit.

After learning how to create their own style guide (Chapter 4), students start Unit Two, "Editing for Correctness," as a grammar refresher. Unit Two is similar to Unit One in that instructors can select the order of what they feel is important for a specific class of students. Chapter 5 provides a basic introduction to grammar, and Chapters 6, 7, 8, and 9 provide more detailed discussion of grammar problems in context and exercises on how editors should address such issues. Exercises for both word-level and sentence-level errors are included in this unit, as well as copious examples of punctuation and mechanics.

We designed Unit Three, "Editing for Readability," to address the increased attention to document design and formatting issues of both online and print documents. Chapter 10 includes an overview of document design theories and applications, and Chapter 11 offers a thorough discussion of types of graphics and how to edit visuals in a variety of contexts.

Unit Four, "Editing for Effectiveness," contains chapters that deal with the more substantive levels of edit. Once students are well versed in the different approaches to learning (Chapter 12), they are then asked to edit for style (Chapter 13), and then to revise documents based on organizational patterns (Chapter 14). The background on different learning theories equips students to more globally identify how a document should be revised to fit the needs of the target audience. Chapter 15 reinforces this approach by encouraging students to recognize and rework errors in reason and logic in order to make the technical document more effective.

Unit Five, "Editing Online Publications," offers students a valuable overview of editing online documents, particularly web pages. Students learn important differences between print and online publications (Chapter 16), as well the editor's role in producing an online document (Chapter 17). In Unit Five, students are introduced to online editing terms (e.g., CSS markup, HTML markup, navigation aids, search engine optimization), major issues to be considered when editing online materials (e.g., page length, organization, readability, and the use of links), and the tools and techniques for actually editing a web page (Chapters 18 and 19).

We created the chapters in Unit Six, "Trends and Issues in Technical Editing," to be taught in any order. Students interested in a career in editing will find Chapters 22, 23, and 24 particularly helpful as they prepare to enter industry. Because Unit Six is designed for a general audience as well as the career-minded, all students will learn how to address ethical and legal issues in technical editing (Chapter 20), as well as how to edit documents so that the text and the visuals in those documents are more suitable for a global audience (Chapter 21).

SUPPLEMENTS

Technical Editing in the 21st Century Student Website

Exercise documents discussed in the book and additional exercise materials are available online to download at www.pearsonhighered.com/amare.

INSTRUCTOR'S MANUAL

The Instructor's Manual includes chapter outlines, teaching notes, additional exercises, and sample testing materials. The Instructor's Manual can be downloaded from our Instructor's Resource Center. To access supplementary materials online, instructors need to request an instructor access code. Go to **www.pearsonhighered.com/irc**, where you can register for an instructor access code. Within 48 hours of registering you will receive a confirming e-mail including an instructor access code. Once you have received your code, locate your text in the online catalog and click on the Instructor Resources button on the left side of the catalog product page. Select a supplement and a log-in page will appear. Once you have logged in, you can access instructor material for all Pearson textbooks.

ACKNOWLEDGMENTS

Special thanks to the reviewers of this text:

Michael J. Albers, University of Memphis
Joyce D. Brotton, Northern Virginia Community College
Roger Engle, Franklin University
Jean Farkas, Bellevue Community College
Marge Freking, Minnesota State University
Douglas K. Gray, Columbus State Community College
Donna Kain, East Carolina University
David A. McMurrey, Austin Community College
Jacqueline S. Palmer, Texas A&M University
Stan Dicks, North Carolina State University
Pamela Ecker, Cincinnati State Technical and Community College
Jim Kovarik, University of Memphis
Mae Laatsch, Madison Area Technical College
Robert Lynch, New Jersey Institute of Technology
Megan Morgan, UNC Charlotte
Scott Sanders, University of New Mexico
Karl Smart, Central Michigan University
Jason Swarts, North Carolina State University

ABOUT THE AUTHORS

The authors of *Technical Editing in the 21st Century* bring a unique blend of experience and expertise to this textbook:

Nicole Amare, Ph.D., is an associate professor of technical writing at the University of South Alabama, where she has taught technical writing, editing, ethics, stylistics, grammar, and composition. She has written *Real Life University*, a college success guide, and has edited *Global Student Entrepreneurs, Beyond the Lemonade Stand*, and *Giving Back*. Some of her research has appeared in *Business Communication Quarterly, IEEE Transactions on Professional Communication, Women & Language, Technical Communication*, and the *Journal of Technical Writing and Communication*. She is currently the associate editor of industry practices for the *IEEE Transactions on Professional Communication*. Her most recent work on editing is the chapter "The Technical Editor as New Media Author: How CMSs Affect Editorial Authority" that appears in *Content Management: Bridging the Gap between Theory and Practice*, edited by George Pullman and Baotong Gu and published by Baywood Publishing (2009).

Barry Nowlin, previously an assistant professor in the English Department at the University of South Alabama (USA), is the current interim chair in the Department of Interdisciplinary Studies at USA. He holds an M.A. in English and Education and earned his Ph.D. in Instructional Design and Development. After teaching high school for nine years, he moved to the college level and has taught courses in composition and in American, British, and Southern literature for the past twenty years. He coedited an anthology of Southern writers called *Mobile Bay Tales* and has published scholarly articles pertaining to composition studies. In addition, he has written articles for trade magazines on topics ranging from local history to sightings of the preternatural.

Jean Hollis Weber has worked as a scientific and technical editor and writer for over thirty years, specializing in software documentation. She has also taught short courses and workshops, lectured in technical writing and editing at several universities, and spoken at professional conferences. Jean has written ten books, including the award-winning *Is the Help Helpful? How to Create Online Help That Meets Your Users' Needs*. Her most recent work on editing is the chapter "Copyediting and Beyond" that appears in *New Perspectives on Technical Editing*, edited by Avon J. Murphy and published by Baywood Publishing (2010). For many years, Jean was active in the Australia Chapter of the Society for Technical Communication and the Australian Society for Technical Communication. Now retired, Jean keeps active as Co-Lead of Documentation at OpenOffice.org, a voluntary position, and she maintains several websites, including one for technical editors: http://www.jeanweber.com/.

Technical Editing in
the 21st Century

The Essentials of Editing

This unit covers three important aspects of editing: the field of technical editing, the editor-author relationship, and the editing tools needed to succeed. While reading about these three essential areas of technical editing, you will learn editing skills and will practice those skills on sample documents. In short, these four chapters provide the backbone of the profession as well as the basic skills needed to get started.

In this unit, you will learn some social strategies and personality traits to help you negotiate the job scene as well as work successfully with the authors of the documents you'll be editing. You will also learn the essentials of hard-copy and soft-copy editing (marking your edits on paper versus using a word processor or other software). We hope this unit will prepare you for the levels of edit that you will learn later in the book as well as leave you curious and enthusiastic about how this knowledge of editing will influence your life for good.

The Technical Editor

At the conclusion of this chapter, you will be able to

- describe the duties of a technical editor;
- recount the history of technical editing;
- define technical editing;
- research technical editing organizations;
- distinguish the different levels of edit.

As technology advances, so does the gap between the people creating the technology—a group called *subject matter experts,* or SMEs, who are frequently but not always engineers—and the users of that technology, called *end users* or simply customers or just lay audiences. This gap has brought about the field of *technical communication,* which is communicating expertise to less informed audience members so that they can understand.

Technical communication often deals with and includes technology, so it is important for technical writers and editors to be technology-literate. Any individuals who oppose technical or technological changes may find themselves struggling to find work. Technical communication includes fields such as medicine, engineering, computer science, government issues, the military, and environmental science, to list a few.

Technical communicators develop oral or written documentation that explains to end users the concepts, products, or programs developed by SMEs. This documentation has many forms, such as manuals, help menus, PowerPoint slides, videos, and web pages. Now you see where technical editors enter the picture: they are editors who work in the field of technical communication. They sometimes write or coauthor technical documentation, but more often they are responsible for editing that documentation.

Technical editors perform some or all of the following duties:

- Copyedit and proofread.
- Revise documentation text based on the levels of edit.
- Revise documentation graphics for technical accuracy, appropriateness, and visual readability.

- Build relationships with the SMEs in order to better understand the technology, and with the writers in order to produce more effective documentation for the end user.
- Research the end users' demographics and characteristics to discover how they will be using the technology.
- Perform or assist in usability testing by end users, to let the SMEs and the technical communicators know what is ineffective about the documentation or about the product itself.
- Create documents that assist in effective future document production, such as policy handbooks and company style guides.
- Protect the end user from harm and the company from lawsuits by enforcing ethics.
- Be aware of global issues, such as language and cultural differences.

As you can see, technical editors do a lot more than just check grammar.

> Cut out all these exclamation points. An exclamation point is like laughing at your own joke.
> —*Author unknown*

A BRIEF HISTORY

Although technical editing grew as a career field during the twentieth century, there were some technical editors before 1900. According to Frederick M. O'Hara Jr., the invention of the printing press by Johannes Gutenberg around 1440 was the catalyst for a culture of print, which created the need for editing. From 1490 to 1515, Aldus Mantinus ran his own publishing house, and he hired Desiderius Erasmus to edit all of the texts that he published. These ranged from religious content to the more technical subjects of the time, such as medicine and science. Erasmus, then, is the first technical editor. Don Jensen reports that over 450 years later, Bill Zielinski became a first on his own terms: in 1962, at the age of fifteen, high school sophomore Zielinski was named the first technical editor at NASA. He edited the column "Tech Trails" for a readership of just seventy-eight members.

From 1515 on, editors were contracted by owners of publishing houses, but authors still largely edited their own work or relied on other writers to edit for them. Then, in the 1800s, the rise in scientific journals created a need for more formal editorial positions. O'Hara states that by 1830 there were more than 500 scientific journals globally, although most of these were centered in Europe and some in the United States. It wasn't until the World Wars of the twentieth century that we began to witness an international need for technical editing beyond the scientific text—namely, weapons, war technology, and government documents. Eventually, technical editing and writing grew into its own field, and today there are dozens of universities that offer programs specializing in technical writing, editing, or both.

> My life needs editing.
> —*Mort Sahl*

Despite its tremendous advancement, technical editing is still misperceived, and technical editors are still stereotyped. Some assume that the "technical" part of technical editing means being a fussy language editor. Anne Enquist defines technical editing much too narrowly:

> If substantive editing looks at the forest, technical editing looks at the trees, branches, leaves, and even the veins on the leaves. Technical editing is what most people imagine when the word *editing* is used; it is the meticulous examination of the smaller details in a piece of writing. With pen in hand, the technical editor slowly moves through the manuscript, line by line, word by word, even space by space.
>
> Given the description in the last sentence, it may sound like the only real trick to good technical editing is to overcome the inevitable tedium that comes with moving through an article so carefully. Actually, there are two other important aspects to effective technical editing: (1) knowing when to make a stylistic recommendation and when to leave the author's style alone; and (2) not overlooking any of the various types of minor writing problems and errors.

Both require that the technical editor know the difference between true errors and stylistic options, and in the case of stylistic options, know when a choice is one of personal preference and when it is a choice between an effective and an ineffective style.

What Is Technical Editing?

To an extent, Dr. Enquist is correct—technical editing does involve word-by-word, meticulous editing for grammar and mechanics. But it also includes substantive editing at times, and technical editors need to understand technology and technical terms in addition to possessing superior language and graphics skills. Technical editors do much more than just edit the document text and graphics—they are responsible for external text issues such as working with management, the production team, the design team, the SMEs, and the end users. Moreover, gone are the days of technical editors merely dragging their pens (or pencils) over documents; most now use electronic devices for editing text and graphics (see Chapter 3). And technical editors have more say about the writing style of a document than ever before, especially because they are being recognized for their expertise.

In essence, technical editing is working with products, documents, texts, graphics, web pages, programs, and media. A technical editor prepares documentation that describes specialized information to a group of people who need to understand or use that knowledge. A simple example explains clearly what technical editors do: In order for you to print your Spanish homework, you need a printer. Technical writers and editors have created and revised the websites you visit when shopping for your printer. The advertisement provides information about the product's "specs," or specifications. Once you find a printer in your price range, you have to check its compatibility with your system. A technical writer or editor is involved in ensuring this compatibility through a process called *usability testing* (see Chapter 23). Finally, the booklet that explains how to assemble the printer and connect it to your computer is developed by technical writers and editors. They use a combination of text and graphics to explain this process in the clearest manner possible. If you have a problem connecting your printer or printing an alignment page, for example, you might want to consult the software CD that came with the printer; the problem-solving instructions were written by a technical writing and editing team.

Do Editors Really Need All That Technical Knowledge?

Unless an organization specifically defines it as an editor's responsibility, the accuracy of a document's data remains primarily the responsibility of the writer or the SME. However, because writers and editors share a common purpose to effectively communicate information to their readership, a technical editor should examine documents for obvious technical errors. These may include basic addition or subtraction errors, transpositions of numbers, misplacement of decimal points, tables and charts with missing data, and logical gaps in reasoning. But just how much technical knowledge and background must an editor have to discover technical inconsistencies? There are two schools of thought.

Some members of the technical community argue that technical editors do not really need to comprehend the technical nature of the material they are editing. In fact, technical editors may actually be better off *not* having technical knowledge because they are less likely to make assumptions about the reader's knowledge than someone who thoroughly understands the technical aspects. If they have less knowledge of the technical subject, editors might remain objective and more closely aligned to readers' needs.

According to Donna Roper, a technical publications editor for NASA, many other members of the technical community disagree with this logic. Instead, they believe technical editors not only must understand as much as possible about the subject but must work with the writer to enhance the technical content. Roper further believes that technical editors who have a certain level of technical knowledge enjoy greater respect within the technical community and therefore wield more influence.

In an in-house study that Roper conducted at the NASA facility in Virginia in 1993, 71% of the writers she surveyed believed technical editors should complete either several technical courses or an undergraduate minor in a technical field. Furthermore, 96% said technical editors need a background in English, and 95% said they also need a background in technical editing. However, only 32% of the respondents believed aeronautics and engineering were requisites for a technical editor at their NASA facility. In general, the survey suggests that editors who possess "technical aptitude" and whose education includes some technical courses contribute positively to the writer-editor relationship and to the editorial process. But most importantly, 94% agree that they need to have editing skills and expertise, 75% believe they need an understanding of the publishing process, and 51% say technical editors must possess the proper personal characteristics to work with people.

Of course, these statistics represent opinions from only one organization. However, from our experience, we believe that these opinions reflect what you will generally find: editors need some technical aptitude, but above all they need a good command of language. Most agencies and organizations require job applicants to take an entrance test to demonstrate their grammar and editing skills. Often the score on this test is the primary factor in determining whether or not an individual is hired.

Education is a progressive discovery of our own ignorance.
—Will Durant

Audience, Accuracy, Author

What matters in technical editing are the three A's: *audience, accuracy,* and *author.* Above all, the document should be correct. This means that an accurate text is grammatically correct as well as accurate in content. It is very difficult to know whether or not the content is accurate unless you as an editor understand how the product or policy is being described. Although you certainly do not have to know *everything* about technological tools (no one could possibly know it all), it does help to possess more than just language and personality skills. Sure, you could edit an online technical manual or help menu that describes how to properly install a wireless router you have never used. But it would be better if you knew not only what a wireless router was but also how to actually install one yourself. Knowing about technological tools is an emergent need in the technical editing professions, and it is necessary when products undergo usability testing, a procedure we discuss in Chapter 23. For now, suffice it to say that you should know something about the product, procedure, or policy as you edit the document.

But what if what you know about a product, like the wireless router we mentioned, disagrees with what the SME has written? What if the SME has written that a router will be functional with a cell phone, but you know that is not the case from either logical reasoning or from personal experience? We'll be honest and tell you that some editors do nothing about it. (We discuss ethical dilemmas for technical editors in Chapter 20.) At the base level, an editor's job is to correct the document. However, even though editors may be labeled as geeky wordsmiths, this textbook shows how they can and should be involved in all areas of document production. Technical writer Janet K. Christian remembers an engineer who labeled all technical writers as just "glorified clerk typist[s]" and a computer programmer who thought technical writers were "those people who flowerize our writing." In "How to Get a Good Job," technical writer and editor Don Bush says that "engineers complain that technical writers are bureaucrats . . . because we demand glossaries, overviews, forewords, abstracts, lists of figures, [and] 'About This Chapter' sections." Editors are good at fixing texts, yes, but they often do more.

Therefore, technical editors need to know something about tools and technology. Knowledge of the appropriate tools and technology will help make you a more competent editor. It could also keep customers from receiving a deficient product or manual and save a company from embarrassment or financial loss. Good editors are vital in today's market. Remember the three A's of good editing: a document should be technically *accurate* as well as grammatically correct, should reflect the *author*'s intended meaning (provided that intention is ethical), and should communicate the necessary information to the target *audience*.

TYPES OF WRITERS

When we think of writers, we may first think of professionals who write novels, poems, and magazine articles, or nonfiction such as biographies, sports books, or entertainment books. Actually, these individuals who write represent only a small portion of professional writers. Think of all the documents we read daily: brochures, pamphlets, manuals, instruction books, newsletters, help and instruction guidelines for computer software, newspaper articles, procedural and government documents, how-to articles, and so on. The writers of these materials have one common goal: to share information with others.

Many technical writers are SMEs. They may be engineers, business executives, scientists, researchers, psychologists, computer programmers, government bureaucrats, lawyers, educators, or anyone having expertise in a given area and a need to disseminate information. For example, researchers who have completed an empirical study of the effects of aspirin on the heart might want to publish their findings in a medical journal like the *New England Journal of Medicine*. Business executives who have advice on how to market a product successfully or retain valuable personnel in a declining economy might publish their advice in a popular magazine like *Forbes*. After studying the data from the Cassini-Huygens probe that landed on Saturn's moon Titan, space geologists may want to publish their conclusions about Titan's geological features for the scientific community. A public relations director might want to create an in-house pamphlet explaining how employees can better serve their customers. All of these professionals are presumably experts in their given areas; not all, though, are good writers.

Organizations employ freelance and in-house professional writers who work with SMEs to communicate information to individuals who do not have technical expertise. These writers act as an *information bridge* between the experts and the lay audience. For example, several mechanical engineers working with an in-house writer might collaborate to develop an instructional manual for auto mechanics. A popular fishing magazine might hire a freelance writer to work with that year's champion bassmaster to create an article on new fishing techniques.

TYPES OF EDITORS

Whereas writers generally know that their primarily responsibility is to *write something,* editors may be involved in any number of publishing activities, from proofreading to testing software. Technical writers understand they will be writing predominantly technical documents; however, technical editors may merely copyedit a writer's draft or actually be asked to write the technical documents themselves.

Job titles alone tell you little about the editorial roles and responsibilities in an organization, because each organization has its own in-house nomenclature. You may encounter copy editors, production editors, managing editors, literary editors, developmental editors, and technical editors. Some organizations even make up their own titles, such as *chief editor*. Below is a general guideline to what editors do, but it is extremely flexible. Just be certain you ascertain the specific job description for any designation of editor when applying for a new job.

- *Proofreader*—checks for grammatical, mechanical, and spelling errors but often has no authority to make changes.
- *Copy editor*—proofreads, checks for conformity to guidelines established by the agency, and generally performs line-by-line editing. Copy editors may work with writers but have little authority to make major changes.
- *Literary editor*—acts as the reader's advocate by recommending changes to the writer. Literary editors may have considerable authority over a manuscript.
- *Developmental editor*—actively participates in the production process by consulting with the writers throughout the planning, organizing, and final production stages of a document.
- *Managing editor*—supervises the production process and manages other editors.

Technical Editing Organizations

One of the best ways to learn about what technical writers and editors do is to visit websites owned by technical communication organizations, such as that of the Society for Technical Communication (STC), www.stc.org. There are literally hundreds of organizations for technical editors, because people with similar interests in writing enjoy communicating with one another. Editors understand the importance of sharing information and networking. Many smaller or regional organizations (e.g., STC has branch chapters around the world) can be found by checking the larger organization's website or by using your favorite search engine. For instance, if you live in the Boston area, STC Boston's website reveals that you are fortunate to have at least two dozen organizations nearby (see Figure 1.1).

In addition to STC, the Professional Communication Society (PCS) also has excellent recourses for technical editors and writers (www.ieeepcs.org). Because PCS is sponsored by the IEEE, members have the benefits of the largest international engineering association, www.ieee.org. If you are interested in technical editing in the medical field, check out the resources available from the American Medical Writers Association (www.amwa.org). Of course, there is the esteemed Council of Science Editors (www.councilscienceeditors.org), if you are primarily interested in editing for the sciences.

Although you can visit these websites for free, we encourage you to consider joining if you are interested in a future in technical editing or writing. Membership to one or more of these organizations will help you learn issues important to technical editors and writers. It will also open the doors to networking and potential employment opportunities. These organizations cater to student populations by offering discounts to students, sometimes as little as $20 per year for a membership, a fee that often includes access to job lists, magazines, journals, and information about online seminars and other educational opportunities.

For those of you interested in a teaching career in technical editing and writing, there are organizations geared toward both academics and practitioners. The membership of PCS consists of academics, engineers, and technical writers/editors. And because IEEE is the parent organization, PCS members have exposure to a wealth of information regarding engineering issues and concerns. If you want an organization that is predominantly focused on teaching technical editing and writing, check out the Association of Teachers of Technical Writing (www.attw.org), or the Council for Programs in Technical and Scientific Communication, (www.cptsc.org).

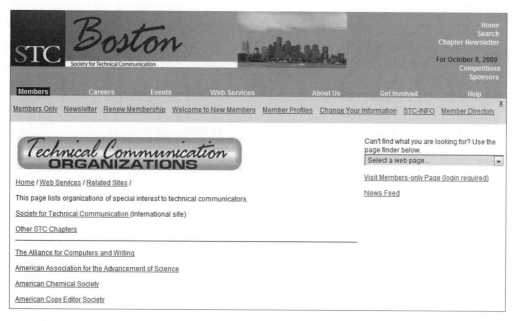

FIGURE 1.1

Page from Boston STC Chapter's Website.

Reprinted with the permission of the Society for Technical Communication, Boston Chapter.

TIP FOR TECHNICAL EDITORS

Besides membership in technical communication organizations, you can also learn a lot about technical editing issues by joining a mailing list or forum. A *mailing list* (also called a *listserv*) is an electronic discussion group that allows users to communicate by means of a communal mailing list. Once you join a list, you may email all members at once (using the list's email address) with questions or responses, or you may choose to email a member individually, or "off list." Mailing lists are usually free, although some require membership to the sponsoring organization. *Forums* (also called *bulletin boards*) are similar to mailing lists, but readers post questions and answers on a website instead of emailing comments to a central address. See Figure 1.2 for some relevant listservs.

TIP FOR TECHNICAL EDITORS

Another way to make yourself more comfortable in the field of technical editing is to read short articles in blogs and newsletters about what issues are relevant to technical editors and writers. *Blogs* may be run by an individual or a group; the owner posts articles and readers can discuss the articles by commenting on them. *Newsletters* were more common before blogging became widespread; they are often less interactive because they may not have an easy way for readers to share comments. Figure 1.3 lists some free blogs and newsletters. These editing resources provide information about everything from grammar wars to dealing with the tab function in Microsoft Word.

ATTW-L	Association of Teachers of Technical Writing: http://www.attw.org
TECHWR-L	Technical Writer's List: http://www.techwr-l.com/techwr-l-list
Copyediting-L	Copyeditor's List: http://www.copyediting-l.info/
STC-LIST	Students of Technical Communication. Email: majordomo@nmt.edu, and in the body of the message, write "subscribe STC-LIST" (without quotation marks)
Stctesig-l	Society for Technical Communication's Technical Editing Special Interest Group: http://www.stc-techedit.org/tiki-index.php?page=discussion list
TCP	Technical Communication Professionals: http://techcommpros.com/mailman/listinfo/tcp_techcommpros.com

FIGURE 1.2
Listservs for Technical Editors and Writers.

Beyond the Elements of Style	http://beyond.jeannettecezanne.com/category/editing/
Bookworm: Newsletter of the Society of Editors	http://www.editorswa.com/submit_bkworm.html
Copyeditors' Knowledge Base	http://www.kokedit.com/library.shtml
Corrigo	http://www.stc-techedit.org/tiki-index.php?page=Newsletter
The Editorium	http://www.editorium.com/euindex.htm
Technical Editors' Eyrie	http://www.jeanweber.com/
Terribly Write	http://terriblywrite.wordpress.com/

FIGURE 1.3
Selected List of Blogs and Newsletters for Editors.

EXERCISE 1.1

Do the following to better familiarize yourself with the field of technical editing.

1. Go to your favorite search engine or directory and type in "technical editing" as a phrase search. You should get at least 25,000 hits! Click on each of the first ten links returned. What did you learn about technical editing that you didn't know before?

2. Join one or more of the mailing lists that you find interesting. (Please note that some lists deliver over 100 messages a day; if your email account cannot handle that many emails in its quota, please subscribe to the *digest* form of the list.)

3. Read at least three articles from the list of blogs and newsletters for editors, and present your findings to the class or in small groups.

THE LEVELS OF EDIT

The most widely accepted method of technical editing is called the *levels of edit*. Originally developed in 1976 by Robert Van Buren and Mary Fran Buehler of the Jet Propulsion Laboratory, the five levels of edit denote "separate and distinct applications of the editorial process." The levels are named simply 1, 2, 3, 4, and 5. Level 1 is the most thorough edit, whereas level 5 is the least thorough.

In addition to the levels of edit, Van Buren and Buehler developed nine types of edit for technical writing at the Jet Propulsion Laboratory. As you can see in Figure 1.4, the five levels of edit intersect with the nine types of edit to describe the amount of editing to be performed on a document.

The types of edit each reflect a different aspect of the technical document that needs attention. For example, a *policy* edit checks for whether or not the document adheres to company policy standards. A technical editor would ensure in a policy edit that if it is company policy that all external documents contain the company logo, the document being edited must also contain that logo.

The Changing Concepts of Editing

In the last three decades, several editors have tried to adapt the levels of edit to their own company or organization. Some have tried to simplify the types and levels. In the late 1980s, Graham Unikel published "The Two-Level Concept of Editing" in an attempt to reduce the five-level edit plan proposed by Van Buren and Buehler. His concept focuses on simplicity. Using a two-level editing concept that he developed from the Lockheed organization, Unikel wanted his technical communication students to have an easy way to categorize editing functions. Our best summary of Unikel is that his first-level editing has to do with a text's content, whereas second-level editing has to do with format.

We agree with Unikel that technical editing novices do better with fewer levels and types of editing. Van Buren and Buehler's levels do reflect industry standards in the main, but unfortunately, every organization tends to have its own editing system, which means you'll have to learn a new system for each organization. Moreover, companies are changing their policies all the time. Judyth K. Prono's "Exploring Editing" discusses ways of reducing her company's four levels of edit down to three: proofreading edit, grammar edit, and full edit. Donald Samson also reduces the levels of edit from five to three in his "Degrees of Edit." Samson refers to his levels as the light edit, the medium edit, and the heavy edit.

Like Prono and Samson, we employ three levels of edit. We find that nine types is too much specification, especially when editing types are usually organization-specific. In today's global economy, freelance and LTE (limited-term employee) work is becoming more

> There are three levels of editorial crime: Level 1: to use the odd wrong word from time to time. Level 2: to fail to correct a wrong word by someone else. Level 3: to change somebody else's right word into a wrong word.
>
> —*Nicholas Hudson*

Type of Edit	Level of Edit				
	Level 1	Level 2	Level 3	Level 4	Level 5
Coordination	x	x	x	x	x
Policy	x	x	x	x	x
Integrity	x	x	x	x	
Screening	x	x	x	x	
Copy clarification	x	x	x		
Format	x	x			
Mechanical style	x	x			
Language	x	x			
Substantive	x				

FIGURE 1.4
Van Buren and Buehler's Types and Levels of Edit.

popular in technical editing because many editors can work and like working from home using their computer and the Internet. This means that technical editing students need to be well versed in how to edit a document effectively, not necessarily with learning what the different levels and types of edit are called. What used to take weeks or even months to edit and mail to international clients now just takes days or hours with electronic editing and email. Because you are often isolated from other people, though, freelance editing takes some special personal and technical skills; we discuss these in Chapters 2 and 3.

A Three-Level Approach to Editing

As much as we would like to reduce all of the editing tasks down to two levels as Unikel did, this is impossible with today's new technical documents, particularly web pages. Content is no longer linked only to paragraphs, and format is much more than just graphics and logos. Moreover, litigation in corporate America is at an all-time high, and technical editors are responsible for helping the home organization avoid lawsuits by producing error-free documentation on a number of levels. Finally, as the number of texts increases—the Internet offers billions of pages daily—the quality of these documents is decreasing. This is why STC is correct that technical editors are more valuable now than at any other point in history.

Our three levels of edit are meant to describe the complex process of editing a document as simply yet as clearly and effectively as possible. While we like Samson's light, medium, and heavy edit levels, these terms may not describe the various functions within those levels. We therefore call the first level *editing for correctness*. The second level requires more knowledge about document design, formatting, and graphics. All of these factors have to do with the reader's visual comfort when viewing the document. This level is called *editing for visual readability*. The third level, *editing for effectiveness,* deals predominantly with content and meaning in a technical document and requires editing large chunks of text. Figure 1.5 shows the three different levels and what is covered in each.

Please note that these levels do not constitute a hierarchy; that is, level one is not necessarily more valuable than level two or three. We can say, though, that level one probably contains the most common types of tasks you will have to do as a technical editor. Because level one, editing for correctness, covers errors that people in industry often find the most egregious (spelling mistakes, incorrect punctuation, etc.), we think students should learn this

Level of Edit	Tasks
Level 1: Editing for correctness	Fixing grammar and mechanics errors within text.
Level 2: Editing for visual readability	Editing for document design issues such as bullets; spacing; font style and emphases; white space; section headings and subheadings; position of headers and footers; footnotes and endnotes; margins and borders; and color issues. Because visual readability involves document formatting and the design of a page, this level also includes all graphics.
Level 3: Editing for effectiveness	Substantive editing for content issues such as organization, sentence structure, style, logic, and meaning.

FIGURE 1.5
A Three-Level Approach to Editing.

level first, although the levels can be learned in any order. Because level three, editing for effectiveness, uses so many of the word-level and sentence-level skills taught in level one, it may be best to learn level three after level one. Level two, on the other hand, can be learned second, first, or third, as it contains advice about the visual rhetoric of a document.

Although this book focuses on editing, most technical editors are also technical writers. The field of technical editing is becoming more lucrative and specialized, but companies still rely heavily on their technical writers to do editing as well. This is mainly due to budgetary concerns. A great technical editor, then, must have knowledge not only of the SME's and the end user's expectations for the document but of the technical writer's as well, especially when the editor may be asked to collaborate with the writing team. This textbook is designed to teach valuable technical writing skills through editing, and many of the best editors are great technical writers. For instance, learning how to *reorganize* a help menu should teach you the necessary skills to *write* an effective, well-organized help menu should the need arise.

A BRIEF OVERVIEW OF THE EDITING PROCESS

In addition to learning the specific duties and responsibilities of the editing staff, you will also need to understand the editing process. A team approach where designated Information Development (ID) leaders manage, plan, and coordinate document production is one process. These managers let everyone know well in advance the writing and editing schedules and the production deadlines. Unfortunately, many organizations do not place enough emphasis on product managers and planners, making editing a "rush job."

At some point during the production process, the editor and writer confer on a number of issues. If the organization does not have an ID manager, it is a good idea for the editor to take the initiative to meet with the writer. Face-to-face meetings are one option, but email and the telephone work also. Writers and editors primarily cooperate on setting due-in and due-out dates. Editors should keep meticulous records of meetings with the writer. The editor should log entries specifying deadlines and the type or level of edit to be performed. (Sometimes the organization has its own specifications and signoff forms.) In addition, the editor should be sure to write down the writer's name, page count, planned dates, scheduled conferences, and an editorial signoff column. It is also a good idea to keep records of all edits until the final version is published.

Although editors may have many responsibilities associated with document production, if handled poorly, the most difficult undertaking may be returning the *markup edit* to the writer. The markup edit is the copy that includes the editor's comments, corrections, and so forth. (Chapter 3 explains the differences between hard-copy and soft-copy markup.) Bad feelings may result if the markup seems too critical or condescending. Therefore, it is a good idea to allow the writer at least a day to review the edit. If the writer is in-house, the editor should schedule a meeting to discuss and negotiate the edits with the writer. This is when good listening and people skills become essential to the success of a well-developed document. If the writer agrees, we recommend "walking" the writer through each page and explaining the rationale for the markup. However, you do not want to sound too preachy; just balance an opportunity for quiet teaching with friendly negotiations. (Chapter 2 addresses the editor-writer relationship in detail.)

No diagram or flowchart can completely demonstrate the editing process. Each corporate culture dictates how things are done within that organization. However, the generic diagram in Figure 1.6 provides a conceptual framework for understanding the editorial process. Remember, though, that variations may occur.

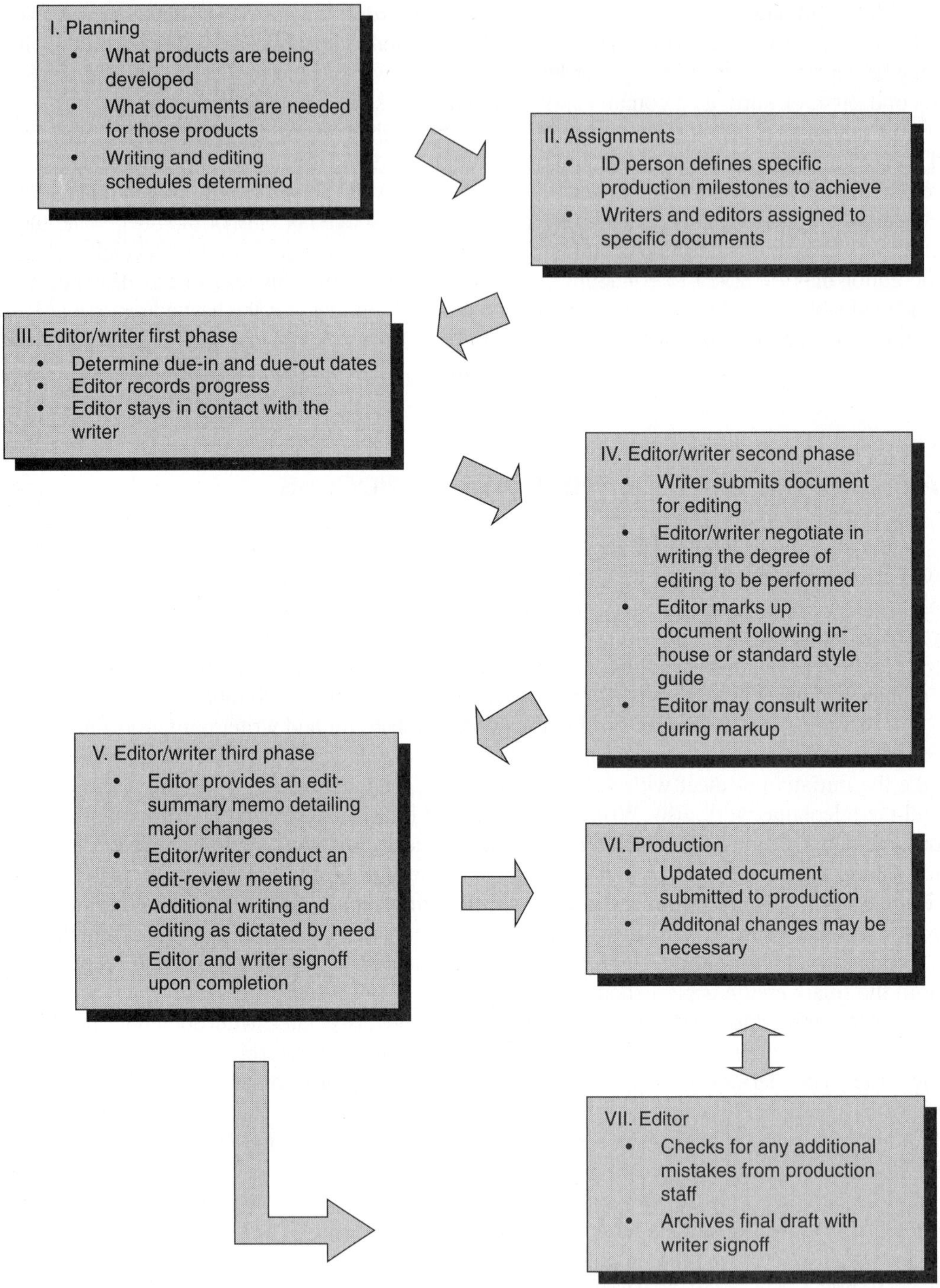

I. Planning
- What products are being developed
- What documents are needed for those products
- Writing and editing schedules determined

II. Assignments
- ID person defines specific production milestones to achieve
- Writers and editors assigned to specific documents

III. Editor/writer first phase
- Determine due-in and due-out dates
- Editor records progress
- Editor stays in contact with the writer

IV. Editor/writer second phase
- Writer submits document for editing
- Editor/writer negotiate in writing the degree of editing to be performed
- Editor marks up document following in-house or standard style guide
- Editor may consult writer during markup

V. Editor/writer third phase
- Editor provides an edit-summary memo detailing major changes
- Editor/writer conduct an edit-review meeting
- Additional writing and editing as dictated by need
- Editor and writer signoff upon completion

VI. Production
- Updated document submitted to production
- Additonal changes may be necessary

VII. Editor
- Checks for any additional mistakes from production staff
- Archives final draft with writer signoff

FIGURE 1.6
The Editing Process.

EXERCISE 1.2

Do the following to familiarize yourself with the field of technical editing.

1. Read at least two articles that discuss the levels of edit from the list of newsletters for editors in Figure 1.3. Report back to the class as to how these levels are similar or different from the ones discussed in this chapter.
2. Search a major job website, such as monster.com, for jobs in technical editing. Report your findings to the class.

SUMMARY

Jobs in technical editing require both linguistic and technical expertise. You can learn a lot about the field of technical editing and improve your chances of finding a job later by joining a listserv, subscribing to organizations, and keeping yourself abreast of issues important to technical editors and SMEs by reading the editing newsletter articles. The next two chapters focus on skills needed to succeed as an editor, a technical editor, and a technical writer.

Even if you are not interested in making a career of technical editing, being able to write and communicate effectively transcends all professions. It's safe to say that individuals who are able to write and edit effectively greatly enhance their chances for career advancement.

Audience and Authors

At the conclusion of this chapter, you will be able to

- identify the basic skills required of a technical editor;
- understand the similarities and differences between technical writing and technical editing;
- conduct an audience analysis;
- understand the importance of maintaining strong writer-editor relationships;
- apply proven strategies for effective editorial feedback.

As we discuss in Chapter 1, the past four decades have seen a dramatic shift within the emerging field of technical editing. The role of the technical editor continues to rapidly transform and expand in new directions, especially as technology increases. The workplace is evolving to accommodate teams of specialists who do everything from providing feedback on user interfaces, to editing graphics, web pages, and tables, to ensuring that a document meets the expectations of the reader. Depending on the agency or business, technical editors' responsibilities range widely; many are active decision-makers and problem-solvers throughout the entire production process. And just as often, technical editors have a range of expertise beyond good communication skills. They may have backgrounds in engineering, science, business, instructional design, computers, software production, or any number of related technical areas. Yet despite their wide-ranging backgrounds, technical editors must possess a common denominator—good oral and written communication skills coupled with the ability to deal with technological concepts.

> The editor's position in technical communication is unique since the editor functions in the center of a series of rhetorical situations, linking the author and the potential reader, and serving the needs of both.
> —*Mary Fran Buehler*

REACHING YOUR TARGET AUDIENCE

Before describing the dynamics between writers and editors, we want to emphasize the editor's most important responsibility: the reader. As we discuss in Chapter 1, the three A's of editing are audience, accuracy, and author. In this chapter, we'll cover a basic audience analysis approach and how to work well with authors. A technical document's readership may be narrow or broad; it may be directed to the general public, or it may target researchers, academicians, policymakers, stockholders, clients, or any number of groups to whom specific technical information must be disseminated clearly and accurately. Generally, editors are in a much better position than writers or SMEs to determine the composition of the target audience because they are usually more detached from the subject matter and can therefore be more objective.

Have you noticed that too often individuals who know something in detail often make assumptions that their audience has the same knowledge or skills, when in reality the audience members are novices? In the past, you may have tried to follow instruction manuals that assumed you knew the names of certain parts or that you possessed intimate diagnostic skills. Most of the problems that arise from technical documents result from the lack of an *audience analysis*.

People who read technical documents are primarily interested in the content; however, the style and form of the document reinforce the content through appropriate word choice, language that is accurate and suited to the reader's level of comprehension, and graphics, diagrams, and tables that support the text. As a technical editor and advocate for the reader, during the production process you assume a role much like the conductor of a symphony orchestra. Just as all the instruments must play together to achieve harmony, the editor must coordinate many different elements to achieve a successful document. Your "symphony," then, begins with a thorough knowledge of the target audience and the continual adaptation of the document to the reader's needs.

Conducting an Audience Analysis

In their rush to meet deadlines, many writers and editors fail to conduct even a cursory audience analysis. Not understanding the nature of one's audience can have serious consequences. For example, suppose your agency has been subcontracted to write a field manual for cleaning the army's new M-17 rifle. Obviously, creating an accurate and readable manual is of the utmost importance. But for whom? Just who is going to be reading this manual? Experts who designed the rifle? Technicians who must repair and maintain the rifle? Administrators who make legal and political decisions governing the rifle's use? Nonspecialists who have little technical knowledge of rifles but as soldiers need practical knowledge of the M-17? Or will the manual be read by a combination of all of these?

At first, you might say the primary audience likely would be the soldiers who need practical knowledge of the weapon but not an engineering degree. However, you could also argue that soldiers are also technicians because they must be able to operate, maintain, and repair the weapon in the field. In fact, soldiers could even be considered experts, especially when historically many have recommended engineering modifications that have improved a weapon's performance. And let's not leave out the generals who make the final administrative decisions about whether or not to deploy the weapons into combat.

How important is audience analysis? Consider that scores of new magazines enter the publishing field each year. Did you know that few survive? Although there could be any number of reasons for failure, the most common is the inability to connect with an audience. For example, why do you think the *Ladies' Home Journal* is the oldest continuously published magazine in the United States? Do you think the editors and writers have a clear sense of the type and interests of women who read their magazine? What about a magazine such as *Bassmaster*? You can be certain thousands of dollars have been spent to determine what type of individuals read *Bassmaster,* including their average age (thirties), average disposable income ($35K–$50K), marital status (married), and so on.

No one expects you or your organization to spend thousands of dollars conducting polls, sending out questionnaires, or completing an extensive demographic analysis to determine the intended audience. However, the writer, in conjunction with the technical editor, should conduct an audience analysis during the planning phase of any new project. Once an initial analysis has been conducted, though, how will you use this information? This is primarily the technical editor's responsibility: to continually assess and adapt successive revisions to meet the readers' needs. This responsibility might mean adding or deleting information, providing more or fewer examples, eliminating jargon and buzzwords, providing more or less background information, modifying the style to conform with the reader's level of expertise, shortening sentences, adding or changing graphics, shortening paragraph length, changing the format, adding or eliminating tables, and more.

Below is an overview of the process editors takes to understand their audience. These steps include determining the purpose of the document, determining the intended audience, and determining the characteristics of each audience type.

THE PROCESS OF AUDIENCE ANALYSIS

Step One: Determine the Purpose of the Document

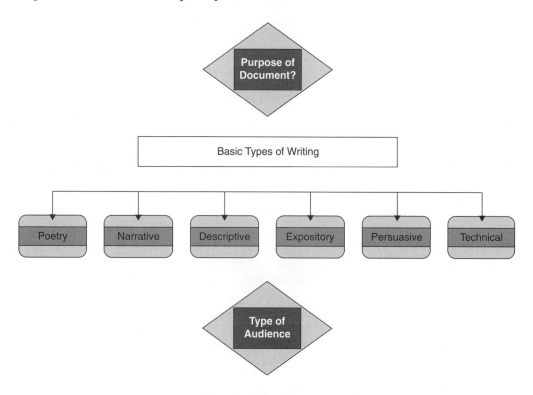

Audience analysis begins by asking, "What is the essential purpose of my document?" Will it attempt to inform the reader through exposition? Perhaps it will describe a step-by-step process, such as a chronological sequencing of the steps someone would take to construct a shortwave radio. Or the writers may desire to persuade the reader to perform some type of action or accept some particular point of view. For example, using a brochure from an environmental organization, a biologist may desire to persuade the reader why the proper mitigation of wetlands is necessary to protect the environment. Often, documents of a technical nature combine description, information, persuasion, and occasionally narration.

Be sure, though, that before you begin to edit the document, the document clearly articulates its primary purpose. As an editor, you should be able to summarize the purpose in a single clear sentence: *The purpose of the brochure is to convince the City of Elibom to convert to propane-powered cars.* Notice that here the primary purpose is *to persuade.* Therefore, the brochure must also provide evidence in the form of facts, information, statistics, examples, or expert testimony as to why the city government should make the conversion. We suggest you practice discerning the purpose underlying everything you read.

Step Two: Determine the Intended Audience

After identifying the primary purpose, the next step is to determine the intended audience. Generally, your audience falls into one of the following four types, or as happens most commonly, a combination of audience types. *Experts* are the individuals who already know the product thoroughly, usually work in research and development areas for the government, business, or industry, and have advanced degrees. Although *technicians* do not usually possess the theoretical knowledge of experts, they maintain, operate, and repair the actual products. Legal, political, and implementation decisions are typically the domain of the *administrators*, who generally do not have a technical background. Finally, the *non-expert,* while similar to the administrator in having little technical knowledge, simply desires to understand the product well enough to use it or to make sound judgments about whether to buy it.

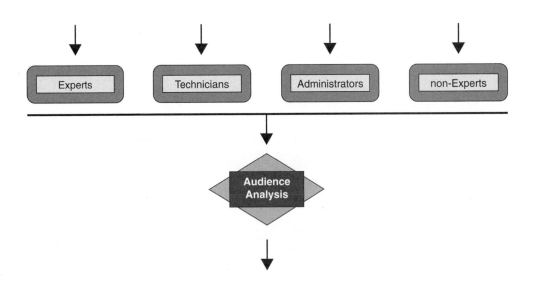

Step Three: Determine the Characteristics of Each Audience Type

In addition to conducting an audience analysis, you will want to ask some standard questions when beginning an editing project. We suggest that you familiarize yourself with these questions and use them as a checklist. As you become more familiar with editing, these questions will become second nature.

Background   How much does this audience type already know about the subject? How much previous experience in this subject area? Will some have more experience than others? Is basic background or preliminary information needed for this audience?

Motivation   Is this audience type interested and motivated toward this subject area? What particular needs or expectations do they have of the document? How will they want to use the document? What can the document contain to address these issues?

Demographics   What is the audience's relative age? Gender? Lifestyle? Education? Disposable income? Political preferences? Other pertinent demographics?

THE EDITOR-WRITER RELATIONSHIP

As in any other field, an individual's ability to work with others is often as important as technical expertise. No doubt you can provide a list from your own experience where the success or failure of a project (or a personal relationship, for that matter) came down to communication between you and another person. Conflicts do arise. The three major problems confronting editors are:

- Failure of clients, coworkers, and management to fully understand the editor's role.
- Failure of editors to be afforded the level of authority appropriate to their responsibilities.
- Failure of project managers to include editors early enough in the document process (or early enough in the production cycle).

These problems can be minimized through good management procedures and active cooperation among the production team. SMEs, writers, and editors must never forget that they share a responsibility to their clients and the organization they work for. Inaccurate, careless, and poorly organized information is a disservice to both. Every member of the production team, especially the writers and editors, must concentrate on delivering a product that effectively represents the intended goal. To this end, then, everyone—regardless of position within the organization—should work in a spirit of partnership.

This may be easier said than done. Unfortunately, some organizations do not establish policies or guidelines that clearly define individual responsibilities. This is particularly true for relatively new organizations or agencies that often infer the need for guidelines or job descriptions only after problems have surfaced. Even in long-established organizations, questions may arise about how much ownership a writer has over a particular document, or how much authority the technical editor has over a given document, or how much value the SMEs and writers place on editing. Ultimately, though, the job tasks, responsibilities, levels of authority, and performance expectations derive from the culture of the organization. And because every organization is different, these change from one to another. So what should newly hired editors do if they have questions about the organization and their responsibilities? Ask a superior.

It is never too early when beginning a new job as an editor to ask about your responsibilities and your role in document production. (You probably want to ask this question during

People ask you for criticism, but they want only praise.
—Somerset Maugham

the job interview.) You will also want to know to what extent you are to be involved in any editorial project, including the timing of your involvement and the type of editing you are asked to perform. After working with the organization for some time, you may be able to negotiate or modify your role and help establish editorial policy and your authority. Remember, though, that you may encounter individuals who are not appreciative of your contributions, and you may need to remind them of your role and value to the organization. Just do it tactfully.

Figure 2.1 provides a general overview of the responsibilities and tasks performed by technical editors. However, remember that each company or organization is structured differently and will vary from our summation here.

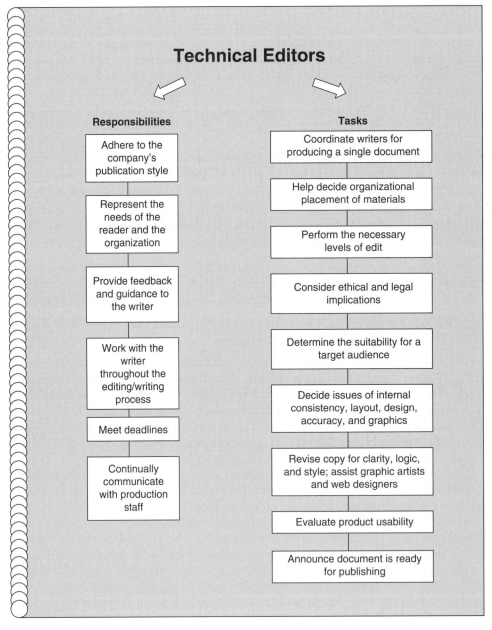

FIGURE 2.1
Technical Editor's Flowchart.

So, if you have never worked before as a technical editor, what can you expect? Regardless of the size and type of organization you work for, you can be certain to work with a wide variety of individuals with just as wide a range of egos. We have found that most people act professionally and desire to produce an excellent product. However, you can expect some individuals to become defensive or even combative when their work is criticized. Unfortunately, when this occurs, working relationships may become permanently damaged. Generally, editors must be sensitive to the tone and type of the feedback they give to the writer. This is where your emotional intelligence can mean the difference between achieving a harmonious, productive relationship or a resentful, acrimonious one.

Editors are in a unique position. Often they are the hub within the document design team wheel. They must work with a variety of individuals and are frequently called upon to assimilate everyone's input. By definition, then, editors must confront what they perceive to be lapses in a document and recommend changes. This may be something as simple as a misspelling or as serious as suggesting a completely different organizational approach. Regardless, recommending changes and persuading writers to make changes may be as much an art form as anything else.

> Writers respect you more when you admit to making a mistake than when you concoct weird reasons and justifications to explain your error.
> —*Author unknown*

The Editor-Writer Conference

Developing a friendly and professional working relationship with the writers and other members of the document design team is essential. Relationships may be established through email, the phone, snail mail, or fax; however, the most productive meetings usually occur when individuals meet face-to-face. Normally, review conferences to plan, discuss changes, establish goals, determine schedules, and assign responsibilities and tasks are routinely scheduled facets of the document design process. These conferences may be *preliminary edits, copy edits,* or *production edits.* Furthermore, these conferences may be between a single writer and technical editor, or they may incorporate the entire team of technical and editorial reviewers. Before such conferences that involve the entire production team take place, though, the writer and editor usually have communicated and resolved most minor editing issues. Larger issues, such as extensive editing, reorganization, changes in layout, or usability testing must be addressed and generally call for a full conference, one where everyone on the team can be present.

Conferences are meetings designed to "confer," not to "instruct." Editors should remember that one of their principal responsibilities is to act collegial, support the writer, and direct the discussions toward achieving the intended goal: user-friendly documentation. Using conference time to "instruct" on grammatical or mechanical issues makes the editor appear arrogant and condescending to everyone sitting around the conference table, not just the writer. Grammatical and mechanical issues should be resolved between the writer and editor before the review conference. Also, editors should avoid the temptation to bring in a heavily edited copy at this stage if possible. Resolve as many editing issues with the writer as you can before the conference and bring in a relatively clean, revised copy.

Remember, editors should not pose as barriers to publication but rather should function as resources to make the writer's work read better. You do not want to earn the reputation as an obstructionist because writers will simply figure out a way to go around you. Whether writers feel that having their work edited by you is like a dentist using a slow-speed drill that grinds their teeth or like an opportunity to completely redecorate their office depends on the rapport that evolves between you. As an editor, you should listen attentively while keeping an open mind regarding the writer's viewpoints. Yet you should also display confidence in your edits. Be sure your markup comments are constructive and reasonable. Most importantly, offer solutions that make the writer's work better.

████████
TIP FOR TECHNICAL EDITORS

████████
TIP FOR TECHNICAL EDITORS

Always practice professionalism by having a positive attitude toward everyone and by delivering work on time.

Editing the Manuscript

An editor can help the writer appear to know more, to look sharp in print. Surface editing gets rid of the distractions, while substantive changes can enhance the author's reputation. The main job of the editor, though, is guarding an author's credibility.
—*Carol Gerich*

Veteran writers will undoubtedly have stories about an "editor from hell." Their horror stories may be attributable to a really difficult editor, or when, the full truth be known, the writer may have been just as difficult. Regardless of what the writers have previously experienced, as a new editor you will have to prove yourself through your use of common sense, good judgment, and qualified editorial skills. You will need to let the writers know that you and they share a common goal. Too often, thoughtlessly worded markups returned to the writer create friction. However, learning a few rules of diplomacy can go a long way in establishing a mutually respectful working relationship.

Every editor is different when approaching a manuscript. Some approaches may seem unorthodox. For example, one editor may skim the beginning, then skip to the end, and then edit the figures and graphics in between to determine continuity. Another may take a more conventional and methodical approach by reading the manuscript linearly, examining the subtopics for organization and development, and then completing an edit for style and surface errors. Regardless of the approach, technical editors have specific preliminary responsibilities, again depending on the organization's culture. For example, if the manuscript is to be published in a scientific journal, the technical editor must first determine if the journal's criteria for submission have been met (e.g., APA style). Another preliminary aspect but one that might be a bit tricky is to certify authorship. That is, does the manuscript have coauthors and to what extent did each contribute? This is very important because the authors often will have to relinquish copyright permission to the journal. Figure 2.2 lists some questions to ask yourself before, during, and after editing a document.

Editors should also create a series of records that follow the manuscript throughout the production process. These should include an identification number, when the manuscript was received for review, when the review was completed, whether or not the manuscript was rejected, accepted, or returned to the writer for a revision, whether or not it passed a technical review, and ultimately, if and when the manuscript was published. Figure 2.3 shows a tracking table that Nicole used when she edited an anthology for the Global Student Entrepreneur Association. She updated the table as submissions arrived. The table served as record-keeping for Nicole, but she also emailed it weekly to Michael, the anthology's compiler, to keep him updated on the submission process.

Technical Editors as Diplomats

Nothing is more critical, sensitive, and potentially explosive than the actual feedback the technical editor gives the writer. Many writers have already had bad experiences with editors whose comments they felt were condescending, authoritarian, or arrogant. So, just how do you "criticize" someone else's work without raising their ire and causing their temper to flare? Can someone such as an editor interpose into the creative process of the writer and still maintain a cooperative, helpful working relationship?

Although many believe writers and editors by definition assume adversarial roles, effective interpersonal relationships can often be established when editors know how to phrase

Questions to Ask Yourself before Editing
- Is the document informative? persuasive? descriptive? a combination?
- What type of expertise, if any, does the reader have on this topic?
- Will my audience already have a position or point of view?
- How much background information do the readers already have regarding the topic?
- What information will be new to them?
- Are there any terms or procedures that may be difficult or require special graphics?
- Is the main point clearly established within the text?

Questions to Ask Yourself during Editing
- Do I need to add or omit information such as key steps in a procedure or definitions for specific terms?
- Is the information that I provide too technical or difficult for this audience; is the vocabulary appropriate?
- Has the intended purpose of the document been clearly established early so there is no confusion?
- Are there sufficient examples, facts, statistics, data, authoritative statements, tables, graphics, and other forms of evidence to support the document's claims?
- Is the information organized logically such as chronological or comparison contrast?
- Do the body paragraphs have topic sentences?
- Do the topic sentences reinforce the main idea of the document?
- Is coherency achieved through the use of transitions and repetition of key words?
- Are the sentences varied and of appropriate lengths?
- Does the writing seem wordy and overdone?
- Is there a good balance of graphics and illustrations?
- Does the document use sufficient headings to break up dense paragraphs?

Questions to Ask Yourself after Editing
- If the document has a wide variability in audience, have you made the necessary adjustment to provide supplemental information for those who may need it?
- After letting the document sit aside, does its readability still seem appropriate to the intended audience?
- Are the findings summarized?
- Does the document need an introductory abstract?
- Does the research exist within a context of related research, and is that context clear?
- Does the overall appearance of the document seem appealing?

FIGURE 2.2
Questions to Ask Yourself.

criticism and suggestions in nonthreatening language. Mackiewicz and Riley in "The Technical Editor as Diplomat: Linguistic Strategies for Balancing Clarity and Politeness" argue that when providing feedback, editors must be both clear and polite—clear in what the writer is obliged to change and polite in the manner the writer is told to make those changes. Editors who plan to be successful, then, must learn to be ambassadors of goodwill and willing to improve their interpersonal skills.

Authors	Bios	Story	Notes
Adrian Bold	Received 1/31	Received 3/02	Still awaiting revised chapter (due April 25)
Russell Hancock	Received 1/30	Received 4/1	Revised chapter received 4/22; still no "giving back" theme used
David Hauser	Received 1/25	Received 2/27	Received revised chapter 4/15
Kristina Kallur	Received 1/26	Received 1/13	Received revised chapter 3/29
Dustin Lindahl	Received 1/26		STILL WAITING FOR FIRST DRAFT OF CHAPTER!
William Palma, Jr.	Received 3/02	Received 2/7	*Done*
Molly Reiling	Received 1/31	Received 4/6	Just had baby; will send revisions by 5/07
Arnaldo Jose Rivera Henriquez	Received 4/7	Received 3/07	Received revised chapter 5/01
Justin Sanders	Received 2/1	Received 2/27	Received revised chapter 4/15
Michael Schoeff	Received 2/14	Received 3/18	*Done*
Michael Stebinger	Received 1/31	Received 2/28	Received revised chapter 4/22

FIGURE 2.3
Sample Record-Keeping Table.

Providing editorial feedback on a manuscript has two principal functions. First, it should provide helpful advice on how the writer could improve the manuscript from the reader's point of view. Second, it should initiate positive interpersonal relationships between everyone on the production team who has made a contribution to the manuscript, especially the writer and editor. Both objectives can be accomplished through a conscious awareness of how the language in which we frame our editorial comments affects those who have contributed to the manuscript.

Just how does written feedback affect the writer-editor relationship? That relationship has gained attention during the past decade; however, much more research needs to be conducted in this area. Despite the lack of extensive research, from our experience in the classroom and as technical editors, we believe there are many correlations between teaching and editing. In short, regardless of the particular situation, people's reaction to criticism of their work is affected by the type, quantity, and tone of the feedback they receive from those judging their efforts. As a technical editor, what you say to the writer is important, but *how you say it* is just as important.

> The very nature of the editor-writer relationship, coupled perhaps with negative associations about past editing experiences, may well put writers on the defensive even before they interact with an editor about their writing.
> —*Mackiewicz and Riley*

TIP FOR TECHNICAL EDITORS

Here are a few additional tips to apply when providing feedback. An editor's comments should be:

- *Specific and precise*, e.g., "Your manuscript clearly develops the topic in an easy-to-follow, direct style," not general, e.g., "Nice job, Bill."
- *Directed toward the manuscript*, e.g., "The description of how the blowout valve operates needs more details in this paragraph," not toward the individual, e.g., "Mary, you did a poor job describing how the blowout valve works."

- **_Informational_**, e.g., "Chapter 1 needs to include more background information on how fuel cells reduce energy costs," not emotionally charged, e.g., "No, don't include this paragraph," or "What were you thinking by including these graphs?"
- **_Useful by recommending solutions_**, e.g., "This sentence is in passive voice and needs an active verb," not cryptic, e.g., "This sentence is awkward."
- **_Honest and sincere_** while directed toward helping the writer create a more effective manuscript.

Remember to focus on the document, not the author. A team approach is much better than assuming an authoritative stance. And, if possible, construct your edits in the form of questions rather than statements that may sound accusatory.

THE LANGUAGE OF EDITING

If you are conducting a substantive edit, generally you will be expected to provide summary remarks either after each section, after each chapter, or at the end of the entire document. These comments may be as short as a few sentences, or they may be several pages in length, depending on the level of edit and the need for corrections.

How you present your comments, though, will often determine the future working relationship with that writer. So how can you provide the necessary criticism yet prevent animosity? How direct can you be with a writer, especially one with whom you have not established a rapport? Balancing directness with language that is courteous and respectful is difficult. For example, which form of feedback is better: "You _should_ create a graphic here," "You _could_ create a graphic here," or "Do you think a graphic placed here would add interest?" Notice the level of directness in each response.

Based on their extensive research, Fiona and Ken Hyland, and Mackiewicz and Riley offer several strategies for providing constructive editorial feedback while reducing the potential friction that can easily occur between those who sit in judgment and those being judged. These strategies include what they term as _paired patterns, hedges, bald-on-record, locution-derivable_, and _interrogative syntax_. Familiarize yourself with the various patterns of feedback presentation.

Strategy 1—Paired Patterns

Summary remarks may take the form of criticism, praise, suggestions, or a combination of the three. The most common patterns include _praise-criticism_ followed by the _criticism-suggestion_, and finally the _praise-criticism-suggestion/solution_ model. Each pattern has its advantages and disadvantages depending on how forceful the editor intends the criticism. Most importantly, though, each pattern contains some form of mitigation that assuages the criticism to some extent.

The _Praise-Criticism_ Pattern

       praise   _criticism_
The grammar and mechanics are well done, but the style of writing makes your key points difficult to follow.

       praise   _criticism_
The organization of the minor claims is logical; however, none of the minor claims is fully developed.

Avoid overusing this pattern, though, because continual praise might sound too insincere or contrived. Also, the emphasis on the criticism becomes subordinated to the praise and may be lost to the writer. Of course, you might reverse the order to criticism-praise, but this too can appear contrived if you are not careful.

Another type of paired pattern commonly used, the *criticism-suggestion/solution*, begins with a negative remark but becomes softened with a suggestion for correction. The criticism-suggestion can be effective because the need for correction is emphasized, and a possible solution is also provided. Notice that the emphasis is on the need for a correction and not directed toward the writer.

The *Criticism-Suggestion/Solution* Pattern

criticism

The reader will have trouble following the rambling style here.

suggestion/solution

Try to eliminate a lot of the verbiage that seems redundant.

criticism   *suggestion/solution*

The introduction begins too abruptly. I suggest you provide more background information for the reader.

Of the three paired patterns, the *praise-criticism-suggestion/solution* is generally the most likely to mitigate the criticism while encouraging changes.

The *Praise-Criticism-Suggestion/Solution* Pattern

praise

The comparison and contrast between widgets A and B are well done—no problem here.

criticism

But your discussion of how they operate using DC current instead of AC seems superfluous.

suggestion/solution

I suggest you leave that information out but elaborate further on why the reader needs to understand how they work.

praise

Presenting the solution before presenting the problem is an excellent idea for this topic.

criticism

However, I'm concerned that the problem is not clearly stated and the reader may miss your entire point.

suggestion/solution

Elaborate more on the problem and be sure to provide a good summary at the end.

EXERCISE 2.1

It is not often that technical editors have the opportunity to actually write fiction. However, here is an exercise that asks you to use your creative talents. Consider each of the three possible scenarios below. After thinking about writing and how you would react in each situation, write a 500- to 750-word dialogue of a plausible discussion between an editor and writer for each scenario. Feel free to incorporate some of the internal thoughts of the editor and writer as the dialogue develops. Be prepared to read your short narrative in class.

Scenario #1: If you are a perfectionist, how would you handle writers who hand you a document and say, "Just a light copyediting is needed on this," but you find substantive organizational and stylistic problems throughout the text? Perhaps the writer says, "Minor proofreading is good enough." How would you react?

Scenario #2: What would you tell a subject matter expert (SME) whose language is too sophisticated for the intended audience? Would you approach the SME diplomatically and offer positive suggestions to change his or her erudite vocabulary?

Scenario #3: Perhaps your role as a technical editor has expanded to include graphics editor. The graphic designer hands you a full-color series of graphics that are beautiful works of art but in your judgment are superfluous, too costly, and do not correspond with the text. What do you say or do?

As we discussed earlier, the team paradigm requiring interaction among coworkers on different levels of production is rapidly replacing the lockstep, systematic model. That is, interactive and cooperative approaches to achieving goals are replacing the static, compartmentalized approaches. In these new corporate cultures, how team members work together toward achieving the desired goal determines the quality of the finished product. Now read the scenario (in Figure 2.4) and, given your particular temperament, consider how you would respond.

As a recent graduate of Mossy Rock State University, you landed your dream job working for the Department of Homeland Security. The starting salary and benefits are good. Sure, you are starting out as a copy editor with minimal responsibilities, but you know that with some hard work, you will be able to advance rapidly to developmental editor before too long.

After a few weeks on the job, a SME with a Ph.D. and seniority comes to you and says, "I need a light editing on this." She plops down a thirty-page document on your desk. "These are the guidelines I've developed for the local port authorities to follow in case of a terrorist attack. Just a quick once-over should do it. I'd like to get it out by the end of the week." She turns and walks out of your cubicle.

Later that afternoon, you direct your attention to the document. Eager to show that you are a team player, you decide to get the document out before week's end. However, as you begin to read it, you discover the document is full of jargon and "bureaucratese." Replete with profundity and long sentences, it is barely readable. Certainly, the local port authorities will have great difficulty understanding the security procedures. They may simply give up and not read past the first page. You know the writer has a reputation for being somewhat arrogant and believes her writing is beyond reproach. What do you do?

FIGURE 2.4
Sample Editor-Writer Scenario.

Should you ...

A. Copyedit the entire document for grammar and punctuation errors and simply return the document to the writer? (After all, this is what you were hired to do.)

B. Only copyedit certain pages, make positive notations on improving the writer's style, and personally visit (or at least phone) the SME with your suggestions for improvement before going further?

C. Copyedit the entire document but also make the substantive changes necessary for readability, talk to a coworker about supporting your position, and then approach the writer?

D. Ask the developmental editor what to do?

If you're somewhat impulsive, your first inclination might be to slash through all the stylistic gobbledygook and insert language more suitable for the intended audience. Or you might want to immediately run to the developmental editor and complain about the writer's bad writing. If you are conservative and believe your best choice is to remain a team player, you might rationalize just to do your job and return the document to the writer after a light edit.

Although there are many possible courses of action, we would first remind you that your editorial authority depends on the culture of the organization where you work. The culture particular to any organization evolves from its traditions and management style. An older company, for example, may maintain a more formal, systematic approach where management makes decisions from the top. New businesses, though, may be more democratic, where teamwork is encouraged and the employees make most of the decisions themselves. So, depending on their *corporate culture*, some organizations hold editors in high regard and grant them considerable authority, even over writers. Other organizations do not. As an editor, though, and regardless of the corporate culture, your primary responsibilities include helping writers communicate information effectively and efficiently to their audience.

Given this context, we recommend that you first personally approach the writer with your concerns using positive, helpful suggestions for improving the style. You might even admit to her that although you are still in the learning phase and that you were asked only to copyedit, you had some additional suggestions you thought could make the document more effective, and you wanted to get her opinion regarding those suggestions. Generally, most people respond positively to individuals who approach them in a nonthreatening way and are sincere about wanting to help. The writer, given the chance, would probably consider your suggestions.

Perhaps, though, the rumors you have heard about this writer are true and after your honest attempt to make positive suggestions fail, you decide to return to your desk and keep your mouth shut. Often this may be the best course of action, depending on the organization's culture, and especially if you are a new employee; however, look again at what is at stake. If it's national security at local ports throughout the United States, you have an additional *ethical* responsibility that overrides all other considerations; homeland security procedures must be clear if they are to be followed. (Chapter 20 discusses ethics in detail.) So if you believe that any document, report, manual, instruction sheet, brochure, and so forth in any way jeopardizes people's lives or health, you must address the issue, even if it means "rocking the boat." Of course, many of the documents you edit may not have such dire consequences; however, always consider the possible ramifications of someone misinterpreting a document you have helped produce. Even if people's lives are not at risk, the organization's as well as your own reputation is at stake.

EXERCISE 2.2

Assume the role of copy editor. Create three of your own examples of feedback for each of the three types of feedback patterns. Consider experimenting with each pattern by varying the order.

Strategy 2—Hedges

Hyland and Hyland offer another mitigation strategy to help maintain a positive relationship with the writer, not necessarily by diminishing the degree of criticism, but by making the criticism seem less disparaging. This strategy calls for *hedges*, which often are indefinite quantifiers such as "occasionally," "often," "a little," "sometimes," "perhaps," and so on. Again, you must judge how direct you want to be and whether hedges should be used to soften the criticism. Also, notice how hedges can be applied to the *feedback pattern* strategy.

> *The first three paragraphs are focused and well developed; however, the fourth and fifth paragraphs are **slightly** (hedge) too long.*
>
> *The summary after chapter six **seems a bit** weak.*
>
> *You **might consider** changing these passive verbs to active.*
>
> *The graphics are too complex for the reader to understand. A **little** less emphasis on the line-diagrams **may** make the graphics workable, though.*

Strategy 3—Bald-on-Record

Mackiewicz and Riley argue that technical editors must function as diplomats by learning when to use direct and indirect strategies to convey feedback. For example, the bald-on-record strategy contains a directive to the writer such as "Insert this table here." This unmitigated imperative clearly states the writer's obligation. Interestingly, some international writers (e.g., Taiwanese) prefer this direct approach simply because it is clear. Others, though, may prefer a more polite approach. In these cases, the bald-on-record strategy might incorporate a *downgrader*—a word or phrase that softens the imperative. Remember, this approach is the most direct and clear, and possesses the potential to appear overly critical if used too often.

> ***bald-on-record***
> *Add another example to complete this section.*

> ***bald-on-record***   ***downgrader***
> *Add another example to complete this section, all right?*

> ***bald-on-record***
> *Eliminate the first two paragraphs.*

> ***bald-on-record***   ***downgrader***
> *Eliminate the first two paragraphs. Don't you agree they do nothing to further the manuscript?*

EXERCISE 2.3

Test your ability to write good effective feedback. Compose at least three statements using hedge strategy and three using the bald-on-record strategy. Exchange your work with another student and compare your answers. After you complete these sentences and receive peer feedback, continue by writing locution-derivable and interrogatives. Exchange answers again with another student.

Strategy 4—Locution-Derivable

This strategy is nearly as direct as the bald-on-record; however, the locution (the directive to be performed) is contained in the modal verb that expresses the writer's obligation. For example, these verbs may include *must, should, ought, will.* Also, these verbs can be expressed either in the active or passive voice, depending on the degree of directness desired.

> *You **ought** to include an additional data table here.* (active voice)
>
> *You **should** provide a glossary so your readers can define the terms.* (active voice)
>
> *A glossary **should** be included before the appendix.* (passive voice)

Notice the difference between the active and passive voice. While both examples of locution-derivables are direct, the passive voice removes the agent *you* and may be considered slightly less direct. However, editors should be careful when using the passive voice; it may seem a bit more polite but may be more difficult for international readers to understand.

Strategy 5—Interrogative Syntax

One of the most common strategies editors employ to mitigate their feedback is to ask a question. Using questions to suggest editorial changes interwoven with more direct strategies such as the bald-on-record provide relief and variety to the markup. Notice, though, that questions are much less direct and do not initiate an obligation on the writer.

> *What do you think about leaving out this paragraph?*
>
> *Could you **possibly** leave out this paragraph?* (question with a downgrader)
>
> *How can we better arrange this table for the reader?*

Choosing which of these strategies to use with a particular writer will take some time, depending on how quickly you develop a relationship with the writer. Initially, we recommend being polite and diplomatic by maintaining good linguistic manners. Some writers will not have a problem with a direct approach, whereas others may be more defensive, and in some cases downright hostile. Just try to remember that an editor's goal should be the same as the writer's—to develop a good manuscript. Keep the feedback goal-oriented and as positive as possible. Also, be sure that your schedule leaves enough time to provide a thorough edit if called for. Keep in contact with the writer throughout the editing process and be adaptable.

Finally, always be positive, display confidence in your recommendations, and encourage the writer's confidence as well.

Other Strategies to Consider

Providing the writer with good, positive feedback in a timely manner is essential but is not the only aspect of maintaining a professional working relationship. Notice that these strategies can and should be used in conjunction with the "Questions to Ask Yourself" figure earlier in this chapter.

Strategies before the Editing Process

- Get involved as early as possible in the project.
- Determine deadlines, level of edit, and your role.
- Help the writer conduct an audience analysis.
- Discuss the organizational structure of all documents by creating a working table of contents.
- Define specific requirements or policies you have regarding how the documents will be edited.
- Discuss the readability level appropriate to the target audience.

Strategies during the Editing Process

- Provide feedback that is constructive and diplomatic.
- Provide solutions for improvement.
- Edit from the perspective of the reader and report your reactions to the writer.
- Meet only as frequently with the writer as necessary to ensure a quality product.
- Make suggestions and offer solutions but don't rewrite.
- Listen actively.
- Be accountable for your decisions and willingly explain any markups if asked.
- Focus on cooperation and not confrontation.

Strategies after the Editing Process

- Provide instruction or help for those who need it.
- Keep abreast of trends and issues in technical writing and editing.
- Become a knowledgeable resource.

Despite all of these rhetorical strategies, though, every once in a while you may encounter a particularly grumpy author, one who takes great umbrage at any corrections. If none of the above strategies work, including providing documentation that establishes your editorial suggestion as desirable or "correct," then you as editor may need to contact someone in management, especially if the author insists on promoting inferior documentation. However, it is best to try to work it out with the author as much as possible before running to management to complain about a coworker.

EXERCISE 2.4

Examine the document in Figure 2.5. Do you feel the style, length of sentences, and word choice are appropriate? Are the paragraphs too long? Review the five strategies discussed in the "Language of Editing" section above and edit the document using diplomatic and constructive remarks in the margins to address these issues. You may want to compare your edit to a classmate's.

---

1. INTRODUCTION

1.1. PURPOSE AND SCOPE OF THE GUIDELINES

These guidelines revise and replace the U.S. Environmental Protection Agency's (EPA's, or the Agency's) *Guidelines for Carcinogen Risk Assessment,* published in 51 FR 33992, September 24, 1986 (U.S. EPA, 1986a) and the 1999 interim final guidelines (U.S. EPA, 1999a; see U.S. EPA 2001b). They provide EPA staff with guidance for developing and using risk assessments. They also provide basic information to the public about the Agency's risk assessment methods.

These cancer guidelines are used with other risk assessment guidelines, such as the *Guidelines for Mutagenicity Risk Assessment* (U.S. EPA, 1986b) and the *Guidelines for Exposure Assessment* (U.S. EPA, 1992a). Consideration of other Agency guidance documents is also important in assessing cancer risks where procedures for evaluating specific target organ effects have been developed (e.g., assessment of thyroid follicular cell tumors, U.S. EPA, 1998a). All of EPA's guidelines should be consulted when conducting a risk assessment in order to ensure that information from studies on carcinogenesis and other health effects are considered together in the overall characterization of risk. This is particularly true in the case in which a precursor effect for a tumor is also a precursor or endpoint of other health effects or when there is a concern for a particular susceptible life-stage for which the Agency has developed guidance, for example, *Guidelines for Developmental Toxicity Risk Assessment* (U.S. EPA, 1991a). The developmental guidelines discuss hazards to children that may result from exposures during preconception and prenatal or postnatal development to sexual maturity. Similar guidelines exist for reproductive toxicant risk assessments (U.S. EPA, 1996a) and for neurotoxicity risk assessment (U.S. EPA, 1998b). The overall characterization of risk is conducted within the context of broader policies and guidance such as Executive Order 13045, "Protection of Children From Environmental Health Risks and Safety Risks" (Executive Order 13045, 1997) which is the primary directive to federal agencies and departments to identify and assess environmental health risks and safety risks that may disproportionately affect children.

The cancer guidelines encourage both consistency in the procedures that support scientific components of Agency decision making and flexibility to allow incorporation of innovations and contemporaneous scientific concepts. In balancing these goals, the Agency relies on established scientific peer review processes (U.S. EPA, 2000a; OMB 2004). The cancer guidelines incorporate basic principles and science policies based on evaluation of the currently available information. The Agency intends to revise these cancer guidelines when substantial changes are necessary. As more information about carcinogenesis develops, the need may arise to make appropriate changes in risk assessment guidance. In the interim, the Agency intends to issue special reports, after appropriate peer review, to supplement and update guidance on single topics (e.g., U.S. EPA, 1991b). One such guidance document, *Supplemental Guidance for Assessing Susceptibility from Early-Life Exposure to Carcinogens* ("Supplemental Guidance"), was developed in conjunction with these cancer guidelines (U.S. EPA., 2005). Because both the methodology and the data in the Supplemental Guidance (see Section 1.3.6) are expected to evolve more

---

FIGURE 2.5

Guidelines for Carcinogen Risk Assessment.

Source: Environmental Protection Agency.

rapidly than the issues addressed in these cancer guidelines, the two were developed as separate documents. The Supplemental Guidance, however, as well as any other relevant (including subsequent) guidance documents, should be considered along with these cancer guidelines as risk assessments for carcinogens are generated. The use of supplemental guidance, such as the Supplemental Guidance for Assessing Cancer Susceptibility from Early-life Exposure to Carcinogens, has the advantage of allowing the Supplemental Guidance to be modified as more data become available. Thus, the consideration of new, peer-reviewed scientific understanding and data in an assessment can always be consistent with the purposes of these cancer guidelines.

These cancer guidelines are intended as guidance only. They do not establish any substantive "rules" under the Administrative Procedure Act or any other law and have no binding effect on EPA or any regulated entity, but instead represent a non-binding statement of policy. EPA believes that the cancer guidelines represent a sound and up-to-date approach to cancer risk assessment, and the cancer guidelines enhance the application of the best available science in EPA's risk assessments. However, EPA cancer risk assessments may be conducted differently than envisioned in the cancer guidelines for many reasons, including (but not limited to) new information, new scientific understanding, or new science policy judgment. The science of risk assessment continues to develop rapidly, and specific components of the cancer guidelines may become outdated or may otherwise require modification in individual settings. Use of the cancer guidelines in future risk assessments will be based on decisions by EPA that the approaches are suitable and appropriate in the context of those particular risk assessments. These judgments will be tested through peer review, and risk assessments will be modified to use different approaches if appropriate.

FIGURE 2.5 (*continued*)   Guidelines for Carcinogen Risk Assessment.
Source: Environmental Protection Agency.

EXERCISE 2.5

Identify the purpose of the document in Figure 2.6, and then determine the potential audience. Do you feel that the document can be edited, or should it be sent back to the writer for a complete revision? Write some explanatory notes in the document's margins explaining how the writing might be improved. Use the strategies we have previously suggested.

Now completely change your position and write a short but tactful paragraph explaining to the author why this document is unacceptable and needs a complete revision. Try not to be offensive, but make it clear the document needs to be clearer for the general public. You might want to edit one or two paragraphs for the author as an example.

The Navy took the lead in identifying and examining potential problems associated with the use of this wire insulation type and in mitigating hazards. In the mid-1980s, when the Navy began experiencing problems with this wire insulation, it enlisted the support of experts from other military services and FAA to further characterize the problems and identify possible solutions. Researchers determined that prolonged exposure of this type of wire insulation to moisture could cause it to deteriorate and that it was susceptible to arc tracking. Arc tracking can occur when two cracks in the insulation are close

FIGURE 2.6
In-flight Fires Traced to Aging Wiring.
Source: Government Accountability Office.

enough together to allow the current to form a conductive path between them at temperatures that can cause the insulation to char and carbonize. This carbonization can turn the insulation into an electrical conductor, and, eventually, can trip a circuit breaker. When a pilot presses the switch to reset a tripped circuit breaker, an entire wire bundle can be disabled and potentially compromise the safety of an aircraft's entire electrical system.

Ultimately, the Navy and the Coast Guard took the most active measures to address potential problems with aromatic polyimide, which some experts attribute to these entities' unique aircraft operations near water. In December 1985, the Navy decided that aromatic polyimide would no longer be its wiring insulator of choice and subsequently removed it selectively from parts of aircraft where it was most problematic (e.g., fore and aft flaps, wheel wells, and around unsecured seals that could leak). The Coast Guard lagged behind the Navy in taking action to address problems with this wire insulation; however, it took the most extensive action by stripping it from its largest fleet of helicopters as a precautionary measure after occurrences of in-flight fires and cockpit smoke and fumes between 1993 and 1996. While no aircraft were destroyed, these incidents led to poor visibility in the cockpit and, in some cases, the loss of all electrically powered flight instruments. A senior Coast Guard safety official said that the Coast Guard completed removal of this wire insulation from its entire fleet of H-65 helicopters in September 2001. In contrast, the Army did not experience similar safety problems. While it independently confirmed the Navy's findings in 1986, the Army concluded that it did not have the same problems with aromatic polyimide. The Army did have durability concerns, however; it found the degree to which aromatic polyimide chafes in Apache and Blackhawk helicopters is unacceptable over time and decided to remove it gradually as it refurbished older aircraft.

In response to the Navy's findings of potential hazards with the use of aromatic polyimide, FAA conducted independent research, tracked related research and operational data from industry and the military services, and decided that mandating the removal of this wire insulation from commercial aircraft was not warranted. However, FAA did issue three Advisory Circulars related to the use of this wire insulation type, in 1987, 1991, and 1998, to provide policy guidance to help prevent electrical problems and potential fires and to describe acceptable practices for aircraft inspection and repair, including wire installation.

Recognizing the need for sustained attention to aircraft wiring issues, FAA has ongoing efforts to assess the health of wire in aging aircraft through its Aging Transport Systems Rulemaking Advisory Committee (ATSRAC). To date, working groups under ATSRAC have conducted visual (nonintrusive) and extensive physical (intrusive) inspections of wiring on aging aircraft. However, according to National Transportation Safety Board (NTSB) officials, it is too soon to determine how well the agency is doing in its assessment. These officials pointed out that ATSRAC has a seven-step objective of reviewing wiring in aging aircraft, and its recent intrusive inspection is only one step in the process.

In August 2001, FAA announced a new initiative, the Enhanced Airworthiness Program for Airplane Systems (EAPAS), a cooperative effort with industry that is intended to (1) enhance the safety of aircraft wiring from design and installation through retirement, (2) increase awareness of wiring degradation, (3) implement better procedures for wiring maintenance and design, and (4) ensure that the aviation community is informed. In the same month, the Transportation Safety Board of Canada announced that, as a result of its investigation of the crash of Swissair flight 111, it (1) concluded wire failure can play an active role in fire initiation and (2) recommended a more stringent certification test regime. FAA officials told us that the agency has not yet responded to the conclusions and recommendations of the Transportation Safety Board of Canada.

FIGURE 2.6 (*continued*)   In-flight Fires Traced to Aging Wiring.
Source: Government Accountability Office.

SUMMARY

Anyone unfamiliar with the recent transformations taking place in the field of technical editing is often amazed to learn of the broad range of skills required of technical editors. Although good communication skills continue to be essential, today that is only the beginning.

One aspect about this field that we have tried to emphasize is the need to establish strong relationships between editors and writers and to understand that their goals are essentially the same. Good editors should be able to adapt to new and demanding situations, continue to acquire new skills to remain current in the field, and, most importantly, possess the emotional intelligence to work with others.

Tools and Technology

At the conclusion of this chapter, you will be able to

- discuss the benefits and drawbacks of hard-copy versus soft-copy editing;
- use editing symbols for marking hard copy;
- use some of the tools in a word processor to help you edit more efficiently and effectively;
- describe the importance of styles and the significance of different file formats.

Effective technical editors have excellent people skills; employ a robust knowledge of grammar, style, organization, and logic; use proficient copymarking skills; are familiar with technical editing tools and software; and understand the products or policies described in the documents they edit. Chapter 2 establishes the polite yet authoritative approach to take with writers and subject matter experts (SMEs). Units Two and Four address grammar and content issues in editing a document. This chapter covers technical editing tools and software as well as basic *copymarking*: the pencil strokes you'll use to mark up a text on paper to show suggested changes.

Two important skills to acquire as a technical editor in the twenty-first century are effective hard-copy *and* soft-copy editing. *Hard copy* refers to documents in printed form, whereas *soft copy* refers to the computer files from which documents (whether published in print or online) are produced. A printed book (hard copy) exists as a file (soft copy) somewhere, and a web page (soft copy) can be printed out (hard copy). Therefore, the form in which a document will be published does not always correspond to the form in which it is edited. An online text can be marked up on paper (hard copy), and a text that will appear only in print can be marked up in soft copy. In many cases, documents are edited and marked up in *both* hard and soft copy at different stages in their development.

One thing is for certain: electronic editing is here to stay. The rush for technology—from cell phone cameras to portable MP3 players to multiple computers and wireless networks in residential homes—has affected the editing world, particularly technical editing. In addition to textual documents, soft-copy editing includes web pages, graphics, and online help.

We have organized our tools and technology discussion based on established tradition: what has been the standard is mentioned first, and newer tools are discussed later in the chapter.

MARKUP

The term "markup" refers to two types of indicators: *change* indicators and *structural* (also called *layout*) indicators. Both types of markup may be used at the same or different times in the document production cycle, and both can be performed either on hard copy (on paper, by hand) or in soft copy. Editors generally use change indicators to communicate with authors, and structural or layout indicators to communicate with people doing the final layout of a document for publication.

Examples of change indicators include most copyediting marks, such as correction of spelling, punctuation, and grammar; questions to the author to clarify points; suggestions for reorganizing material; and any other items related to the content of the document.

Examples of structural indicators include marking heading levels; marking paragraphs to be set as quotations, callouts, examples, lists, table contents, or other elements where the indentation, fonts, borders, or other attributes vary from normal text; placement of graphics; and any other items related to the structure of the document, which at some point affects its visual appearance.

Most documents today are developed in electronic form and can be output into different media for publication and distribution. For example, documents may be placed on a website in HTML as well as printed or distributed as PDFs. Any document that may be reused (in whole or in part) should be marked up electronically to show its structure; when markup is properly done, a variety of computer programs can interpret and translate it as needed for presenting the information visually. Structural markup makes extensive use of *styles*. (See the section on the importance of styles later in this chapter.)

In addition, many documents are stored in databases from which portions of the document (for example, the title, the headings, the abstract, the summary) can be extracted for use in ways other than as a full individual document. Again, structural markup is essential for this type of retrieval and reuse to be successful.

HARD-COPY VERSUS SOFT-COPY EDITING

Most technical editors today edit electronically: some for the ease of use, others because their job requires electronic editing. Research conducted more than five years ago frequently refers to resistance from editors related to problems with the hardware and software of the previous ten or fifteen years. For example, in "Electronic Editing in Technical Communication: The Compelling Logics of Local Contexts," David Dayton explains that some editors still find editing on paper the easiest, quickest, and most thorough means of catching errors. Advocates of hard-copy editing also assert that editing on the computer causes eyestrain and other injuries or inconveniences.

According to Dayton's study, electronic editing is growing among technical editors, but technology companies have been slow to create user-friendly programs for technical editors and writers. In an earlier article titled "Technical Editing Online: The Quest for Transparent Technology," Dayton summarizes a study by David Farkas about software designed for electronic writing and editing that failed because electronically encoding the copymarkers was difficult and time-consuming for the technical editor. Newer technologies are making the process more user-friendly, although problems remain in some cases, such as when editing web pages. Students can expect to see major developments in editing technologies over the next few years.

Even we as writers have different editing preferences: Nicole prefers electronic editing, whereas Barry prefers hard-copy editing, and Jean prefers a mixture of both.

For many years, Nicole did all of her editing on hard copy using copymarking symbols, then submitted the document in person, faxed it, or sent the manuscript by mail or courier to the writers. This process was expensive for the writers as well as time-consuming for Nicole as an editor. Then, in the 1990s, she moved to electronic editing because it was faster and cheaper for the writers, although it was still slow for Nicole because she was still editing hard copy, transferring the hard-copy edits to the screen, and then emailing the changes to the writer. Today, she is used to reading documents online, so she edits almost all documents electronically using Microsoft Word's Track Changes feature (discussed later in this chapter). Because most technical editing students today study by reading texts electronically, Nicole believes that the next generation of editors will be much more accustomed to writing, reading, and editing electronically.

Barry prefers to edit hard copy for reasons of convenience and accuracy. He finds it much easier to bring paper with him to edit as opposed to editing on a laptop screen. Many people argue that paper is simply more portable than a computer. Computers can't be used during a portion of air time when taking a trip by plane and don't neatly fit in a pocket. However, all of this is changing with the influx of new hardware, and soon documents on a computer may be just as portable as paper ones.

Jean uses a mixture of electronic and hard-copy editing, but she sends all of her edits to her clients and collaborators in soft copy. (Jean lives in Australia and works on projects with people scattered around the globe; it's impractical and expensive for her to send hard copy.) She uses the computer to do repetitive tasks that she can partially automate because the computer can find many errors that she might miss when reading the document. In addition, she often prints out documents to read on paper because she finds words on paper easier to read than words onscreen, and she spots some errors more easily on paper. After reading and marking up a paper copy, she then transfers her edits to the electronic document. Jean doesn't find this method inefficient because her scribbles on hard copy are often just notes to herself of things to be corrected in soft copy, and she finds that the final result is much better than she could achieve using only one method or the other.

> The most corrected copies are commonly the least correct.
> —*Francis Bacon*

Benefits of Hard-Copy Editing

Portability

As we mentioned earlier, sometimes being able to bring your document with you is the most important element of your editing process. As an editor, you can hard-copy edit a document almost anywhere: while sitting in an airport, waiting in a doctor's office, or riding the subway, train, or ferry to and from work. Modern laptops are becoming smaller and lighter, decreasing the inconvenience of lugging around an extra piece of equipment when you're running to make your connecting flight or train, but many people still find paper more portable.

Ergonomics

Engineers have tried to make using computers more comfortable through *ergonomics,* or design factors to maximize productivity by minimizing operator fatigue and discomfort.

But let's face it: working on a PC for long periods of time can cause eyestrain, fatigue, as well as stiffness and pain in the arms, hands, neck, and back. As the editors in the Dayton study mentioned, editing hard copy eliminates many of the physical discomforts of working on a computer.

Speed

Those editors accustomed to hard-copy editing argue that it is faster than electronic editing, especially if they need to use copymarking symbols instead of a word processor's change tracking markup.

Quality

Advocates of hard-copy editing in Dayton's study contend that it allows the editor to see more errors. While this accuracy based on editing medium is still out for debate, some argue that a hard copy of a document shows the document as a whole, making it easier to find all the mistakes.

Drawbacks of Hard-Copy Editing

Handwriting

One of the biggest complaints about hard-copy editing quality is penmanship. Editors with poor or flamboyant handwriting make copymarking symbols almost impossible to decipher. Some use colored pens and pencils that are difficult to read. It is important to edit using neat and clear copymarking symbols and to write your comments clearly. Use a sharp pencil, write with a strong stroke, and make sure that the lead isn't too faint to be visible on the page.

Ergonomics

Although working long hours on a computer can cause problems for some people, writing on paper has its own problems. Poor desk height and poor lighting may cause back strain or eyestrain, and some people (Jean is one) get more hand, arm, and wrist discomfort from handwriting than from typing.

Clutter

There's always something new by looking at the same thing over and over.
—*John Updike*

Copymarking symbols are typically placed within the lines of the text. Editorial comments are usually listed in the left-hand margin. Although many hard-copy texts tend to be double-spaced with one-inch margins so as to leave enough room for the editor, sometimes the space just isn't enough. Every other word in a sentence may require copyediting, or a single word may need three copymarking symbols. Squeezing in all of the symbols in such a small space makes it very difficult to read. These symbols become particularly difficult to read if the document has to be faxed to its recipient.

Comments in the margin also create clutter, which is why some editors use sticky notes for extensive comments. But sticky notes may fall off or, if they stay put, clutter the page, and removing them to read the text often means wasted time trying to correspond the comment with the appropriate word/line/paragraph. Increasing the font size, triple-spacing the text, and creating two-inch margins around the text are some ways to create more room for hard-copy editing.

Time

Although the editors in David Dayton's study mentioned that hard-copy editing was quicker for them than electronic editing, some editors who are fast typists find hard-copy editing drudgery. Also, if you as the editor are required to submit the document via email, you may have to eventually enter the hard-copy edit into an electronic format, which makes hard-copy editing a time-consuming step. Finally, documents that are edited on hard copy must be submitted, faxed, mailed, or couriered to the recipients, whereas an electronic document can be emailed in a matter of seconds.

Benefits of Soft-Copy Editing

Portability

Many people find a laptop computer more portable and convenient than a large volume of paper. For people like Jean who often travel for extended periods (several weeks at a time), the bulk and weight of thousands of pages of manuscript can be a major problem in her luggage; her travel computer weighs less and takes up far less space. Jean also finds that in dim lighting conditions, she can read a document onscreen much more easily than on paper—and she can make the type size larger, and thus easier to read, as well.

Speed

Many editors can type their changes and comments faster than they can write them by hand. Many repetitive, time-consuming tasks can be semi-automated and done must faster (and with more accuracy) than when relying on a human to see each individual error. Editors who soft-copy edit also have the advantage of saving successive versions, which is a timesaver.

Quality

The biggest advantage of editing in soft copy is the ability to use the word processor's built-in tools to do some of the repetitive, time-consuming tasks that can be difficult to do with complete accuracy when relying on the eyes of an editor. By combining what a computer does best with what a human editor does best, you can improve your effectiveness as well as your efficiency. Some of the most useful features for an editor are listed below. We'll look at each of these in turn later in this chapter.

- Spelling checker, combined with custom dictionaries
- Grammar checker
- Find and Replace
- Automatic table of contents

- Automatic cross-references
- Outlining
- Styles
- Change tracking

Change Tracking

Using a word processor's change tracking feature eliminates the need for copymarking symbols that writers may not understand or handwriting that is hard to read. Change tracking has other advantages as well: reducing clutter on the page and minimizing double handling of changes.

Clutter Reduction

If editors use the word processor's change tracking feature, writers and editors can switch between viewing the changes and viewing the result of the changes. Many authors like to read the edited version without seeing all the changes yet still be able to view the changes if they wish.

Minimizing Double Handling

If change tracking has been used, authors can simply accept or reject the editorial changes; they don't need to retype them and potentially make more mistakes while doing so. This can save authors hours of work on a long manuscript. They can then concentrate on editorial questions and other matters of substance rather than copyediting corrections.

Drawbacks of Soft-Copy Editing

Ergonomics

As mentioned earlier, ergonomics is often a key issue for editors who prefer hard-copy over soft-copy editing. Staring at a monitor for a length of time can cause eyestrain and fatigue, potentially leading to poor editing.

Quality

Most editors find that reading onscreen makes seeing the "big picture" more difficult. Comparing one part of a manuscript to another is generally much easier to do if you can put the two pages side by side. It's possible to do this in most word processors but generally less convenient than with hard copy.

Writer Resistance

Some writers are uncomfortable working with electronically edited copy, although this problem is less common among technical writers than others. Some professional technical writers do not want an editor working on their files—even on a copy of the file.

Incompatible Software

You may not have a copy of the software used to create the document you are editing. Importing a document into a different program—or even into a different version of the same program (e.g., Microsoft Word)—can cause problems, although that depends on the programs involved and the complexity of the manuscript. Unit Five specifically discusses how to edit online documents such as web pages.

HARD-COPY EDITING: COPYMARKING SYMBOLS

Fortunately, the rules for editing on paper are fairly standard. Most editors use the same copymarking symbols found on the front or back cover of any hardback dictionary. Some companies, however, have a "house style," so they use their own copymarks.

Whatever system is used, the copymarks are probably some extension of the symbols in Figure 3.1. Editors and others use these symbols to indicate text changes. The symbols are universally understood among writers and editors.

The writers or companies you work with may have this same copymarking table or one with slight variations and some additions, depending on what type of documents you are editing. If you work predominantly as a freelance editor, you will have to conform to the company's house style. If you work with independent writers using hard-copy markup, you can include your own copymarking table on top of the document that you submit to the writers. No matter what symbols you decide upon, though, you will also need to determine how detailed you want each edited page to be. For example, some editors will correct errors in the text using the editing symbol and then write the symbol again in the margin (usually the left margin) so that the author doesn't miss the correction. However, others find this redundant as it makes for a very messy page when there are numerous errors to be corrected in the text.

At first, the copymarking symbols in Figure 3.1 may seem illogical to you. For example, you might wonder why there are two lines for letters that need to be capitalized and a slash for those that need to be lowercased, when it seems more "logical" to you to write "caps" in the margin, just as you may have seen your English professors do on your term papers. However, "caps" does not tell the author if you want lowercase or uppercase letters. It also doesn't indicate whether you want the entire word or just the first letter of the word capitalized or lowercased.

> To finish is both a relief and a release from an extraordinarily pleasant prison.
> —Robert Burchfield, on completing *The Oxford English Dictionary*

To compensate for some of the ambiguous marks, editors may use words or abbreviations in the margin to help the writer understand the suggested edit. Again, if your author is not familiar with your editing symbols, you may want to include your editing symbol guide at the top of the document. Finally, please note that the copymarking symbols will change depending on the company you work for.

Copymarking symbols constitute technical editors' *discourse* with other editors, designers, users, and writers. It is helpful to think of this conversation almost as a foreign language. Again, what seems "logical" at first is not always the best means of communication. When some native English speakers first learn a foreign language like Spanish, their instinct may be to simply add a vowel to the American term: "house" becomes *el houso*; "toilet" becomes *el toileto,* etc. This approach is severely incorrect, and no Spanish-speaking person would understand you because "house" in Spanish is *la casa* and "toilet" is *el baño*. This co-optation of Spanish—referred to in English slang as Spanglish—can create some pretty embarrassing translation issues, such as confusing the Spanish infinitive *embarazar* as "to embarrass" when it actually means "to impregnate." (We discuss global editing and translation issues in Unit Six.)

> Many learned persons have read themselves stupid.
> —*Arthur Schopenhauer*

Symbol	Explanation of Meaning	Example of Use
^	Caret: Insert letters or words at point indicated	Six^point^elements
ℰ	Delete: Eliminate words or punctuation indicated	Temperature (°C)
⌒	Close up, no space	mock⌒up
⌣	Delete and close up	non⌣destructible
stet	Stet: Latin for "Let it stand." Restore elements deleted, or retain elements indicated	The power was
#	Insert space	U#O Mo
∼	Transpose	hTe increase of power
/	Make lower case	Sodium Reactor
=	Make upper case (capitalize)	Janet greyston
//	Initial caps	INITIAL CAPS
ITAL	Italicize (underscope in typewriter composition)	an off position ITAL
⊣	Move right to the position indicated	The final design ⊣
⊢	Move left to the position indicated	⊢There has been
⊓	Raise to the position indicated	the word
⊔	Lower to the position indicated	the word
¶	New paragraph	Completed. Work has
⊙	Insert period	test loop. One run has
ℐ	Move to point indicated	to determine
○	Write out or abbreviate	3 6-in pipes
][	Center	]Heading[
∨	Superscript	106/255U
∧	Subscript	H2O

FIGURE 3.1

Copymarking Symbols for Hard-Copy Markup.

Source: http://www.pnl.gov/ag/usage/editsyms.html. Material not copyrighted.

   Similarly, you are required to learn symbols to compute simple math. For example, from an early age, you understood that + meant to add two numbers together, × meant to multiply, and = meant that the two sides of an equation are of the same value. In much the same way, copymarking symbols are a foreign language you must learn in order to communicate effectively as a technical editor. At first, you may need to memorize symbols and their meaning, but after you have used these symbols enough in your hard-copy edits, this "foreign language" will become second nature, and you will become a fluent member of the discourse community of technical editing. Figure 3.2 shows some sample editing using copymarking symbols of a document given during Prentice Hall's editing test.

   Of course, not everyone wants to or needs to know editing symbols. For example, does a software engineer whose manual you've edited need to know that *WW* means *wrong word*? Probably not. In fact, you will probably alienate yourself from your engineer-authors if you insist that they learn your editing lingo just to understand your suggested changes. Chances are that the writer will not even read your comments or heed your suggested edits. So unless you work for a company that requires all employees to memorize and adhere to standards of editing symbols, comment on the document in a way that the author can comprehend without having to memorize a complex table of editing symbols. (Strategies for dealing with writers in a polite yet authoritative manner are discussed in Chapter 2.) This is a good reason for tracking changes using a word processor: no need for special copymarking symbols.

Abbreviation	Meaning	Example
Ab	a faulty abbreviation	She had earned a Phd along with her M.D.
Agr See also P/A and S/V	agreement problem: subject/verb *or* pronoun/antecedent	The piano as well as the guitar need tuning. The student lost their book.
Awk	awkward expression or construction	The storm had the effect of causing millions of dollars in damage.
Cap	faulty capitalization	We spent the Fall in Southern spain.
CS	comma splice	Raoul tried his best, this time that wasn't good enough
DICT	faulty diction	Due to the fact that we were wondering as to whether it would rain, we stayed home.
Dgl	dangling construction	Working harder than ever, this job proved to be too much for him to handle.
- ed	problem with final *-ed*	Last summer he walk all the way to Birmingham.
Frag	fragment	Depending on the amount of snow we get this winter and whether the towns buy new trucks.
‖	problem in parallel form	My income is bigger than my wife.
P/A	pronoun-antecedent agreement	A student in accounting would be wise to see their advisor this month.
Pron	problem with pronoun	My aunt and my mother have wrecked her car The committee has lost their chance to change things. You'll have to do this on one's own time.
Rep	unnecessary repetition	The car was blue in color.
R-O	run-on sentence	Raoul tried his best this time that wasn't good enough.
Sp	spelling error	This sentence is flaude with two mispellings.
- s	problem with final *-s*	He wonder what these teacher think of him.
STET	Let it stand	The proofreader uses this Latin term to indicate that proof- reading marks calling for a change should be ignored and the text as originally written should be "let stand."
S/V	subject-verb agreement	The problem with these cities are leadership.
T	verb tense problem	He comes into the room, and he pulled his gun.
Wdy	wordy	Seldom have we perused a document so verbose, so os- tentatious in phrasing, so burdened with too many words.
WW	wrong word	What affect did the movie have on Sheila? She tried to hard to analyze its conclusion.

FIGURE 3.2
Common Copymarking Abbreviations.

TIP FOR TECHNICAL EDITORS

Editors tend to choose pencils over pens when marking hard copy. Why? Pencil marks can be erased when editors make mistakes! Just be sure to press down firmly enough, and choose a pencil with a high enough lead number so that the author can distinguish your marks from the text. The pencil color is up to you, but recognize that some color—blue, red, or green, preferably—will be needed to contrast with the black text. Use different colors if your document is not the traditional black type on a white background or if you are editing colored graphics.

EXERCISE 3.1

Now it's your turn to try your hand using editing symbols. Copyedit the document below using the copymarking symbols found in Figure 3.1. Try to make your symbols and marginal edits as legible as possible. If you make a mistake, erase fully. Press hard enough on the pencil to make the symbol visible without smudging the copy.

Introduction to Feasibility Study

Purpose

The purpose of this feasibility study on the Santa Rosa College Dining Services is to identify problem areas and provide ample measures to improve those areas. Improving those areas will create a more enjoyable dining experience for all visitors, faculty, and students, at the Santa Rosa College.

Problems

Value

Students at Santa Rosa that have less than 24 credit hours and live on campus are required to purchase a traditional meal plan. 4 traditional meal memberships are offered that range from 16 meals per week to 12 meals per week with prices ranging from $1049 to $935 per semester.

Students are not offered an unlimited meal plan. In addition, these required students are not given the option of any meal plans that offer less than 12 meals per week at a lower price. Table 1 shows the offered traditional meal memberships (<u>Santa Rosa Dining Services</u>; <u>USM Dining Services</u>; <u>UAB Dining Services</u>).

Table 1   Santa Rosa Tradition Meal Memberships (student receives more bonus bucks)

	Meals/Week	Price/Semester
Rosa	16	$1049*
Rosa 16	16	$999
Rosa 14	14	$972
Rosa 12	12	$935

58% of the 114 surveyed students rated the value of Santa Rosa's Dining Services below average or poor. Chart 1 shows this is in detail.

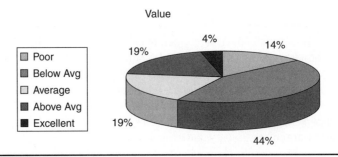

Value

- Poor
- Below Avg
- Average
- Above Avg
- Excellent

4% 14% 19% 19% 44%

SOFT-COPY EDITING

Most soft-copy editing of documents intended to be printed is done using a word processor, but new tools and technologies are being developed rapidly. These new tools are particularly important when the document will be delivered as a web page, help system, or in some other online form.

Effective soft-copy editing involves much more than just reading a document onscreen and tracking changes as you type them in. This section describes some of the tools in a word processor that you can use to help you edit more efficiently and effectively. Later in this chapter, we'll look at the use of Adobe Acrobat for editing, and Unit Five covers editing materials intended for online delivery.

Checking Spelling Using Custom Dictionaries

Everyone knows that spelling checkers have one major limitation: they can't tell whether a correctly spelled word is being used incorrectly in a sentence. Although grammar checkers are becoming better at finding some of the more common mistakes, you might think that the task of ensuring correct word use cannot be automated. Human judgment is essential here, but you can certainly use the computer to help you find at least some of the words you need to verify in context. Jean uses this technique to speed up this repetitive task and ensure that no instance of certain words is overlooked.

Here's the trick: in word processors such as Microsoft Word and OpenOffice.org Writer, you can define your own lists of words ("custom dictionaries") to supplement those provided with the program or available as specialist dictionaries from other sources (for example, medical or legal terms). One type of custom dictionary is known as an *exclude* or *exception dictionary*; this is a list of words that you don't want the spelling checker to recognize as correctly spelled, even if they are in the main dictionary. Here are some examples:

- If an author frequently uses a wrong word, such as *affect* instead of *effect,* put the word *affect* in the exclude dictionary. When you next run the spelling checker, it will flag each use of *affect* as an unknown or incorrect word, and you can decide whether to change it to *effect.* (If the words your author misuses aren't in the grammar checker's list, or it doesn't always catch them when they are misused, try the exclude dictionary technique.)
- If authors frequently mix up *there, their,* and *they're,* or *its* and *it's* or other pairs or groups of words, put all of them in the exclude dictionary. Again, your word processor may provide this function through the grammar checker.
- If your style guide says to use words like *click, press, select,* and *check* in specific situations, and not to use other words such as *hit* or *tick,* you can put them in the exclude dictionary to bring them to your attention.

Consult the documentation for your word processor to learn how to set up an exclude dictionary.

Checking Grammar

Grammar checkers have a bad reputation with many technical editors because they often offer incorrect advice concerning either what a problem is or how to fix it. But you can put a grammar checker to very good use if you carefully choose what to look for and then use your judgment to decide whether something needs changing and what the change should be. Here are some examples.

- The author uses a lot of passive voice in a document where active voice is supposed to be used. You can change the text as you read through the document, but you'll want to check that you didn't miss anything. Use the grammar checker to locate any remaining uses of passive voice and use your judgment to decide whether to change them. Or you can use the grammar checker at the beginning to bring common problems to your attention.
- Remember, a grammar checker may have options for checking "commonly confused words" and "misused words"; if so, you may find these checks useful in speeding up your work.

Using Find and Replace

The Find and Replace feature can be used for a variety of editorial tasks. You can use Find and Replace to

- Locate each instance of specific words. However, you will need to use Find numerous times, once for each word you want to check, and you will probably need to keep a list of words to find. Use Find for phrases, which a dictionary doesn't handle.
- Remove excess spaces, for example after periods or between words. (Some programs provide another way to check and correct the spaces after periods; for example, Microsoft Word has this function available in the grammar checker.)
- Remove blank paragraphs. Many authors press Enter twice to space out text, but publishers generally don't want blank paragraphs, as they use styles to create space before or after paragraphs. (Some programs provide another way to remove blank paragraphs; for example, OpenOffice.org Writer has this function available under Auto-Correct.)
- Change styles or find everything tagged with a particular style. Did the author use a style named "Body text" when it should have been "Text body"? Change all the instances at once using Find and Replace.
- Find special characters. Some publishers want editors to replace special characters such as em dashes with three hyphens. Use Find and Replace to do this quickly.

TRACKING EDITORIAL CHANGES IN SOFT COPY

Word processors such as Microsoft Word and OpenOffice.org Writer have a change tracking or change recording (also known as "redlining") feature that allows both the editor and author to track editorial changes in a text using underlining, new font colors (red for first editor, blue for second editor, etc.), and other features such as ~~strikethrough~~.

Figure 3.3 shows an example from Microsoft Word. Word uses different colors to represent changes made by each editor. This feature allows the author to send the text to more than one reader and be able to distinguish changes among editors, or allows one editor to go through more than one edit with the same document.

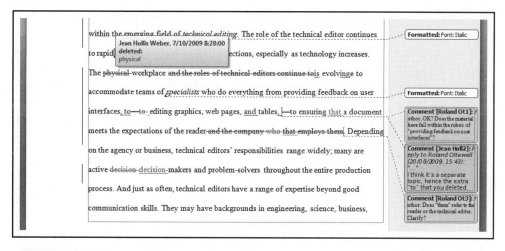

FIGURE 3.3

Tracked Changes and Comments in Microsoft Word.

Credit: Microsoft product screen shot(s) reprinted with permission from Microsoft Corporation.

Once the editor has edited the document using change tracking, the author can choose to accept or reject the changes made using the "accept/reject changes" feature. Some writers use this feature to make corrections, whereas others prefer to retype the changes into the original, unedited document. The second method is slower, but some writers prefer the control it gives them.

Writers can switch between viewing the changes and viewing the result of the changes by "showing" or "hiding" the change markup.

In addition to marking inserted and deleted text and changed formatting, most word processors provide a comment feature. At any point in the document, the editor can insert a comment about the text being corrected or a question about the text. Depending on the software being used, this comment may be placed in the right margin, or it may appear when the author hovers the mouse pointer over the comment marker. See the example in Figure 3.3.

Some technical editors prefer the comment feature because it allows the editor to be less invasive by making suggestions to changing the text as opposed to merely striking through text and replacing it with what the editor prefers. Some editors use different methods for different writers, depending on the personality of the writer and the relationship between the editor and the writer (see Chapter 2), or they may use a combination of the two. Comments are particularly useful for asking the writer questions.

> Originality is nothing but judicious imagination.
> —*Voltaire*

THE IMPORTANCE OF USING STYLES

A style is a set of formats that you can apply to selected elements (such as pages, paragraphs, headings, graphics, and others) in your document. When you apply a style, you apply a whole group of formats with one click. Using styles can dramatically improve visual consistency in a document. In addition, styles are essential for making full use of some other functions in a word processor. We'll look at both uses of styles in this section.

Word processors support several types of styles, although not all programs support all of them.

- Page styles control page margins, headers and footers, borders, and backgrounds.
- Paragraph styles control all aspects of a paragraph's appearance, such as text alignment, tab stops, line spacing, borders, backgrounds, and fonts.

- Character styles affect selected text within a paragraph, such as the typeface and size of text, and bold and italic formats, overriding the style of the paragraph as a whole.
- Frame styles are used to format graphic and text frames, including wrapping type, borders, backgrounds, and columns.
- Numbering styles control alignment, numbering or bullet characters, and fonts in numbered or bulleted lists.
- Table styles apply borders, background colors, spacing and alignment of contents, fonts, and other attributes of tables.

Styles as an Essential Formatting Aid

Most people are used to formatting documents according to *physical* attributes. For example, you might specify the font family, font size, and weight for a second-level heading as Helvetica, 14-point, bold, centered; and you might apply those attributes manually to each second-level heading.

Styles are *logical* attributes. Using styles means that you shift the emphasis from what the text looks like to what the text is: a title, a third-level heading, a paragraph, a list item, a figure caption, a table cell, and so on. So you stop saying, "The font is 14-point Helvetica, bold, centered," and you start saying, "Heading2"—and you mark those second-level headings with the "Heading2" style.

TIP FOR TECHNICAL EDITORS

We use the equivalent of styles every day. For example, there are two main types of computers: desktop and laptop. Each has its own distinctive set of properties. You never say, "My computer is a low-weight, one-piece unit with an LCD screen attached to a rectangular casing containing the computing components and the keyboard." Instead, you say, "I have a laptop." Styles are a way to do the same thing for a document.

Professional writers and editors should use styles, not manual formatting. There are several advantages to using styles:

- Styles are faster and easier to apply than manual formatting.
- Styles help ensure consistency of formatting, especially when several people are working on the same document (for example, writing different sections of a report). Manual formatting can result in a mess when all the parts are combined. Cleaning up the mess can be time-consuming; avoiding it is a better strategy.
- Styles make formatting changes easy. For example, if you need to change the font, type size, or alignment of a book element such as chapter titles, third-level headings, or quotations, you only need to change the style; you do not need to find and change every paragraph of that type.
- Styles greatly assist in producing output in more than one form (print, web page), or when moving a file from one program to another, a common occurrence in the publishing industry, where drafts are often done in a word processor such as Microsoft Word but the final layout is done in a desktop publishing program such as Adobe InDesign or QuarkXPress.

Using Styles for Other Purposes

As described earlier under "Benefits of Soft-Copy Editing," styles are important not only for formatting but also for making full use of other functions in a word processor. For example, if headings are consistently marked with styles (such as Heading1, Heading2, Heading3), a word processor can generate a table of contents complete with page numbers. If someone adds or deletes a few pages, you can regenerate the table of contents with one click—much more efficient and accurate than doing it manually.

Running headers and footers in books often include information such as the name of the chapter or section, which changes from one section or chapter to the next. If chapter titles and section headings are marked with styles, you can put a field in the header or footer that will automatically display the correct information for each page in the document.

Word processors use styles to determine which items to display in a list of items (such as headings, figures, or tables) for use in automatic cross-references to other parts of a document, a powerful tool to ensure that the text and page number of a cross-reference is always up to date.

Headings are used for outlining; most word processors have some way for you to easily display only the headings, and by moving the headings in outline view you can reorganize a document without having to cut and paste the material.

Using Templates

Templates are files that can store styles, text, graphics, and other information. They are extremely useful for ensuring consistency from one document to another. A simple example of a template is a letterhead containing a company's name, street address, phone number, website address, and logo, all carefully placed on a page—the electronic equivalent of preprinted letterhead paper.

If all chapters of a book (written in separate chapters, as is commonly done) are based on the same template, they will be consistent in design and layout without the author needing to redefine the margins, styles, and so on in each new chapter.

As with individual styles, you can quickly change the appearance of a document by changing the template, or by applying a different template. If two templates have been set up to use the same style names, and you have applied styles correctly in the document, the process of changing from one to the other should be easy.

Automating a Table of Contents

Word processors and desktop publishing programs can compile a table of contents from text marked as headings (or other elements specified by the writer or editor). By using an automatic table of contents and updating it when you make changes in the document, you can ensure that it always contains the most up-to-date changes in the text of headings and their corresponding page numbers.

An automatic table of contents can be quite useful to an editor even if it won't be included in the final published document. Compile one for your own use, so you can see at a glance if the headings are parallel, if any heading levels have been skipped, and if the document's structure is logical and its various parts are in the correct order. (If the program you use has a outline view, you can use the outline to find the same information.)

The method for inserting a table of contents varies a bit from one program to another. In Microsoft Word, place the insertion point (cursor) where you want the table of contents to appear, and then choose Insert > Index and Tables and follow the onscreen instructions (consult the help for details).

Styles are the most common method used to indicate headings so the table of contents generator will recognize and include them. In some programs, using styles may be the only method that works.

Automating Cross-References

If the author has typed in cross-references to other parts of the document, those references can easily get out of date if you reword a heading, add or remove figures or tables, or reorganize the sequence of topics. Replace any typed cross-references with automatic ones and when you update fields, all the references will update automatically to show the current wording or page numbers.

In Microsoft Word and OpenOffice.org Writer, all headings indicated by heading styles and all figure and table captions (created using Insert > Caption) are listed in the cross-reference dialog so you can easily choose the one you need. You can also mark any text or page using a bookmark. Other programs have a similar function.

If necessary, you can change all the automated cross-references into plain text at the end of the editing. This may be necessary if the final manuscript will go into another program for page layout and publishing.

Outlining

Some programs (such as Microsoft Word) have an outline view in which you can show or hide different levels of headings and text, and easily move portion of text by dragging and dropping a heading; all the text associated with the heading moves with it. This feature can make reorganizing a document easier than cutting and pasting large amounts of text.

Outline view depends on the headings being marked with styles.

SOFT-COPY EDITING USING ADOBE ACROBAT AND ADOBE READER

Sometimes you will be required to provide edits in soft copy, but you won't be able to edit the document (or a copy of it), either because you don't have the software that produced the document or because the procedures at your workplace don't allow you to work on the document. Examples include editing a website or a technical drawing or production editing of the final layout of a book. In these cases, it's quite common for the editor to be provided with a PDF of the document. You could, of course, print the PDF and write on the hard copy, but you may be required to mark your edits on the PDF soft copy itself. Fortunately, recent versions of Adobe Acrobat have powerful editing tools that you can use to mark your changes, although it is a bit more cumbersome to use than a word processor. Even better, Acrobat Professional version 8 or later can be used to enable commenting rights in a PDF, so the editor can mark it up using similar tools in the free Adobe Reader, version 7 or later. Adobe Acrobat also provides tools for managing reviews, including email-based and browser-based reviews.

Adobe Acrobat (both the Standard and Professional versions) can be used to make some last-minute changes to a PDF just before sending it to the printer or publishing it on a website. It's always better to make the changes in the original document and recreate the PDF, if you're able to do that. However, sometimes you're up against a strict deadline and don't have access to the original document, so you are forced to edit the PDF directly.

Adobe Acrobat (but not the free Adobe Reader) provides some tools for doing the following:

- Minor changes to text or graphics
- Adding, deleting, or changing bookmarks
- Cropping graphics
- Inserting hyperlinks
- Adding, deleting, and replacing pages
- Rotating pages
- Cropping pages
- Combining two or more PDFs into one
- Splitting one PDF into two or more
- Amending the PDF's page numbering to match the page numbers in the document itself

FILE FORMATS

A *file format* is a way to encode information for storage in a computer file. Different kinds of information—for example text, graphics and photographs, video, audio, archives (compressed files), databases, spreadsheets, presentations, or programs—use different formats.

File formats are generally designated by the *extension* in a file's name. Figure 3.4 lists some common file extensions; there are many more than shown in this brief list.

Within any format type, such as word processor documents, there are several different formats. For example, the file format used by Microsoft Word is different from the ones used by Corel WordPerfect or OpenOffice.org Writer. In addition, file formats change as technology develops, so early versions of (for example) Microsoft Word used a different format from more recent versions.

Type of File	Extensions
Text, including word processor	.txt, .doc, .docx, .wpd, .rtf, .sxw, .odt
Graphics and photos	.bmp, .jpg, .gif, .png, .exif, .svg
Video	.mpeg, .ogg
Audio	.mp3, .wav
Archives (compressed files)	.zip, .jar, .cab
Databases	.dbf, .mdb, .odb
Spreadsheets	.xls, .xlsx, .wks, .sxc, .ods
Presentations	.ppt, .pptx, .sxi, .odp
Programs	.exe, .hlp
Web pages	.html, .xhtml, .xml, .asp, .php

FIGURE 3.4
Common Extensions for File Formats.

Editors often need to deal with documents that will be converted from one file format to another. For example, a manuscript written and edited in Microsoft Word may be converted into the format used by QuarkXPress or Adobe InDesign for page layout and publishing. Some conversions can lose formatting or other information; other conversions have few problems. Technical editors may need to learn how to clean up manuscripts to minimize conversion problems. A detailed discussion of this topic is beyond the scope of this book, but the editor's checklist shown in Figure 3.5 gives some indication of the types of cleanup that may be required.

(Note: This checklist has been adapted from guidelines provided by a major commercial publisher for freelance editors who are preparing material in Microsoft Word to be imported into QuarkXPress for page layout. People accustomed to editing files to be published in Word will notice considerable differences in the techniques used here, which include stripping out most of the formatting that Word puts in.)

Preparation

- Make sure you have a printout of the original manuscript before you start editing.
- Save the files onto your hard disk in a folder.
- Make sure each file is saved as a separate, logically named document; for example, ch01uned.doc, ch02uned.doc, and so on.
- Keep each level of work (original files, first edit, second edit) in a separate folder. Name edited files ch01ed1.doc, ch02ed1.doc, and so on for the first edit, and ch01ed2.doc and so on for the second edit.
- Create a template (if one is not supplied) in which you define the Normal or Default paragraph style as left-aligned, with all tabs cleared. Define styles for everything needed in the text, using (when possible) style names that correspond to the style names used in the page layout program. If you cannot use the same style names or need to define extra styles, keep a style sheet showing the exact style names you have used so that the layout person can use the same names.

At the beginning of the each file in a book

- Select the whole document (Ctrl-A). Choose Normal (or Default) from the style list. Now your whole document should be in the Normal or Default style, and any styles that the author has applied will no longer be in effect. Bold, italic, and bullets will be retained, but all tabs and alignments, different fonts, styles, and indents will be gone.
- Open a new blank document based on the template you defined earlier. This new document will have available in it all the styles you have defined. Copy all the contents of the original book file into this blank document and save it under a new name.
- Repeat for each chapter in the book.

When editing

- Assign the appropriate heading level style to all headings.
- Replace double spaces with single spaces.
- Use three hyphens to indicate an em dash and two hyphens to indicate an en dash.
- Delete all hard page breaks.
- Turn on paragraph marks and ensure there are no spaces at the end of a line and no hard or soft returns within a paragraph.
- Take out all double hard returns that have been used to create space between paragraphs.
- Do not use automatic numbering because it does not usually import correctly.

FIGURE 3.5

Sample Checklist for Working with Files to Be Converted to Another Format.

- Do use automatic footnotes.
- Change tables to text. (Usually it's easier for the layout person to turn the text back into a table within the layout program than to cope with an imported table.)

Check with the editorial supervisor to determine

- What editing changes to mark; this will depend on what the author wants to see and how heavy the editing is expected to be.
- How to insert author queries; for example, they could be inserted as comments directly into the text in a different color and style, or preceded by "AQ" or some other indication that the material is not final.

FIGURE 3.5 (*continued*)   Sample Checklist for Working with Files to Be Converted to Another Format.

As this book is being written, major changes are taking place in the world of file formats. The most important is the rise of *open* formats such as OpenDocument. An open format is a published specification for storing digital data, usually maintained by a nonproprietary standards organization, and free of legal restrictions on use. In contrast to open formats, *proprietary* formats are controlled and defined by private interests. The primary goal of open formats is to guarantee long-term access to data without current or future uncertainty with regard to legal rights or technical specifications. A common secondary goal is to enable competition instead of allowing a vendor's control over a proprietary format to inhibit use of competing products. Governments are increasingly showing an interest in open format issues.

OTHER TECHNOLOGY

Being familiar with other technology and products will increase your credibility as a technical editor. As we discuss in Chapter 1, knowing a product will help make your editing of that product's technical manual or other documentation that much more correct and valuable. Moreover, you will garner the respect of your writers and gain even more authority.

SUMMARY

This chapter introduced the budding technical editor to important editing skills as well as some of the software tools needed to successfully edit documents electronically. Technical editors will make themselves more marketable if they become familiar with the tools and technology used in producing technical documents.

CHAPTER 4

Style Guides: Reinforcing Consistency

At the conclusion of this chapter, you will be able to

- distinguish between static and dynamic style guides;
- create a style sheet;
- recognize the credibility an excellent style guide brings to technical editors;
- understand the need for a global style guide.

In Chapter 13, we discuss editing for style. There we treat style as a combination of grammar, tone, and effective writing (including concision). The term, however, has several definitions, one being the "customary manner of presenting material, including usage, punctuation, spelling, typography, and arrangement." This is a fancy way of saying that style is sometimes a verb as opposed to a noun, such as when a technical editor makes a document consistent with rules of style: he or she might *style a manuscript*. You have probably heard us use the term *house style* more than once, and although this could technically signify a mansion, a ranch, a clay abode, a teepee, or a three-story Tudor, clearly we mean *house* figuratively, as in the company where you work. Figure 4.1 shows a sample style guide outline. Many companies use the inclusive pronoun "our" to refer to their ownership of a house style. Note that the style guide can be titled a "style sheet"; this terminology is merely a matter of preference. Most style sheets are shorter than traditional style guides (hence the term "sheet" to imply one page), but that is not always the case.

As you can see from Figure 4.1, style guides cover the detailed format in which the company wants the material presented. This level of detail ranges from typography to abbreviations to how to style a bulleted list. Why is it so important that authors and editors follow the house style? Liability is one answer. As we discuss in Chapter 20, companies are responsible for making sure that their materials, products, and documents have a uniform consistency that follows the company's guidelines. A break in the house style may misrepresent the company or mislead customers or users. At the very least, it could be embarrassing or unprofessional.

This document accompanies "Developing a Departmental Style Guide" (Figure 4.2). Some of the sections include detailed sample text; others do not. The examples shown here are not necessarily the "correct" choices, or the "preferred" choices, or the "best" choices; they are simply **examples** of things to include. Your project may require additional items, especially if your writing will be used on a Web site.

Style Guide for XYZ Corporation

Follow this style guide when writing or editing materials to be published by XYZ Corporation, to ensure that documents conform to corporate image and policy, including legal requirements, and to improve consistency within and among our publications. Use this style guide as part of the specifications for your writing projects, along with

- XYZ Corporation Technical Writing Process Guide
- XYZ Corporation Document Design Guide

References

This style guide lists decisions we have made for this company. It supplements several standard style guides, dictionaries, and other reference material. If you can't find something in our style guide, look in these references or refer your question to the departmental editor.

Dictionaries

Merriam-Webster's Collegiate Dictionary, Tenth Edition. Springfield, MA: Merriam-Webster, Inc., 1994, is the preferred source.

Style Manuals

This guide takes precedence over all other sources.

The Chicago Manual of Style, 14th Edition. Chicago, IL: University of Chicago Press, 1993.

Read Me First! A Style Guide for the Computer Industry. Mountain View, CA. Sun Microsystems, Inc. 1996.

Grammar/Usage Guides

For questions about grammar, consult any of the following:

Webster's Dictionary of English Usage. Springfield, MA: Merriam-Webster, Inc., 1989.

Skillin, Marjorie E., and Robert M. Gay. *Words into Type, 3rd Edition.* Englewood Cliffs, NJ: Prentice-Hall, 1974.

Strunk, William, Jr., and E.B. White. *The Elements of Style, 3rd Edition.* New York: Macmillan, 1979.

Online References

[Fill in as appropriate]

Mechanics

Graphic Elements

Include drawings, figures, tables, and screen shots, whenever it seems useful or otherwise appropriate. Give each graphic element an informative caption.

Refer to each graphic element at an appropriate place in the running text.

For graphic elements used on Web pages, such as navigation aids, icons, photos, and other images, always include alternate text within the IMAGE tags. For example . . .

Hyphenation

Punctuation

Use standard American punctuation. [If some of your writers are more used to British punctuation, you may need to add an example or two here.]

FIGURE 4.1
Sample Style Guide.

Use of Language

    General

        Use short, simple, easy-to-understand words and sentences.

        Avoid the passive voice, except where appropriate. [You may want to include an example of an appropriate use.]

        In general, use the present tense and, where appropriate, the imperative mood ("Do this.").

        Use strong subject-verb constructions. Avoid weak constructions such as "There are."

        Be concise; avoid wordy phrases.

        Use gender-neutral language.

    Capitalization

        Use sentence-style capitalization for headings.

        Capitalize and spell screen element names to match their appearance on the user interface. To avoid ambiguity, capitalize the first letter of each word (including articles, prepositions, and so forth) in the names of menus, dialog options, commands, fields, and other such elements, regardless of their capitalization on the user interface.

    Spelling

        Use American, not British, spelling; consult the preferred dictionaries.

        See the Terminology section for spelling of computer terms not found in the preferred dictionaries.

    Articles

        When using a term that includes special leading characters, choose the article that agrees with the way the term is pronounced. For example, the term "#include" is usually pronounced "include," so the correct article is "an" ("an #include statement").

    Units of Measurement

        Use American units of measurement.

    Numbers (in tables and text)

    Lists [capitalization and punctuation; treatment of nested lists]

Terminology

    Abbreviations and Acronyms

    Generally Preferred Words

    Computer Terms

        Keyboard Key Names

        Choosing and Selecting

    Special Characters

Content of topic types [particularly for online help, but also useful in any situation where multiple writers are involved]

    Overview topics

    Conceptual topics

    Reference topics

    Problem-solving topics

    Frequently-asked question topics

    Field-level help topics

    Procedural ("how to") topics

FIGURE 4.1 (*continued*)   Sample Style Guide.

Procedural topics provide step-by-step instructions on how to complete a user task. User tasks often involve the use of more than one window or dialog box.

Users access procedural topics through the index, contents page, or links from window- or dialog-level topics.

Procedural topics contain

- Task title (briefly identifies and describes the task procedure)
  - Phrased using terms that are familiar to users; users should be able to predict whether the topic matches their task goals just by reading the title
  - Begins with the gerund form of a verb (ending in –ing; for example, Submitting), followed by an object

- Purpose of task or procedure from the user's perspective (a sentence or two that explains the task purpose, its usefulness, and the expected outcome or result)
  - Explains, in the users' own language, why they would want to perform this task and how it relates to their work; focuses on user needs, not on how the application works
  - Answers questions like "What user problem does the procedure solve?" and "How does the procedure fit into the user's work?"

- Prerequisite conditions or tasks that users must perform before beginning this task
  - If the prerequisite tasks have procedures of their own, link to help topics for those tasks

- Step-by-step instructions or procedure (numbered steps that describe how to complete the task)
  - Begins with an infinitive tag (a short phrase beginning with "To"; for example, To submit the form:, To change an emergency contact:). The infinitive explains the purpose of the steps that follow.
  - Each step describes a single action, such as clicking on a button, selecting an item, choosing a menu item, or typing text in a field. It is written as a verb followed by a noun phrase.
  - If there are multiple ways to complete an action, document a single approach and choose the approach that users will easily understand and learn. Provide cross-references to other topics that describe alternative ways to complete the action, but don't provide cross-references for common actions.

FIGURE 4.1 (*continued*)   Sample Style Guide.

THE IMPORTANCE OF STYLE GUIDES

Style guides are crucial in maintaining a consistent corporate image. They also act as an important reference tool for technical writers, editors, graphic designers, managers, production assistants, and others. Although many groups of individuals may consult a corporate style guide, the creator of the guide is often the technical editor. Therefore, style guides are important to the corporation's image, but they are also a way that the technical editor or editing team can establish credibility and build rapport with other employees in the corporation. For example, style guides may be used to train new employees, a process that reaffirms the expertise of the technical editor. Because of the potential impact of corporate style guides, technical editors should try to persuade in-house SMEs to support the standards explained in the guide. In addition, technical editors should try to create the style guide with the production cycle as the backbone; that is, the style guide should integrate product design, development, and marketing and serve as the definitive resource for document guidelines. This will help streamline any production process.

- Procedural topics typically contain about 5 or 6 steps. If a procedure starts to exceed 8 steps, consider breaking it into two procedural topics.

- What happens now? What happens after a user performs the task steps (outcome; results and follow up information)

- Related topics list

Sample procedural topic

Title of help topic

Purpose

Steps

What happens now?

Related topics

Adding a customer

To add a customer record:

1. Open the **Customer Maintenance** dialog by clicking a **Customer** button from any dialog.
2. Type the relevant details about the customer in the **Customer Contact, Customer Address** and **Account Manager** sections.

   You must fill in the following fields:

   - Customer Name (the full company name of the customer)
   - Customer Contact section: First Name, Last Name, Phone
   - Customer Address section: Street Number, Name, Type, Town, State, Post Code

You may also fill in any other details in the relevant fields.

4. Save the record by clicking the **Save** button on the toolbar.

When you save the record, [product name] automatically assigns the **Cust ID** to the customer.

Related topics
Changing customer records
Deleting customer records

FIGURE 4.1 (*continued*) Sample Style Guide.

Many different style guides are available. You may already be familiar with some that you have had to reference with your own academic writing, such as the *Publication Manual of the American Psychological Association* and *The Chicago Manual of Style.* These style guides, along with a few other manuals like the *MLA Handbook for Writers of Research Papers,* provide guidelines for students and professors who are writing and citing academic work. They serve as definitive resources specific to an academic discipline or line of work. Corporate style guides are not much different in this regard.

Some argue that having so many different style guides confuses the writer, editor, and user, whereas others argue that we need style guides to be different in order to address the specific needs of audiences. However, having department-specific style guides within the same corporation may be counterproductive. Jennifer O'Neill advocates a global style guide for corporations. For example, if a large company has divisions in Europe, Australia, and Africa, all divisions should share a common style guide. Of course, the style guide will need to be translated (see Chapter 21 for global and translation issues), but having only one style guide helps streamline the production process for all divisions. O'Neill says this streamlining is particularly important today because so many companies are going global and are acquiring and merging with other companies. Because a global style guide would be a huge and ever-changing project as the company progresses, O'Neill recommends setting up websites, teleconferencing, and chat sites so that technical editors and SMEs in different divisions and continents can communicate effectively with one another about the standards and criteria discussed in the style guide.

EXERCISE 4.1

Compare and contrast two print style guides you have used with your own work. Which one did you prefer and why? Is one more effective than the other? How are they different, and are their differences dependent upon the target audience? Explain.

So far, we have discussed the different kinds of style guides based on academic discipline or industry. There are also different *types* of style guides based on genre and format. Although technical communicators have several different names for the types of style guides, we have simplified the categories into two major areas, *static* and *dynamic*.

Static Style Guides

A static style guide is one that summarizes the criteria for writing and editing corporate materials. The term *static* here does not mean that the style guide never changes; in fact, it is best to change the style guide as often as needed. A static style guide is one that serves as a constant resource for writers, editors, and SMEs. It is sometimes called a *custom* style guide. This type of style guide is traditionally in print form (usually a three-ring binder), but more often than not companies make this style guide available on a website or intranet for easy access and quick searching. For instance, it is much easier and quicker to do a keyword term search online than it is to find the term in a 1,000-page manual, even if you are able to find the term listed in the print index. Technical editors may even want to incorporate the revision of a company style guide into an annual schedule to confirm that elements of the guide are not out of date. For example, we have seen the name of electronically sent messages evolve from *electronic mail* to *e-mail* to *email*. Generally, revisions progress toward concision, but sometimes highly technical terms or processes require more detail or specificity.

EXERCISE 4.2

Go to www.google.com or some other search engine (or directory) and type in "style guides" as a phrase search. From the list, choose two or three online style guides to analyze. In what ways are the style guides static? Which one would you be most or least likely to use in your chosen or future profession and why?

Dynamic Style Guides

Unlike a static style guide, a dynamic style guide takes the standards and rules in the static style guides and puts them to work in a hands-on way. For example, you as a technical editor could work by yourself or with someone to create soft-copy templates that match the standards found in the style guide. If all feasibility reports, for example, need to be in a particular font, and the font size changes for headings and captions, you could create a template that allows writers to just type the documentation and—voilà!—the template already has the styles set to the correct fonts and type sizes.

In addition to templates, Geoffrey Hart advocates the benefit of macros and reference tools in creating dynamic style guides. Although the static style guide is clearly not "dead" and still serves its function, dynamic style guides have gained tremendous popularity in the past decade or so because of their ability to help users follow the rules almost transparently. For example, macros allow you to embed instructions into a template to help users even more, such as putting "type title here" in the space where you need the document title to go. Macros function as template "wizards" and guide the author through "all kinds of repetitive steps that you can automate for them." Hart gives the example of creating a macro that automatically formats the document (double-spaces it, checks the spelling, etc.) each time the file is saved.

Reference tools such as dictionaries and spelling checkers can be created to match the SME's needs and, again, to save time. Nicole recalls one instance when tweaking a reference tool also helped to save face. When she was working with a class on some new writing software, the installed spelling checker automatically changed the first name of her professor colleague to *moron* when the automatic spelling checker was used (the first in the suggested list of alternatives for *Myron*). This change caused the professor much chagrin when students used the automatic spelling checker and did not manually proofread their document! The software company adapted the reference tool by removing the suggested *moron* substitution. The company also added to the dictionary some jargon common to the professor's classes so that the students (and the professor) wouldn't have to see a red squiggly line under *Myron* or check the spelling every time they correctly used a term that was not listed in the software's dictionary.

CREATING A STYLE GUIDE

Chances are, your company has some sort of style guide already in use. If so, you will want to review it and see if it can be improved. If the company is functioning without a working style guide, it is probably going to be your responsibility as a technical editor to create one.

Of course, a style guide will only work well if people read it and actually use it. If you find yourself in a corporate situation where your style guide isn't being utilized, try some of the other communication strategies that you have learned on your road to becoming an effective technical editor. For example, if your SMEs claim they don't have time to read your "style encyclopedia," maybe pass out one-page style sheets instead. Better yet, try a different mode of communication. Arrange a once-a-month lunch meeting (brown bag) where you go over a different aspect each month of the guide that is guaranteed to help your SMEs produce better documentation.

You can address your authors' questions individually, by phone, face-to-face, or by email, or you can encourage more group participation through a chat list, threaded discussion, intranet, mailing list, or responses to a blog. Use the communication tools best suited for your rhetorical situation and knowledge base. Remember that your style guide users are going to treat a static style guide like a phone book, using it only when they need to find an answer and probably only *after* they have already called you first to see if you have the answer (to avoid hunting it down in the print or online version!). This is why dynamic style guides have become so popular, although static style guides still perform their necessary function as providing consistent standards for the look and feel of corporate documentation. Figure 4.2 is an explanation of how to create an effective style guide.

This article provides information that will help you in planning and developing a style guide, including guidelines for what should (and should not) be included, whether to develop one or more style guides, and how detailed the style guide should be.

What Is a Style Guide, and Why Use One?

A *style guide* is a reference document that includes rules and suggestions for writing style and document presentation. Style guides often specify which option to use when several options exist, and they include items that are specific to the company or industry and items for which a "standard" or example does not exist through commercial style guides. The specific content in the style guide is not usually a matter of "correct" or "incorrect" grammar or style, but rather the decisions you or your employer or client have made from among the many possibilities.

More specifically, style guides can serve several purposes:

- To ensure that documents conform to corporate image and policy, including legal requirements.
- To inform new writers and editors of existing style and presentation decisions and solutions.
- To define which style issues are negotiable and which are not.
- To improve consistency within and among documents, especially when more than one writer is involved or when a document will be translated.
- To remove the necessity to reinvent the wheel for every new project.
- To remind the writer of style decisions for each project, when one writer works on several projects that have different style requirements.
- To serve as part of the specifications for the deliverables, when writing for clients outside your company or when outsourcing writing projects.

A style guide contains both rules (nonnegotiable) and suggestions or recommendations (negotiable). Which items should be rules and which should be suggestions is a matter of opinion and corporate policy, though items that result from audience analysis and usability testing are more objective and thus more likely to be rules.

Keep in mind as you're planning and developing a style guide that it should be an evolving document. You don't need to include everything on the first pass; add items as questions arise and decisions are made, or alter items as you make new decisions to deal with changing situations.

Additionally, be aware that developing a style guide can often turn into a major power struggle within an organization, if someone attempts to impose it on a group of people who are accustomed to working without one. If the writers and editors are involved at all stages, and if the development can be seen as a cooperative effort with clear benefits to everyone, then developing a style guide can be a productive experience and the document can take less time to produce.

FIGURE 4.2
Developing a Departmental Style Guide.

What Topics Should *Not* Be in a Style Guide?

Before determining what you should include in the style guide, consider topics that should *not* be included. Although style guides can include a range of topics, the following are often best included in separate documents:

Process information (how we do things in this company or this department). Process information does not belong in a style guide, but it often ends up there because you need to have it written down and no one in your company knows where else to put it. Style guides are intended to help writers to write and editors to edit; process information could go in a documentation plan, project specifications, or other project management document.

What do I mean by process information? Here are some examples:

- Who is responsible for what and when (writer, editor, reviewer, manager, subject matter expert, graphic artist, and anyone else who might be involved)?
- How many reviews are required, by whom, and at what stage of a document's development?
- Which tools are to be used for specific purposes (page layout, help authoring, web page development, graphics development, and so on)?
- Where are documents stored (in a directory on the network, for example)? What file-naming conventions, document numbering, document version numbering, and so on are used?

Design information (what our documents should look like). Design information has traditionally been an important part of a style guide, but it is best provided in a separate document. Here's why:

- Information content is increasingly being reused in a variety of situations and media, both hardcopy and online. The presentation of information is again becoming separated from the creation and maintenance of the content, as it was in the days of typesetters and layout artists—but with the addition of markup languages such as XML providing easier multiple-use capabilities.
- Design decisions may be made at a corporate level, with writers and editors required to follow those design decisions; or writers' and editors' work may be turned into books or web pages by some other part of the company or an external organization.

Because many technical writers continue to be responsible for layout as well as content, and deal with both at the same time (rather than sequentially), you might prefer to put some or all design information in a series of templates, rather than in a checklist-style document.

Design decisions that belong in a design guide or in document templates, but not in a style guide, include

- Page size and margins, number of columns, offset style (if used), typefaces and sizes.
- Bullet characters, including whether and when to use nonstandard bullet styles or more than one bullet style.
- Whether list numbers in procedures have a period after the number.
- Use of horizontal and vertical rules in tables of data.
- Use of spot color for headings, bullet characters, and other navigation aids.
- What file formats and resolutions to use for graphics.
- Numerous other issues related to the appearance of the final product.

Design decisions that directly affect writing and editing *should* go in the style guide; for example

- Chapter and section numbering conventions.
- Capitalization style for headings, vertical lists, figure and table captions, and other situations.

FIGURE 4.2 (*continued*)   Developing a Departmental Style Guide.

- When to use various types of highlighting (e.g., bold or italic type).
- Explicit information on what to do about writing for single-sourcing.
- Which template to use for which type of document, if writers are also responsible for layout.

Grammar and writing tutorials. Too many style guides get turned into tutorials on grammar, spelling, and punctuation. When the style guide is intended to be used by people who are not professional writers, this emphasis is understandable, but still misplaced. It's often a matter of trade-offs between brevity (including only what's needed for consistency) and completeness (when you know that the audience does not know the basics). I generally try to solve this problem by having a separate document for the writing tutorial, if one is needed. Remember that style guides are references, consulted when a question or problem arises, rather than books to be read as a training tool.

Rationale for decisions. I recommend including only as much information as is required, and leaving out the rationale for why specific choices were made. If necessary, include the rationale in a supplementary document, or separate the rationale from the specific style guide points. The less writers have to read and remember, the more likely they will read and remember the important points.

What Topics *Should* Be in a Style Guide?

What topics, then, should a style guide include? Remember, the choices you record are usually not a matter of "correct" or "incorrect" grammar or style, but rather the decisions you or your employer or client have made from among the many possibilities.

A style guide should include the answers to questions such as these:

- The version of English to use (American? British? Other?), specifying any variations. For example, in Australia, when writing computer software documentation for the local market, it's common to use Australian or British English but spell computer-specific terms (e.g., *program* or *disk*) in the industry-standard way rather than the vernacular way (*programme, disc*). If your employer or client has a list of nonstandard words, include them in the style guide.
- The system of measurement to use (American? Imperial? Metric?), specifying any variations (e.g., "dots per inch" in a metric guide) and whether conversions should be included in parentheses. If conversions are included, make sure to confirm them with an appropriate expert and to indicate how many decimal places the conversion requires for acceptable accuracy.
- Any reference materials (such as an industry style guide, a particular dictionary, the company's design and process guides) to use, specifying any variations. Don't reinvent the wheel when perfectly suitable wheels already exist; focus on the things that are unique to your company.
- Which template to use for each type of document.
- What document elements (e.g., title page, preface, table of contents, glossary, index, summary of changes, copyright information) are required, and what to include in them.
- Content of headers and footers, and what pages they appear on.
- Chapter and section numbering conventions to use, if any.
- What legal elements are required (copyright, trademark information), and what goes in them.
- Index style: How many levels? Page numbers (for books) on main entries if they have sub-entries, or not? Page ranges: when to use. General guidelines on deciding what goes into an index.
- Glossaries, bibliographies, footnotes, and references: what style, and when to use.
- Caution, danger, and warning notices: wording and usage.
- The style of capitalization to use for headings, vertical lists, figure and table captions, and other situations.
- The style of punctuation to use for running lists and vertical lists.

FIGURE 4.2 (*continued*) Developing a Departmental Style Guide.

- The minimum level of information to include in a particular type of document (e.g., conceptual information or procedural information).
- Content templates or outlines, where appropriate (e.g., corporate policy and procedures or online help topics).
- What information to include in specific topic types in online help, perhaps with examples.
- The style to use for cross-references or clickable links, both when cross-referencing (or linking) within a document and to other documents.
- Whether to use within-document navigational features such as a clickable contents lists at the beginning of a chapter or long web page.
- Whether illustrations and tables always need captions, or under what circumstances captions are required. Whether captions should always be referenced in the text; if not, under what circumstances references are required.
- When to use various types of highlighting (e.g., bold or italic type).
- When to spell out numbers and when to use numerals; the use of commas, spaces, or other punctuation in numbers over 999.
- Word use (e.g., company-specific or product-specific terms; acronyms and abbreviations, preferred words, and words to be avoided).
- Writing style and preferred wording for common phrases.
- Acceptable jargon, and the spelling, capitalization, and hyphenation of names and terms.
- Abbreviations used, including in measurements.
- Revision bars: what style, and when to use.
- How pages are numbered (sequentially throughout, or by chapter).
- Special requirements for the audience (e.g., language or knowledge levels). For example, whether the singular "they" is acceptable; any terms to be avoided, such as non-English abbreviations or words, or terms that your audience might find confusing or unacceptable.

Do You Need One Style Guide or More Than One?

Some organizations may need only one style guide that covers all of their publications, both hard-copy and online. Most decisions about writing style will probably be the same for all the work done by a publications department, but some details may vary. Most differences will probably be design issues. Product-, publication-, or client-specific style sheets can supplement the main company style guide by recording any decisions made for a specific situation. For example

- Your company style might specify sequential page numbering throughout a document, but a particular project might require numbering by chapter.
- Some documents (user guides, for example) might not have numbered headings, while others do (engineering guidelines, for example).
- Some jargon might be acceptable (or required) in documents for one audience, but not allowed in documents for another audience. These differences should be clearly spelled out.
- Hard-copy and online documents may have some different requirements.

How Detailed Should a Style Guide Be?

A style guide can be as short as a single page listing variations from a commercially available guide or the main company style guide, or quite lengthy if it contains detailed specifications of topic content. Many companies adopt a commercially available style guide such as *The Chicago Manual of Style,* and only note any additions or changes in the company style guide. Other companies summarize the most relevant points from a major style guide in the company style guide, because a small guide is more likely to be read.

FIGURE 4.2 (*continued*)   Developing a Departmental Style Guide.

Exactly how detailed your company's style guide should be depends on how much the styles deviate from those included commercial style guides, the types of information products your company delivers, and how many different elements those information products include. In developing a style guide, begin by exploring these aspects, and then plan what details the style guide should include.

Summary

Style guides include rules and suggestions for writing style and document presentation. They often specify which option to use when several exist, and they include items that are specific to the company or industry and items for which a "standard" or example does not exist through commercial style guides. As you develop a style guide, keep in mind that the specific content in the style guide is not usually a matter of "correct" or "incorrect" grammar or style; instead, it's a compilation of decisions that you, your employer, or your client have made from among the many possibilities.

Style guides can be of any length and level of detail; however, they should exclude process and design information, tutorials, and decision rationales that are best included in separate documents. Instead, the style guide should include only as much information as is needed to meet the particular goals of the style guide at your organization.

FIGURE 4.2 (*continued*)   Developing a Departmental Style Guide.

EXERCISE 4.3

Using this chapter as an example, create a style guide or style sheet for publishing textbook chapters on technology-related issues with Prentice Hall. Work with your classmates to create criteria and standards for margins, headings, font choices, grammar, abbreviations, etc.

EXERCISE 4.4

Create a style guide for the class in which you are currently enrolled. Revise the guide as the course progresses.

SUMMARY

In this chapter, we discuss the different types of style guides as well as the importance of creating an effective style guide. If your corporation is large and has divisions in different countries, you will want a global style guide. Using the features available in a dynamic style guide may help you streamline the document production process for all employees.

Editing for Correctness

2

Unit One discusses one of the most common approaches to editing documents: the levels of edit. The levels constitute the different areas of the document that need editing attention. However, editors and scholars cannot agree on a standard definition of levels. Just as many companies or disciplines subscribe to different citation styles, most editors use different levels and types of edit.

Because your career as a technical editor will involve learning more than one house style, that is, the preferred way a company edits and cites documents, we use only three levels of edit: *editing for correctness*, *editing for visual readability*, and *editing for effectiveness*. We feel confident that any preferred style of editing you encounter later will easily adapt to these three levels; besides, we prefer to keep things uncomplicated.

In this unit, you will learn the necessary skills to *edit for correctness* and how to apply that knowledge by locating errors with subject-verb agreement, punctuation, mechanics, and spelling. These are errors that the readers will see, even when just skimming the text. By discovering and correcting them, you establish your own, the company's, and the writer's reputations. In addition, you create better documents for readers. Learning how to edit for correctness becomes the basis for higher levels of editing that call for critical thinking and making sound judgments.

CHAPTER

An Introduction to Grammar: What Technical Editors Need to Know

At the conclusion of this chapter, you will be able to

- explain why we need standards for writing;
- summarize why technical editors need to know grammar;
- justify why conventions of usage are necessary.

L ike most of us, you probably graduated from high school armed by English teachers with a repertoire of grammatical rules, usage skills, and strategies for writing the perfect essay. Throughout those adolescent years of diagramming sentences, completing grammatical exercises, and memorizing rules, you learned to identify the parts of speech, to write compound-complex sentences, and, on good days, to insert commas flawlessly within sentences and write paragraphs without any major grammatical errors. Still, terms such as *gerund*, *transitive verb*, *predicate nominative*, and *infinitive* may have remained somewhat elusive.

No one argues that English is a complex and challenging language, not just for non-native speakers but also for those raised speaking and writing the language all their lives. Yet despite its complexities, English has become the international standard; you can travel almost anywhere in the world and find millions who read and write English fluently. Why? Sociologists will quickly point to economic, political, and historical influences during the twentieth century. However, linguists will argue that because the English language evolved through the blending of many other languages, including Celtic, Latin, Norman-French, and Anglo-Saxon, it emerged as a hybrid language rich in description, nuance, and expression. Today, comprising over 500,000 words, English continues as a dynamic language with no fewer than 1,000 words entering or disappearing annually.

> I am returning this otherwise good typing paper to you because someone has printed gibberish all over it and put your name at the top.
> —*English professor, Ohio University*

Noam Chomsky, a renowned linguist, theorizes that our brains contain a "language-grasping structure" that enables us to comprehend meaning from complex sentence constructions. Even before we could talk around the age of two, many of us could follow a command to "Go get your *Green Eggs and Ham* book off the bed. Bring it here for us to read." Although computers can calculate millions of bits of information a minute, the computer has not yet approached the complexity of the human brain. For example, you surely have noted that spelling checkers and grammar checkers are frequently wrong! Unlike you, they cannot discern the fine nuances and associations of words and phrases in context with their meanings. Some argue that creating a computer that can comprehend, much less appreciate, language as can humans is next to impossible.

So, with our innate ability to read, write, and speak our language, is it really necessary to know all its constructs? Can't a person write a coherent document or forceful piece of prose without being able to identify a direct object or name all the parts of speech? No doubt many professional writers and editors would be hard pressed to diagram a sentence or conjugate the verb *to lay*. They may not even know all the parts of speech. However, through practice and experience, many have become familiar with grammar. Still, it's doubtful that you will find even the most experienced writer or editor more than an arm's length from a well-worn grammar handbook and dictionary or without a least a handful of grammar and usage URLs stored in his or her "Favorite Websites" folder.

WHY DO WE NEED STANDARDS?

No one expects any individual to memorize all the rules of grammar, mechanics, and usage (although you probably concluded that your high school English teachers had other ideas). Despite what you have been told, grammar *rules*, per se, don't exist. What you have memorized (and probably forgotten) as grammar and usage rules are actually descriptions of how educated and successful writers apply the conventions of the English language that have emerged through centuries of use. For example, you have learned rules that dictate when you should capitalize a word. Yet if you examine Jonathan Swift's writing from the eighteenth century, you will discover that he arbitrarily capitalized words and punctuated whenever he desired emphasis. In fact, the paragraph as we know it today—a topic sentence subsequently followed by supporting details—didn't exist before Alexander Bain created it in 1866. Writing prior to this time frequently lacked unity and was difficult to read. Today, the *rules* of grammar actually *describe* conventions of usage accepted by successful writers as standard. No doubt, though, many of these rules will continue to change as the English language matures. Who knows, one day your grandchildren might be using the word "ain't" with their English teacher's blessing.

But why is conformity to any standard important? Perhaps you recall from early-nineteenth-century American history how the first trains couldn't travel large distances because railroad lines used different gauges of track. Traveling across the state of New York by rail from Albany to Buffalo required no fewer than four transfers to different rail systems. In 1869, the United States changed from fifty-six time zones to the present four to eliminate the inevitable confusion that resulted when people traveled across the country. The North won the Civil War, in part, because it standardized parts for the Springfield rifles and made them interchangeable. You may remember a few years ago when a probe sent to land on Mars crashed because the European scientists calculated the lander's trajectory using the metric system while the American scientists based their calculations on the British imperial system.

> Grammar is the logic of speech, even as logic is the grammar of reason.
> —*Richard C. Trench*

Without *standards* or *rules* as they now have become to be called, communicating in a technological world would be impossible. Just as one standard gauge of track permits uninterrupted travel over long distances, standards of usage and grammar allow us to communicate effectively with a minimum of confusion. For example, read the following sentence; before reading any further, decide what it means. Then insert the necessary commas to reflect how you interpret this sentence.

Woman without her man is nothing.

When asked, some individuals punctuate this sentence as *Woman, **without her man**, is nothing.* Notice the gender bias here, suggesting that a woman is helpless unless she has a man. However, suppose you move one comma, insert a colon, and punctuate the sentence as *Woman: **without her**, man is nothing.* Notice how the meaning completely changes simply by shifting one comma! This sentence now contends that a man is helpless without a woman.

WHY DO TECHNICAL EDITORS NEED TO KNOW GRAMMAR?

Few individuals get excited over the subject of grammar, but just as the previous example demonstrates, knowing grammar and how to use it properly can determine the difference between effective communication and misunderstanding. The editor's job, in part, is to know how to use grammar effectively and prevent confusion between writers and their audience. Even beyond misinterpretation, grammatical and mechanical errors create a poor impression and suggest little care was taken in preparing the document. On the other hand, a well-crafted, grammatically correct document intimates professionalism, intelligence, and competence.

Simply stated, successful editors develop an excellent command of grammar. Telling a writer to reword a phrase or clause because it "sounds better this way" without being able to explain why may anger the writer and hinder cooperation. For example, if a writer has written, "Each of the documents have been sent back for proofreading," you may need to explain that the word "each" (not "documents") is the subject of the sentence and requires the singular verb *has*.

Finally, acquiring a command of grammar and the conventions of usage instills confidence. No one picks up a violin and begins playing it, but through practice and repetition, the musician gradually gains confidence and takes command of the instrument. Chances are, you already have a fairly good background in grammar from your earlier schooling and now only need to review what you may have forgotten. Although it may seem intimidating at first, especially if it has been some years since you studied grammar, take advantage of opportunities to attend grammar workshops, enroll in additional college grammar courses, or work with an experienced individual as your "grammar mentor."

CONVENTIONS OF USAGE: A LONG TRADITION

Although distinct from rhetoric, grammar as a discipline can be traced back before 350 BC to the Greek educational system called the *trivium*. Until recently, teaching grammar was never questioned; it was assumed to be a prerequisite to good writing. However, some contemporary composition theorists have challenged this 2,000-year-old tradition, arguing that no relationship exists between knowing grammar and writing well. In fact, during the 1980s, many university writing programs completely abandoned the teaching of grammar in favor of what was termed the *process approach* to writing. This approach concentrated on the invention of ideas and free-writing activities while creating a generation of "grammatically challenged" students.

Our purpose here is not to argue theories of writing. We ask, though, without knowing how words, phrases, clauses, and sentences interplay to make meaning, how can a person create a document that communicates an idea effectively, especially when technical accuracy is imperative? The potential for misunderstanding or misdirection increases. Successful editors

Everywhere I go, I'm asked if the universities stifle writers. My opinion is that they don't stifle enough of them.
—*Flannery O'Connor*

understand this because they *know* grammar. Grammar is their tool for making language work. For example, they know that changing a passive verb to active can create a new mental image for the reader, that adding a descriptive adjective before a noun can clarify the writer's intention, and that adding a comma in the right place can avoid a serious misreading.

Similar to a good automobile mechanic who can diagnose the problem with your car and fix it the first time, grammar is an editor's diagnostic tool for repairing confusing or imprecise language. Unfortunately, some believe that editing involves only "cleaning up" a writer's document, which is a misfortunate assumption. Editors engage in problem-solving. Not only must they know how words relate within sentences but also under what conditions words and phrases are considered acceptable. Being able to comprehend the dynamics and skillfully apply them to the conventions of grammar, spelling, mechanics, and punctuation in a carefully articulated document is both challenging and cognitively taxing.

Chapter 6 through 9 review the major conventions of grammar and usage. To help you evaluate your progress, we have provided two self-diagnostic tests, one below and the other at the end of Chapter 9. Each will help you determine your level of mastery and indicate where you may need additional work. Most importantly, though, we have provided two additional types of exercises: traditional grammatical exercises and application exercises. First review the grammar; then complete the traditional application exercises. If you have difficulty, review the material again. Also, if you haven't already, acquire a good grammar handbook and consult it often. Your professor may recommend online grammar handouts for you, such as those found at Purdue's Online Writing Lab (http://owl.english.purdue.edu/).

SELF-DIAGNOSTIC: HOW MUCH DO YOU ALREADY KNOW?

Now take the time to evaluate what you already know about grammar and usage. The first diagnostic spans a wide range of grammar, mechanics, and usage skills. The second will help you determine if you can apply those skills in the context of a copy editor. Neither of the diagnostic exercises covers all the skills, but working through them should provide you with an indication as to how much you need to review. Remember, locating writing errors comes from considerable practice. If you are unsure how to answer many of these questions, study the following sections closely, complete the exercises, and take the final diagnostic to determine your improvement.

There are more English speakers in India and the Philippines than in England and the United States.
—*Author unknown*

Directions: Below are twenty-five sentences that contain at least one grammatical or usage error. Determine the error and then check your answers against the key that follows.

1. Brandon said it is alright if I borrow his style manual if I promise to return it tomorrow.
2. "I assure you, Davis. It is not I whom is to blame."
3. Due to the fact that the economy has turned downward, we decided not to hire another copy editor.
4. If I had known that you needed a ride to the office, I would of been glad to drive by your house and pick you up.
5. Jamie is one of the students who is going to apply for the copyediting job next week.
6. Neither the class or the professor wanted to have the test next Friday.
7. All of you completed the assignments without hardly any errors.
8. Mike wants to go to the concert tonight, however, his girlfriend insists they study for their technical editing quiz tomorrow.
9. Please loan me a couple of dollars for lunch.

10. Although the professor's pen had leaked into his shirt pocket, everyone continued to work at their desk as if nothing happened.
11. Her attempts to dramatize the affects of driving while intoxicated were not lost on the class.
12. "Why did you bring your boss along?" Marty asked, "we cannot squeeze anyone else into the car."
13. Riding with the top down, the convertible energized the young sales clerk.
14. This is one of the prizes that are going to be given to the outstanding technical editor.
15. When they realized the mechanic was not going to show up, they decided to repair the car theirselves.
16. I'm confident that everybody who works diligently on these exercises will improve their grammatical acumen significantly.
17. Its time for all the students to return to their dorms and study their lessons.
18. Either you or the interns is capable of handling the editing assignments.
19. "Why don't all of you sit down and be quiet" screamed the professor!
20. After you recieve your degree from the university, do you plan to attend graduate school?
21. If I was expected to have my PowerPoint presentation ready for today, you should have notified me earlier.
22. My children got alot of toys for Christmas that made too much noise, contained small pieces that hurt when I stepped on them, or broke too easily.
23. The reason I copied his paper is because I ran out of time and needed to turn in something.
24. Learning grammar is time-consuming, sometimes difficult, but is definitely rewarding.
25. He was concerned that the last check he wrote would overdraw his account; but fortunately, he had just enough money to cover it.

Answer Key

(Some sentences may be corrected more than one way.)

1. *all right*, not *alright*
2. *who*, not *whom*
3. *because*, not *due to the fact that*
4. *would have*, not *would of*
5. *are*, not *is*
6. *nor*, not *or*
7. *~~hardly~~*
8. . . . *tonight; however* . . .
9. *lend*, not *loan*
10. *~~everyone...their desk~~* all the students. . . *their desks*
11. *effects*, not *affects*
12. . . . *asked. "We cannot . . ."*
13. *~~, the convertible~~*
14. *is*, not *are*
15. *~~theirselves~~ themselves*
16. *~~everybody who works~~* all students who work
17. *It's*, not *Its*
18. *are*, not *is*
19. *quiet!" screamed the professor.*
20. *receive*
21. *were*, not *was*
22. *a lot,* not *alot*
23. *~~The reason, is~~*
24. cross out the second *is*
25. *account,*

SUMMARY

Although we are often encouraged by others to think creatively or to think outside the box, there are times when conforming to accepted standards and conventions is necessary. Conforming does not mean we cannot or should not think independently, but it does mean we recognize that playing by the rules promotes an important end: ensuring clear and effective communication.

CHAPTER 6

Parts of Speech: Eight Basic Elements of Grammar

At the conclusion of this chapter, you will be able to

- identify the major parts of speech;
- recognize and label parts of speech in a sentence;
- explain the function of each part of speech.

The most fundamental elements of grammar are the eight parts of speech: nouns, pronouns, verbs, adjectives, adverbs, prepositions, conjunctions, and interjections. Each has a different function in a sentence and, by virtue of its function, affects the meaning of each sentence. For example, notice how the addition of an adjective or adverb enhances the meaning of these sentences and conveys the author's opinion.

The Internet is a medium.
*The Internet is an **inexhaustible, wonderful** medium.*
(adjectives)

Barry designed his web page.
*Barry designed his web page **quickly** but **carefully**.*
(adverbs)

Writers as well as editors have over half a million words at their command. Each word belongs to one of the eight parts of speech. With the knowledge of how each part functions and the skill to apply that knowledge, editors can substantially improve the quality of a document and avoid potential misunderstandings.

Perhaps you already have an extensive background in grammar and only need to quickly review the parts of speech. However, if it has been some time since you learned grammar or if you have a weak background, work through the following sections carefully.

EXERCISE 6.1

Direction: Identify the parts of speech in the following sentence.

Important online text should be placed in a single column because two-column text reduces reading speed and is poor document design.

Answer:

adj	*adj*	*n*	*v*	*v*	*v*	*prep*	*adj*	*adj*	*n*	*conj*
Important	online	text	should	be	placed	in	a	single	column	because

adj	*n*	*v*	*adj*	*n*	*conj*	*v*	*adj*	*adj*	*n*
two-column	text	reduces	reading	speed	and	is	poor	document	design.

NOUNS: WORDS THAT NAME

Functions of the Noun

The noun, which comes from the Latin word *nomen*, **names**:

- Persons (*wrestler, undertaker, editor*). The *copy editor* corrected the grammar in the *technical writer*'s document.
- Places (*lake, classroom, darkroom*). The *classroom* was too small for all of the editors to meet.
- Things (*umbrella, automobile, computer*). His new *computer* came with *software* capable of designing web *pages*.
- Abstract concepts (*loyalty, sadness, honor, peace*). I never questioned his *loyalty* to the firm; however, his condescending *attitude* toward clients resulted in Bill's firing.

Types of Nouns

- Proper nouns: specific names for persons, places, things, or concepts (*Taundra, Empire State Building, Catholicism*). *Susan* suggested that the *Randolph Company* investigate the complaints of poor sales.
- Collective nouns: names for a type of group (*covey, herd, team, audience, crowd*). Our graphics *team* won an award for originality. Note: When a collective noun is the subject of the sentence, make certain that it agrees with the verb. Generally, collective nouns are considered a unit and require a singular verb. *The herd of elephants is ~~are~~ running across the savanna.* However, when the context of the sentences clearly indicates that members of the collective are to be considered individually, use a plural noun. *The jury (members) agree ~~agrees~~ that the defendant is guilty.* Of particular

relevance to technical editors is the word *data,* which is used as a collective noun (taking a singular verb) in the computer industry, but in mathematics is considered to be plural (the singular being *datum*).

- Mass (noncount) versus count nouns. Probably the easiest way to distinguish mass nouns and count nouns is to know that you can count all count nouns. For example, units of money can be counted: *dollars, pennies, quarters*. These are each individual, countable items of money. However, if you are talking about money in general, that is a mass noun because it can't be counted; you wouldn't say *I have one money* but instead *I have less money* or *a little money*. Often, writers will make the mistake of forgetting quantifier adjectives with count nouns. For example, a recent beer commercial advertised, "Great taste. Less calories." However, because *calories* can be counted individually, this count noun requires the quantifier adjective *fewer: Great taste. Fewer calories*.

Singular or Plural Nouns?

Two other aspects of noun usage need to be mentioned here. First, nouns are either **singular,** meaning one (*computer, street, document, horse, potato*), or **plural**, meaning more than one (*computers, streets, documents, horses, potatoes*). Notice that the plural form of a noun is created by adding either *s* or *es* to the end of the noun. Some nouns, though, are irregular when shifting from singular to plural: *man* becomes *men*; *deer* remains *deer*; *goose* becomes *geese*; and *octopus* becomes either *octopuses* or *octopi*.

The Tricky Business of Possessive Case

Nouns also express ownership, usually by adding an apostrophe plus *s* to the noun. This is called the **possessive case**. Even experienced writers and editors occasionally have trouble with possessive case, but in most instances you can follow these guidelines: If the noun is singular, or is a plural noun that does not end in *s*, add *'s* (*Pepito's, horse's, car's, child's, children's, women's*); but if the noun already ends in *s*, add an apostrophe (*horses', cars', students', Puritans'*).

Be aware, though, of special cases. For example, if a singular noun ends in *s,* you can simply add an apostrophe or you can add an apostrophe and *s*. For example *James'* or *James's* is correct. However, some style guides will tell you to make the possessive *'s* when the *'s* is pronounced. For example, the possessive case of *Francis* is *Francis's* because the *'s* is pronounced. Now try to create a sentence using the possessive case of *Francis's* without pronouncing the *'s*. Do you see why the second *s* is necessary?

You should also remember that compounds add the apostrophe and *s* only to the last word: her *mother-in-law's* dress, the *secretary of defense's* speech, *someone else's* homework. When you have a case to indicate joint ownership, most style guides allow for either option: *Mary's and Adam's* house or *Mary and Adam's* house.

Probably the most problematic case facing editors and writers involves the word *it*. The possessive case (*Its engine is too large for the housing*) does not require the apostrophe; however, the apostrophe is used to form a contraction (*It's time for writers to return the proofs*). And finally, remember that apostrophes are not used with pronouns (*his, hers, ours, yours, theirs, whose,* etc.) You would write *The radio is **ours** to keep.* Reminder: Be careful with *it's,* which means *it is* or *it has*.

TIP FOR TECHNICAL EDITORS

1. Be certain to look at the context of the sentence. For example, the *engineers' department* belongs to all engineers (plural), while the *engineer's room* belongs to a single engineer and the *engineers' room* belongs to two or more.

2. Some nouns have irregular plural forms, making the possessive case even more tricky. For example, it's the *children's* school and *women's* rights.

Possessive-case nouns (*Uyen's, oxen's*) are functioning as adjectives in the phrase *Uyen's blue purse* (*blue* and *Uyen's* are adjectives modifying *purse*) and the sentence *The oxen's harnesses are loose, making plowing painful for them* (*oxen's* modifies *harnesses*).

Quick Reference Glossary

Noun....................................a word that names a person, place, thing, or abstract idea
Collective noun...................names a type of group (*covey, herd, team, audience*)
Proper noun.......................names a specific person, place, or thing (*Chris, the South*)

EXERCISE 6.2

Correct the following paragraph for the possessive case and contractions.

Sometimes its difficult to know all the responsibilities that must be performed by the technical editor. For example, one of the technical editors jobs may be to suggest stylistic changes to the writer. Often the editors and writers responsibilities overlap, and its important to clarify whos doing what before any editing begins. Usually the Editor in Chiefs job is to make certain things proceed smoothly.

The Tech Editor's Cubicle

Although we discuss this point in greater detail in Chapter 13, now is a good opportunity to mention how some scientific and technical writers abuse nouns. Frequently these writers forget their audience's lack of technical expertise and attempt to "objectify" their writing style. For example, *The officer issued an **announcement** that the hurricane was the source that **caused the destruction of** the building* is much more clearly written as *The officer announced the hurricane destroyed the building*.

The first example, rather than using verbs to convey most of the meaning, *nominalizes* the sentence by creating abstract nouns from the active verbs *announce* and *destroy*. As a result, the nominalized sentence is longer, less direct, and not as forceful as the second. Nominalized sentences may be technically correct, but they should be edited to be more straightforward, especially for a nontechnical audience.

Which sentence is more forceful?

*This study is an investigation of the role the female rat plays in the contraction of
disease.*

or

This study investigates how female rats cause disease.

*The collection of specimens called for taking an analysis of their DNA in order
for the jury to be more informed.*

or

The specimens were analyzed for their DNA.

Nominalized	Edited for Clarity
John made a decision . . .	John decided . . .
The data resulted in a decrease . . .	The data decreased . . .
The earthquake caused the annihilation . . .	The earthquake annihilated . . .
Mary's intention was to rescue . . .	Mary intended to rescue . . .

EXERCISE 6.3

Although some technical writers believe nominalizations (often a root verb + *tion*) make
their writing seem more objective, in reality nominalizations often make the meaning less
clear. Figure 6.1 has been adapted from a National Park Service brochure. Mark up the
brochure by locating the nominalizations and substituting an active verb.

The diversity of National Parks and other public lands can be said to mirror the diversity of the nation
from which these lands are drawn. As a result, global warming will have the same types of impacts on
these lands as those that occur in areas that are not owned by the government. Sea level rise will tend
to promote erosion and affect the beaches through inundation of the National Seashores and the
wetlands of various National Wildlife Refuges and National Parks in coastal areas. Regional climate
change combined with the fertilization effect of CO_2 in the atmosphere will have the same effect on
forests within National Parks and National Forests as occur in other forests. The intensification of
evaporation and precipitation will tend to increase the frequency during which wild and scenic rivers
experience either extreme floods or extremely low flows of water.

Nevertheless, the impacts of climate change on public lands differ from the implications elsewhere in
two fundamental respects. First, they are often unique. Yellowstone, Yosemite, Everglades, and many
other National Parks were created because previous generations reached a national consensus that it
was important to preserve these unusual areas in their natural state forever. Blackwater, Edmund
Forsythe (formerly Brigantine), Audubon, and other National Wildlife Refuges were once typical of
the natural environments in their respective regions, but today these refuges provide unique habit
within their regions because the surrounding areas are the results of subjection to agricultural and
urban development. EPA, in cooperation with the National Park Service, has prepared a series of case
studies on the potential impacts of climate change on selected national parks and other woodlands in
the western mountains and plans, the Great Lakes region, the Chesapeake Bay area, and South Florida.

FIGURE 6.1
Climate Change on Public Land.
Source: Environmental Protection Agency.

PRONOUNS: WORDS THAT SUBSTITUTE

Functions of the Pronoun

Pronouns act as substitutes for specific nouns. Without pronouns, sentences would sound redundant. For example, *Xavier lost Xavier's handbook so Xavier had to buy Xavier another one.* Using pronouns solves ineffective repetition: *Xavier lost his handbook so he had to buy himself another one.*

However, using pronouns to avoid repetition is not always a good solution. One of the most common sources of confusion results when it is not clear to the reader to whom or what the pronoun refers. For example, *After we visited Alecia's mother, we went over to Aunt Sally's restaurant. **She** is really a great woman.* Is the writer referring to Alecia's mother or Aunt Sally here?

TIP FOR TECHNICAL EDITORS

When a pronoun refers to a specific noun, that specific noun is said to be the *antecedent* of that particular pronoun. Good use of pronouns and their antecedents keeps writing clear and coherent.

> Technical communication today often involves editors and lawyers **who** work collaboratively.

Pronouns can be categorized according to their specific function in a sentence. Figure 6.2 is a quick reference chart describing the types and their functions.

TIP FOR TECHNICAL EDITORS

Other types of pronouns exist, but these are the principal ones. Although you don't need to memorize their names, we encourage you to be familiar with the various types and how they function.

EXERCISE 6.4

One common problem occurs when writers use pronouns but do not make clear the pronoun's antecedent: the noun to which it refers. This can cause confusion and misinterpretation. Distinguish the pronouns here and determine the antecedents. Also, find the sentences with pronouns that do not have an antecedent.

1. *The Chicago Manual of Style* is one of the most comprehensive style guides that we use in our department.
2. Gil was pleased with himself but thanked everyone who worked on it.

3. She only has herself to blame for the technical error.

4. Who shouted "Hasta la vista, baby!" when Harry announced he had been fired?

5. So far, our research project continues to be on time, but because someone failed to complete the graphics, we may be over by a couple of weeks.

Type of Pronouns	Function	Examples
1. Personal Pronouns	refer back to specific person or thing	*I, you, he, she, it, we, us* they, them, etc.

Example: After the designers consulted the original drawing, *they* added more color.

2. Demonstrative Pronouns	identifies specific nouns	*this, that, those, these*

Example: *These* are the web pages I was telling you about.

3. Indefinite Pronouns	refer to quantity but not necessarily to a specific person or thing	*all, anybody, somebody each few, most, everything, nothing, none, any, etc.*

Example: *All* of the engineers gathered in the drafting room for the announcement.

When indefinite pronouns are used as the subject of a sentence, they must agree with the verb in number. Clearly, *all* is plural and takes a plural verb. *All of the copy editors* were *present at the workshop*. However, indefinite pronouns such as *anybody*, *somebody*, *each*, and *none* usually take a singular verb depending on the context of the sentence. Why? For example, *none* means "not one" and considers one person or thing at a time. *None of the copy editors* was *present at the workshop*. *Was* is correct because it refers to one individual at a time. Indefinite pronouns that contain *body* as in "somebody" almost always take a singular verb; just be certain to understand the context of the sentence.

4. Relative Pronouns	begin dependent clauses but refer to a noun in an independent clause	*who, which, what, whom, whatever, whichever, that*

Example: She is the engineer *who* caused all the trouble

(*She is the engineer* can stand alone as a independent sentence, whereas *who caused all the trouble* cannot.)

5. Interrogative Pronouns	ask questions but should not be confused with relative pronouns	*who, which, what,*

Example: *What* is the name of the new person in the drafting department?

6. Reflective Pronouns	refer back to the subject of the sentence or the clause. these pronouns end in self or selves	*himself, herself, itself, myself, yourself, etc.*

Example: Matthew thought highly of *himself* when he got the promotion.

FIGURE 6.2

Types of Pronouns and Their Functions.

Quick Reference Glossary

Pronoun.............................a word that substitutes for a noun
Antecedent........................the specific noun the pronoun refers to
Personal pronoun...............a pronoun that refers to a specific person or thing
Indefinite pronoun...........a pronoun that does not refer to a specific noun
Interrogative pronoun........a pronoun that asks a question
Relative pronoun................a pronoun that introduces a dependent clause

---

The Tech Editor's Cubicle

Who is bold enough to correct William Shakespeare when he writes, "God send everyone their heart's desire"? Well, if Shakespeare were writing today for publication, undoubtedly his copy editor might respond, "Bill, don't you know the pronoun *their* must agree with the antecedent *everyone* in number? I know you're trying to be gender-neutral here, but you must be grammatically correct. Try this, Bill: 'God send *all people* their heart's desire.'" (Or maybe you're like us and simply decide to leave grammatical issues involving Shakespeare alone.)

Gender-specific pronouns that refer to indefinite antecedents, especially third-person masculine pronouns such as *he, him, his,* and *himself,* are no longer acceptable in technical or business writing. To say *A student left his books on the table* implies that all students are male. Nor should you say *A student left their books on the table,* because *their* is plural and does not agree with its singular antecedent *student.* You might say *A student left his or her books on the table,* but it sounds clunky. So what is an editor to do?

Read the following excerpt from the X-Ray Software Company:

Excerpt One

Each software employee is required to download his data onto the mainframe each night. However, if the data files are less than one gigabyte including graphics, the employee may store his data temporarily on his own hard drive, but no employee is to keep more than two days of data on his hard drive. If data storage becomes a problem, then he should immediately notify his supervisor.

Although using the pronoun *his* may have been grammatically correct forty years ago, using the masculine form is no longer acceptable. When the sex of the subject is indeterminate, as in this excerpt, the editor must make certain the writing is gender-neutral. Now examine the next example.

Excerpt Two

Each software employee is required to download his or her data onto the mainframe each night. However, if the data files are less than one gigabyte including graphics, the employee may store his or her data temporarily on his or her own hard drive, but no employee is to keep more than two days of data on his or her hard drive. If data storage becomes a problem, then he or she should immediately notify his or her supervisor.

This revision has satisfied the grammatical requirement to maintain gender neutrality; however, the repetition of compound pronouns *he or she* and *his or her* is awkward and distracting. You will find that many technical writers are sensitive to gender-neutral language but will use this awkward form because they frequently do not know the alternatives. Let's look at three solutions.

Solution One

Software employees are required to download their data onto the mainframe each night. However, if the data files are less than one gigabyte including graphics, employees may store their data temporarily on their own hard drive, but employees are not to keep more than two days of data on their hard drive. If data storage becomes a problem, then employees should immediately notify their supervisor.

Revising the manual in the plural avoids sexist language. Rather than refer to a single *employee,* the text uses the plural *employees.* Also, notice that the gender-neutral pronoun *their* (in this sentence, functioning as a possessive adjective, modifying *supervisor*) is frequently used to refer back to the antecedent *employees.*

Solution Two

As a software employee, you are required to download your data onto the mainframe each night. However, if the data files are less than one gigabyte including graphics, you may store them temporarily on your own hard drive, but do not keep more than two days of data on your personal hard drive. If data storage becomes a problem, then you should immediately notify your supervisor.

Using second-person pronouns (*you, your, yourself, yours*) can be an effective way to directly communicate with your audience, especially if you are editing an instruction manual or a policy handbook. Using the second person gives the writing more immediacy and eliminates the compound pronoun problem of using *his or her.* Be certain, though, that before you recommend using the second person to the client, you ascertain that the familiar "you" is appropriate. And try to avoid overusing the words *you, your,* and so on in your rewrite; often you can make the paragraph more readable by leaving some of them out.

Solution Three

Data files larger than one gigabyte including graphics are to be downloaded onto the mainframe each night by every software employee. However, if the files are less than one gigabyte including graphics, they may be temporarily stored on the employee's hard drive, but no employee is to keep data more than two days. If storage becomes a problem, then that employee should immediately notify the supervisor.

This solution to the gender issue simply eliminates pronouns altogether. The excerpt is clear and coherent. However, if you choose to use this solution, be careful not to overuse the primary noun. After all, pronouns substitute for nouns and add a little variety.

TIP FOR TECHNICAL EDITORS

The use of *they, their,* and *them* as singular pronouns is considered acceptable by many people today, but to prevent annoying those people who do not consider such usage acceptable, it is best to avoid it.

EXERCISE 6.5

Figure 6.3 is a draft of an inclement weather policy we adapted from a state's employee handbook. Imagine that the state is one of your company's clients. You have been asked to perform a quick copy edit. After examining the handbook, you notice it has numerous pronoun errors, especially with gender neutrality. Copyedit this draft using one of the three solutions we provided earlier by changing the wording to (1) third-person plural, (2) second person, or (3) by eliminating all pronouns. First consider your audience and be prepared to justify the choice that you believe is best.

A nonessential state employee will receive inclement weather pay when:

1. Adverse weather results prevent him or her from traveling or the local worksite is closed. This applies only to those employees when travel advisories, including local advisories, and the closure of interstates and secondary highways are considered to be a prohibition on local travel. If travel is not prohibited but if the employee is unable to reach the worksite or wants to leave the office early due to inclement weather, he or she must use vacation leave.
2. The worksite is closed because the state does not control access to the worksite.

An essential employee will always be expected to deliver service. He or she is an employee who is requested by management to work during a period of inclement weather. We will do whatever possible to accommodate him or her to the best of our ability.

FIGURE 6.3
Inclement Weather Policy.

This policy does not mean an essential employee should take undue risks during inclement weather. If he or she believes he or she cannot safely reach the worksite or travel home at the end of the scheduled workday, he or she should make arrangements with their supervisor to be absent from work.

Eligibility for Inclement Weather Pay

- An employee who did not work during the period their office was closed will be paid for his normally scheduled shift.
- An employee who worked during the time their office was closed will be paid for his or her hours worked, but will not receive both inclement weather pay and hours worked. Sufficient hours to make up the normally scheduled workday, however, will be paid to him or her. For example, if an employee worked five hours, and the employee's normal workday was eight hours, he or she would receive pay for five hours worked and three hours of inclement weather pay.
- Only a permanent employee is eligible for their inclement weather pay.
- Only an employee who is scheduled to work during the inclement weather period is eligible for inclement weather pay. If he or she was on any type of approved leave, he or she may not change hours to inclement weather pay.
- Inclement weather pay will be granted only for the number of hours the work location was closed, not to exceed the number of hours for which they were scheduled to work.
- Permanent employees with hire dates on the date of office closings will be paid they're inclement weather pay.

Inclement Weather Pay for Salaried Employees

- A salaried employee who works during a period of office closings must record their hours worked on the proper day. Only hours worked should be recorded on the time sheet if she actually worked. Salaried employees will not be allowed to adjust hours worked during periods of office closings or take the hours off on an alternate day.
- A salaried employee who was absent for their entire scheduled shift should record the number of hours of inclement weather pay they were granted for their shift.

FIGURE 6.3 (*continued*)   Inclement Weather Policy.

VERBS: WORDS OF ACTION

Functions of the Verb

Whether a sentence "lives" or needs resuscitating by a seasoned editor often depends on the writer's choice of verbs. Verbs maintain the pulse of a sentence. They create the action that drives the author's intended idea.

Perhaps this description of the verb's role seems a bit dramatic, but if you examine a piece of good writing, you'll discover that whenever possible, the writer uses well-chosen, active verbs. And because English is a hybrid language derived historically from no fewer than six other languages, it has emerged with verb forms that are flexible yet exact. Just think of how you might describe getting from one place to another on foot. You might walk or run, but you could also stroll, sprint, dash, wander, or slink. Each verb gives the reader a different, and more vivid, mental image of your action.

Verbs also complete the second part of the sentence, the predicate. Without the predicate it would be impossible to know what the subject of the sentence intended to do. For example, suppose someone said, "The new employee" or "You and I." Neither of these subjects alone makes sense. However, when you add a good predicate, the sentence might read, "The new employee *wasted no time hanging her diploma on the wall*." Or, "You and I *should drive over and pick up some burgers*." Therefore, a subject that states the topic needs a predicate that comments on the subject.

Verbal Abuse?

Can you guess what the most commonly used verb is? President Clinton apparently didn't know the answer when he said, "Whatever is, is." Yes, *is*, which *is* a form of the verb *to be, is* the most widely used verb and *is* used to denote existence or **state of being**. Writing a paragraph without using *is* or one of its other forms (*am, are, was, were*) *is* very difficult, as evidenced by the writing in this paragraph. However, no one *is* saying that writers should not use a verb form of *to be*; many situations require it. Yet, depending on the level of editing requested by the client, a good editor should recognize the overuse of *is, was, am, are,* and *were* and be prepared to substitute active verbs. Generally, writing containing active verbs appears more professional, accurate, and intelligent.

EXERCISE 6.6

Select at least four pieces of writing: one from a magazine, one from a newspaper, one from a technical manual, and one from a novel written by a famous author such as William Faulkner, Nathaniel Hawthorne, Toni Morrison, or Stephen King. Locate the verbs in each sentence, and then count how many times the author uses a *state-of-being verb* per paragraph. On the average, which type of writing uses the most active verbs per paragraph? What can you infer about using active verbs versus state-of-being verbs?

Authors' note: How do you locate the verb in a sentence? First, ask, "What action is taking place, or is a form of *to be* present?" When you think you have located the verb, ask "who" or "what" and then state the verb. If you have located the verb, these questions will lead you to the subject.

Forms of the Verb *To Be*

Present tense (*am, is, are*)

	Singular	Plural
First person	I *am*	we *are*
Second person	you *are*	you *are*
Third person	he/she/it *is*	they *are*

Past tense (*was, were*)

	Singular	Plural
First person	I *was*	we *were*
Second person	you *were*	you *were*
Third person	he/she/it *was*	they *were*

Other Verb Forms

We have discussed the two types of verbs in the English language: standard verbs that generally show action, and state-of-being verbs (*is, are, am, was, were*). All verbs have five forms, as listed below. Be familiar with the five forms in order to recognize them in a document.

Standard form	Singular (s, es)	Past tense	Past participle	Present participle
carry	carries	carried	carried	carrying
race	races	raced	raced	racing
walk	walks	walked	walked	walking
serve	serves	served	served	serving
swim	swims	swam	swum	swimming
be	is	was	been	being

Notice that two forms are called *participles*. **Past participles** are verb forms ending in *ed* (or an irregular verb ending form; see examples in verb chart), but often they function as adjectives or the passive voice. **Present participles** also are verb forms, but they end in *ing* and may function as an adjective, a noun (*gerund*), or a verb (with an auxiliary). To designate continuous action, both participle forms require auxiliary verbs such as *have, be, do, can, could, may, might, must, ought to, should,* and *would*. Let's see how the participle operates in everyday sentences.

> *The team **is designing** a new website for the company.*
> (The present participle form *designing* combined with the auxiliary verb *is* shows continuous action.)

> *They **have worked** on that website for months.*
> (The past participle form *worked* combined with the auxiliary verb *have* to show continuous action.)

Lack of agreement between the subject and verb within a sentence is one of the most common errors in writing. The subject might call for a singular verb, but the verb form was mistakenly written as a plural. It's important that you recognize this. Follow these general guidelines:

- Find the subject first; then determine if it calls for a singular or plural verb.
- Singular verbs in the present tense are usually created by adding *s* or *es* to their ending.
- Auxiliary verbs generally do not require endings.
- Nouns often form the plural by adding *s* and *es*; verbs, however, are usually singular when they end in *s* or *es*.

> *The young apprentice decides the format for the page*. (present tense)
> (The subject *apprentice* takes the singular verb *decides*.)

> *Stunned by the grammar errors, Warner shook his head in despair*. (past tense)
> (*Shook* is the past tense form of the verb *to shake* and agrees with the singular subject *Warner*. Notice the past participle form *stunned* here acting as an adjective by modifying the noun *Warner*.)

Formatting templates make editors' work easier. (present tense)
(The principal verb is *make* and agrees with the plural subject *templates*. The present participle form of the verb *formatting* acts as an adjective describing the noun *templates* in this sentence.)

Remember, though, the English language has many irregular verbs that may require special endings. Always consult a dictionary when in doubt.

EXERCISE 6.7

Editors should be able to identify and correctly label the parts of speech, not just for their own sake or to show off in the office, but to understand the complexities of the English language and how words interact to form meaning. Locate the nouns, pronouns, and verbs in the following sentences; look for auxiliary verbs as well. Determine if the verb form is standard, past tense, past participle, present participle, or the (*s*) plural form.

1. She brought the computer programs for us to examine.
2. Why did you choose technical editing as a vocation?
3. Occasional writers often feel defensive about their writing because they work alone.
4. Technical editors need to educate writers about the editing process.
5. She seemed to have accomplished the impossible by editing the entire book in one hour.
6. When you make extensive changes in a writer's manuscript, you will want to have it retyped.
7. With our pencils sharpened, we began hard-copy editing the AIDS education pamphlet.
8. When Lamar leaked toner on Stacy's completed manuscript, she quickly exited the building.
9. Determining the audience is one of the most important considerations when editing.
10. The software we received from the company did not boot properly.

Quick Reference Glossary

Verb............................a word that expresses action or state of being in a sentence
Standard verb..............a verb form listed in the dictionary in the present tense
Past tense.....................a verb form that indicates action took place in the past
Past participle..............a verb form that creates the perfect tense, passive voice, or adjectives. Past-participle verbs cannot form a predicate without an auxiliary verb such as *have* or *be*.
Present participle.........a verb form with an *ing* ending. It must be used with an auxiliary verb and also functions as an adjective or noun (gerund).
Auxiliary verb..............a verb that is used with a standard verb to form a past or present participle
Irregular verb...............a verb form that does not follow the typical pattern of conjugation
Notional agreement....the agreement of verbs with their subjects and of pronouns with their antecedent nouns based on meaning
Nominalization..........creating abstract nouns from verbs (verb plus *tion*)
Linking verb..............a verb that links a subject with its complement. It indicates a "state" or the "result" of a process.
State-of-being verb.....a verb that demonstrates existence or time: *is, am, are, were, was, been, being, be*. It can also be used as a linking verb.

The Tech Editor's Cubicle

English verbs, with all their complexities, present difficulty for native and non-native speakers alike. When someone misunderstands a sentence, often a misplaced or misused verb is involved. So, what can go wrong? Plenty. A verb may disagree in number with its subject, the tense of the verbs may shift in time, or the verb, acting as a participle, may "dangle."

Generally, the participle functions in the present tense when it ends in *ing,* and the past tense when it ends in *ed.* As a technical editor, you will need to correct *dangling participles,* meaning those participles where what the participle modifies is unclear. This error occurs when the subject of the sentence does not agree with the participle. Dangling participles distract and slow the reader. Let's look at a few examples.

Hurrying to finish the lawn, my mower broke.
(The subject of the sentence is *mower,* but it does not agree with the present participle, *hurrying.* I am the one hurrying, not the mower!)

Trained in the basics of first aid, the Boy Scout's bandage secured the wound.
(Was the bandage trained in first aid?)

After being boiled, the waitress brought the eggs to her customers.
(Had the waitress been boiled?)

After winning the Battle of Mobile Bay, the Confederate forces surrendered to the Union.
(Notice the misunderstanding here. The Confederates lost the battle; however, the sentence suggests that they won it.)

At first glance, dangling participles do not seem to be problematic—until you read the last two examples above. But technical documents, especially, must be clear. No one wants to become confused when reading an electronic wiring manual, first aid instructions, or any other type of sensitive material. Confusion can really result with international audiences who are unfamiliar with informal usage.

EXERCISE 6.8

Your agency has been contracted to edit technical documents from a well-known international software firm. Copyedit the page in Figure 6.4, which comes from a management manual. It contains dangling participles, nominalizations, and sexist language. See how many errors you can find, and then continue on to the section that discusses subject-verb agreement. After completing this task, list at least five generalizations you might make about the intended audience for this manual.

As you know, the subject of a sentence and its verb must agree in person and number. If the subject is singular, the verb must also be singular, and if the subject is plural, then the verb must be plural. Fortunately, most verbs in the English language, with the exception of *to be,* are regular. Notice, though, that as a very general rule (with many exceptions), verbs that end in *s* or *es* are actually singular, just the opposite of nouns that end in *s* or *es,* which are usually plural. *She plays. They work. Pablo asks. We answer.*

Although to address all the vagaries that affect subject-verb agreement is far beyond the scope of this text, we provide a few important points here but expect that you will consult a handbook when you encounter difficulty.

AFD CHAPTER 14 612-004
Policy #243
Employees:

1. Completion of the Work Designation Schedule is mandatory before you sign off on Report of Credit Hours, (AFD #25-MM6).

2. Recording of the Report of Credit Hours form, the Compensatory Time or Overtime Earned/Used or Leave Used forms (AFD #29-MM7) must not show deviation from the normal tour of duty. If an employee fails to submit this form properly and in a timely manner, he may be subject to loss of pay.

3. By signing all forms, accuracy is attested to by the employee.

4. You should maintain the Report of Credit Hours and the Compensatory Time or Overtime Earned/Used or Leave Used form during the current pay period and provide accessibility for your supervisor to review the forms upon their completion at the end of each month.

5. If the employee obtains advance approval for leave, he/she must receive written approval from his or her supervisor at least 8 hours prior to the granting of such leave.

6. Submitting at the close-out day of each pay period, the Report of Credit Hours must be signed by the employee's supervisor. He must then submit the Report of Credit Hours to the Accreditation Dept. for validation. If he is unable to complete transmission of the form to the Accreditation Dept., the employee must make arrangements with his or her supervisor to ensure the form is completed in a timely manner.

FIGURE 6.4
Human Resources Management Manual.

Occasionally, verbs do not grammatically agree with their subjects, but this case is rare. Sometimes, verbs must agree with their subjects in meaning, which may not be grammatical. This is called *notional agreement*. For example, *The mumps is a childhood disease,* or *Five thousand dollars is the cost of the car*. Here, the subject and verb do not agree grammatically but do agree notionally. *Mumps* and *five thousand dollars* are considered collectively as single units and therefore use a singular verb. Handbooks often provide lists of troublesome subject-verb combinations, so be sure to keep one handy.

Compound subjects require plural verbs, but be careful. Do not confuse the use of *and* with *or. Xealius and Jigar want to go with us* has two subjects, *Xealius* and *Jigar*. Two subjects joined with *and* are always plural. However, *Xealius or Jigar wants to go with us* still has two subjects, but the subject nearest the verb in this case decides whether or not the verb is plural. For example, *Xealius or the boys want to travel in the bus*. The plural noun *boys* is nearest the verb *want* and therefore dictates that the verb be plural. Here are a few more examples.

Neither the boys nor the girl has time to help with preparing lunch.
Miguel and Hammad swim the channel regularly.
The house, the barn, and the workshop were included with the property.
Either the house, the barn, or the workshops were included with the property.
Either the house, the barn, or the workshop was included with the property.

The most difficult aspect of editing a document for subject-verb problems centers on finding the subject; generally, verbs are fairly simple to find by locating the word or words that express action or state of being. If you cannot locate the subject, find the verb and say "who" or "what" and then state the verb. For example, *Everyone was in class today*. Who or what *was* in class today—*everyone*. Finally, here is one final tip about locating the subject of a sentence; it will not be in a prepositional phrase. *Every one of the boys was in class today*. The noun *boys* is the object of the preposition *of* and therefore cannot be the subject of the sentence.

EXERCISE 6.9

Many, but not all, of the verbs in the NOAA document in Figure 6.5 have been altered. This is a press release to the general public. Determine which verbs have been altered and correct them. Also, if you believe you can clarify or simplify any of the sentences within the document, make the changes.

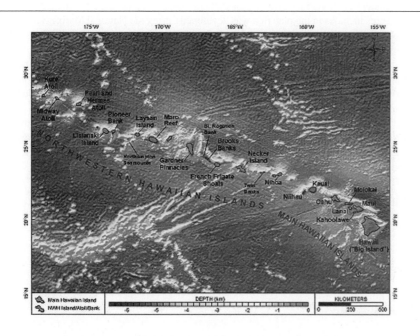

The Northwestern Hawaiian Islands Marine National Monument encompass nearly 140,000 square miles of U.S. waters, including 4,500 square miles of relatively undisturbed coral reef habitat that is home to more than 7,000 species. The monument will be manages by the Department of the Interior's U.S. Fish and Wildlife Service and the Commerce Department's National Oceanic and Atmospheric Administration, in close coordination with the State of Hawaii.

"Along with the two national wildlife refuges already in the area, this national monument provide permanent protection and conservation for the extraordinary natural resources and wildlife of the Northwestern Hawaiian Islands," said Interior Secretary Dirk Kempthorne. "Relatively untouched by human activities, these isolated waters and coral reefs affords vital habitat for the endangered Hawaiian monk seal, the threatened Hawaiian green sea turtle and other rare marine species."

FIGURE 6.5   National Oceanic and Atmospheric Administration press release.

Source: Commerce Department National Oceanic and Atmospheric Administration.

"This is a landmark achievement for conservation, protection and enhancement of the Northwestern Hawaiian Islands," said Commerce Secretary Carlos Gutierrez. "Approximately one quarter of the species here is found nowhere else in the world and a marine national monument provide comprehensive, permanent protection to this region."

The national monument located in waters off the Hawaiian Islands Reservation established by President Theodore Roosevelt in 1909, the Hawaiian Islands National Wildlife Refuge and Midway Atoll National Wildlife Refuge, site of the key World War II sea battle and the Battle of Midway National Memorial.

Permits shall be required for activities related to research, education, conservation and management, native Hawaiian practices and non-extractive special ocean uses. The commercial and recreational harvest of precious coral, crustaceans and coral reef species will be prohibited in monument waters and commercial fishing in monument waters shall be phases out over a five-year period. Oil, gas and mineral exploration and extraction will not be allowed anywhere in the monument.

Prior to today's designation, this unique region had been part of a five-year study under a National Marine Sanctuary designation process, during which federal and state entities, native Hawaiian leaders and the public did participate in strong collaboration with significant amounts of testimony and input to develop a plan with broad-based consensus. Since 2000, more than 52,000 public comments were received, most supporting strong protection.

The President's action today means immediate protection, immediate implementation of the management measures includes in the plan that was developed during the National Marine Sanctuary designation process, and immediate starting of the "seamless" federal/state management process that will be including ongoing consultation and involvement with the public.

Secretary Kempthorne noted that Hawaii Governor Linda Lingle recently approval the establishment of a marine refuge in the Northwestern Hawaiian Islands. "States and federal partners, national and local conservation organizations and thousands of interested individuals have made possible the protection of this national monument. This is collaborating at its best," Kempthorne said.

Under the Antiquities Act of 1906, which is celebrating its 100th year of enactment, the President of the United States authorizes to declare by public proclamation, historic landmarks, historic and prehistoric structures, and other objects of historic or scientific interest that is situated upon the lands owned or controlled by the Government of the United States to be national monuments.

FIGURE 6.5 (*continued*)   National Oceanic and Atmospheric Administration press release.

Source: Commerce Department National Oceanic and Atmospheric Administration. Map from http://www.hawaiianatolls.org/maps/index.php.

ADJECTIVES: WORDS THAT DESCRIBE

Functions of the Adjective

If verbs are the pulse of a sentence, then adjectives (and adverbs, which we will discuss next) maintain the pulse and keep it healthy. Adjectives accomplish this by describing, identifying, or limiting the meaning of nouns or pronouns and by answering the questions *which, what kind,* or *how many.* For example, notice how adding adjectives directs the revised sentence by identifying and limiting the nouns they modify.

The design took editors weeks and dollars to develop.
*The **final** design took **many graphic** editors **three** weeks and **$60,000** to develop.*

Without these adjectives, we would not know which design, how many and what kind of editors, or how many weeks and dollars were involved.

TIP FOR TECHNICAL EDITORS

Technical documents by their very nature must be accurate. Use adjectives to your advantage when editing. First, ask if the writer uses any adjectives that might mislead the reader because they are too imprecise. For example, just what does "many," "a few," "some," or "a lot of" mean? However, given the context of the document, these indefinite adjectives may be sufficient; you'll have to decide. Or you may want to select a good synonym that is more precise and clear. Regardless, you need to pay attention to adjectives that may mislead the reader.

Types of Adjectives

- Descriptive: adjectives that describe qualities and attributes. *Beautiful* flower, *pleasant* afternoon, *old* DVDs.
- Limiting: adjectives that do not describe qualities and attributes but simply identify the noun or pronoun they modify. *My* place, *these* windows, *our* CDs. Note: personal, relative, demonstrative, and indefinite pronouns often function as limiting adjectives. *My* hair, *Verndon's* book, *That* church.
- Articles: *a, an,* and *the* are another form of limiting adjectives. Use *a* or *an* before a singular count noun when the identity is unknown (*An instructional designer who . . .*) and *the* before a singular count noun when the identity is known (*The pharmacist who filled my prescription . . .*).

EXERCISE 6.10

Underline the adjectives in the following sentences. After you have found the adjectives, locate the verbs, pronouns, and nouns as well.

1. A good technical editor can edit onscreen as well as on paper.
2. Editing for audience appropriateness may require the editor to sit down with the author of the document.
3. Sitting in front of a computer for long periods can be exhausting.
4. Most technical editors are also technical writers.
5. Editing requires a good working relationship with the writer and the design team.

ADVERBS: WORDS THAT MODIFY

Functions of the Adverb

Similar to adjectives, adverbs modify other parts of speech by describing or limiting them. However, instead of modifying nouns or pronouns, adverbs modify adjectives, verbs, or other adverbs. Adverbs answer the questions *how, when, where, why, how much,* or *how often*. Entire phrases and clauses—that is, words that combine as a grammatical unit—can also function as adverbs.

*The Internet connection booted **up slowly**.*
Up and *slowly* are adverbs modifying the same verb, *booted*. Which questions do each of the adverbs answer?)

Be careful not to confuse adverbs with adjectives. Adjectives are used when they follow linking verbs, including all forms of *be, look, appear, seem, sound, feel, smell, taste, become, grow, prove,* and *turn*. For example, in the sentence *The technician looked **sleepy,*** the word *sleepy* functions as an adjective because it follows the linking verb, *looked,* and is describing the noun *technician*. However, look at the example *On Fridays, we dress **casually** for work. Casually,* in this sentence, functions as an adverb modifying the verb *dress* and answers the question *how* we dress.

EXERCISE 6.11

Find the adjectives and adverbs in the following sentences. Also, determine if the sentence contains a linking verb.

1. Technical editors often specialize in a particular area or field.
2. *The Chicago Manual of Style* is one of the most commonly used style guides.
3. Editing for correctness includes editing for spelling, grammar, and punctuation.
4. Quickly email this final graphic to the designers before the deadline.
5. Informal tables look good when they break up long paragraphs.

Quick Reference Glossary

Adjective..............................a word that modifies a noun or pronoun and answers the question *which, what kind,* or *how many*
Adverb.................................a word that modifies an adjective, verb, or another adverb and answers the question *how, when, where, why, how much,* or *how often*
Article.................................a limiting adjective such as *a, an,* and *the*
Descriptive adjective.............an adjective that describes qualities and attributes
Limiting adjective.................an adjective that does not describe qualities and attributes but simply identifies the noun or pronoun it modifies

---

The Tech Editor's Cubicle

After you navigate through a stylistically difficult technical document, any problems you encounter with adjectives might seem relatively simple to correct. Unfortunately, adjectives can also be confusing, especially when writers mistakenly substitute them for adverbs. For example, do you know when to use *bad* or *badly, good* or *well, sure* or *surely, real* or *really, near* or *nearly,* or the many other possible word combinations that can function as an adjective or an adverb? Learn the following and you will have little difficulty choosing the correct form when editing.

First, remember that both adjectives and adverbs modify other words. Adjectives modify nouns, whereas adverbs modify verbs, adjectives, and other adverbs. You can recognize an adverb easily. Adverbs often end with the suffix *ly,* such as *carelessly, quietly, extremely,* and so on. Also, adverbs answer the questions "who," "what," "when," "how," "where," "how much," or "how often." Adjectives, on the other hand will modify a noun and answer the question "what kind of."

Now determine the correct form for the following sentences:

William speaks (careless or *carelessly).*

William speaks carelessly is correct. *Carelessly* is an adverb modifying the verb speaks and answer the question how William speaks.

Tipper is careless.

Careless acts as an adjective here because it modifies the noun *Tipper* and answers the question *what kind of* person Tipper is. You would not say *Tipper is carelessly,* or *Mohammad has a happily hamster,* or *I was nervously.* But you might say *Mohammad's hamster happily performed his tricks,* or *I nervously awaited my blind date.* Always look at what the modifying word modifies: a noun or a verb, adjective, or other adverb.

Second, remember that adjectives follow a form of the verb to be when they modify a noun before the verb. Examine these examples:

I am nervous.
They were tired and hungry.
Samantha was sick yesterday.

Finally, one other method you can use to determine the difference between an adjective and an adverb is to remember that an adjective follows a sense verb such as *feel, smell, look, seem, taste, sound, appear,* and so on when it modifies the noun before the verb:

The noise coming from the computer sounds bad.

Here, *bad* functions as an adjective clarifying what kind of noise the computer is making. If you mistakenly used the adverb *badly* you would be saying that the sound itself is not sounding properly.

The mountain air smells sweet.

The adjective *sweet* modifies the noun *air* after the sense verb, *smell.* If you used *sweetly,* an adverb, you would be saying how the air is smelling something, which doesn't make any sense because air doesn't have a nose!

Look at the following sentences and determine the proper form—adjective or adverb.

The copy editor in the next cubicle seems (unhappy or *unhappily) today.*
The Chinese food tastes (divine or *divinely).*
The cat smells (careful or *carefully)*

In the last example, ask yourself if the noun *cat* or the verb *smell* is being modified. Here, *carefully* must be used because it modifies the verb *smell.* The cat does not give off a *careful* odor; that simply doesn't make sense.

Bad or Badly?

Use the adjective *bad* to describe how you feel, because *feel* is a sense verb. *I feel bad.* Suppose, though, you said *I feel badly.* Because badly is an adverb and modifies the verb *feel,* you are saying that you are unable to feel; that is, your hands or your emotions have lost their feeling function!

Good or Well?

Well is an adverb; *good* is an adjective. So, you live *well* and you do *well,* but remember the tricky exception. You say you feel, look, or smell good because the adjective *good* follows a sense verb. If you say *I smell well,* it means your nose is doing a great job!

Sure or Surely?

As you might expect, *sure* is an adjective, *surely* an adverb. Look at our examples:

She was sure that she had given him the memo.
I am surely ready to begin the exam.
Surely, Isaac Newton's prediction for the end of the world in 2060 is wrong.

Real or Really?

The same is true here; *real* is an adjective and *really* is an adverb.

I did really well on my computer exam.
You can't really be going out with that nerd.
That is a real good reason to leave the circuit breaker alone.

Near or Nearly?

Determining the difference here is not as easy as the previous examples. *Near* can function as an adjective, a verb, or an adverb, while *nearly* means "almost but not quite."

As we neared the deadline, we felt our exhaustion.
The lioness circled near.
She is more nearly related to Igor than I am.
They are nearly finished with the audit.

EXERCISE 6.12

Select the correct form of adjective or adverb. Be prepared to defend your answer.

1. I'm (real, really) angry how the challenge match turned out.
2. My dog feels (real, really) (bad, badly).
3. It's (good, well) that the time for the test has come.

4. I'm not (near, nearly) ready for the copy editor to see this.

5. That was a (careless, carelessly) thrown dart at the target.

TIP FOR TECHNICAL EDITORS

Most words that end in *ly* are adverbs, which makes them easy to identify (*really, unfortunately, usually, excitedly*). One exception is *lovely,* which is an adjective: *That **lovely** hat makes my head itch.*

CONJUNCTIONS: WORDS THAT JOIN

Functions of the Conjunction

Similar to verbs, adverbs, and adjectives, which all contribute to the dynamics of a sentence, conjunctions play an important role. However, unlike these three other parts of speech that either place the subject in action or otherwise modify and refine the sentence's meaning, conjunctions join elements together. When conjunctions join elements, they show important relationships between ideas.

For example, in the following sentence the conjunction *and* demonstrates that two elements are equal. *The technical writers **and** the engineers brought their laptops to the conference.* However, conjunctions also join elements together that are unequal, such as ***Although** we waited patiently for the markup copy to arrive, we finally had to quit and go home.* Notice how the clause beginning with *although* establishes a contrast to the independent clause that follows.

Types of Conjunctions

Coordinate Conjunctions

The coordinate conjunctions (*and, but, or, nor, for, so, yet*) join corresponding elements.

*Many students first learn to edit for correctness **and**, after they gain confidence, move on to more complex tasks.*

TIP FOR TECHNICAL EDITORS

Want an easy way to remember all of the coordinating conjunctions? The first letter of each conjunction spells out the acronym FAN BOYS: *for, and, nor, but, or, yet,* and *so.*

Correlative Conjunctions

Correlative conjunctions (*either/or, neither/nor, both/and, not only/but also*) provide greater emphasis between coordinated elements.

__Either__ you finish the markups by five, __or__ you will need to work until they are completed.

__Neither__ the writers __nor__ the editors have completed their assignments on time.

Subordinate Conjunctions

Subordinate conjunctions (*after, although, as, as if, because, before, even though, if, in order that, once, since, so that, than, through, unless, until, when, where, while,* etc.) introduce subordinate clauses.

__Although__ we tried to edit the documents onscreen, we decided to resort to hard copies instead.

__If__ you delete those emails from your files, Mr. Naybob says he is going to issue a summons for your arrest.

She left our employment __because__ she found a higher-paying job in London.

Conjunctive Adverbs

Conjunctive adverbs (*also, besides, consequently, furthermore, however, in addition, instead, moreover, nevertheless, otherwise, therefore,* etc.) help connected independent clauses transition more smoothly.

The graphics were not printed in color; __consequently__, we had to revise our contract.

We worked for hours on our PowerPoint presentation; __however__, when the bulb blew in the LCD, we had to resort to handouts.

TIP FOR TECHNICAL EDITORS

Please do not put a comma before the subordinating conjunction *because*. Usually, no comma is needed before a subordinating conjunction if the dependent clause immediately follows the independent clause. For example, *We were late for class because our car had a flat tire.*

EXERCISE 6.13

It's time for you to create sentences using the various types of conjunctions. This will give you good practice learning how to combine sentences and improve readability. On a sheet of paper, write at least four sentences for each type of conjunction. After you have completed your sentences, exchange your responses with another student and peer-edit each other's papers. Be sure to edit for correct punctuation too.

TIP FOR TECHNICAL EDITORS

By skillfully incorporating conjunctions into a manuscript, an editor can create coherence between ideas and make the writing much easier to follow. When editing, determine whether the manuscript would read more fluently by combining ideas using conjunctions. However, be aware that conjunctions can be overused as well. Stringing too many ideas together with conjunctions will create a run-on effect.

PREPOSITIONS: WORDS THAT SHOW RELATIONSHIPS

Functions of the Preposition

Prepositions communicate relationships in time and space, often in the form of prepositional phrases. For example, *We decided to look underneath the staircase for the old manuscripts.* This sentence contains two prepositional phrases: *underneath the staircase* and *for the old manuscripts.* Notice how the addition of these phrases expands the amount of information. Prepositional phrases can also emphasize important details. Remember, prepositional phrases begin with a preposition and end with either a noun or pronoun.

TIP FOR TECHNICAL EDITORS

Please don't confuse phrases with clauses. A phrase is simply a group of grammatically related words that do not contain a subject or predicate. Phrases can function as nouns and various types of modifiers. A clause, however, is a group of grammatically related words that does contain a subject and a predicate and, like a phrase, can be used as a noun or modifier.

Some Commonly Used Prepositions

about	at	except	near	since
above	below	during	of	through
across	before	except	off	toward
after	beneath	from	on	under
among	between	in	onto	up
around	beyond	inside	out	with
as	by	into	over	without

Be sure not to overdo a good thing. We encourage you to use prepositional phrases judiciously. You will discover that many inexperienced writers string too many prepositional phrases together in a single sentence. If necessary, create two separate sentences when this occurs. Moreover, when determining if a subject agrees with its verb, don't look for the subject within a prepositional phrase. For example, *None of the copy editors is ~~are~~ here today* is correct because the subject is *none,* not *copy editors,* which is the object of the preposition *of.* Study the examples below where the prepositional phrases are double-underlined. Identify the subject of the sentence. Then identify each noun or pronoun that is the object of the prepositional phrase.

Headings and subheadings help organize information <u>within a document</u>.

Although editors must be excellent wordsmiths, they must also be able to evaluate the effectiveness <u>of a document on screen</u> or <u>in the pages of a book</u>.

Being able to think <u>outside the box</u> is an important attribute <u>for graphic designers</u>.

Jamel tried to install the new hard drive <u>without the manual</u>; however, <u>after he</u> removed many components, he decided to read the directions.

EXERCISE 6.14

Circle the prepositional phrases in the following sentences.

1. Finally finished, Abigail ran the report across the room, down the stairs, and into the hands of her boss.
2. Although it was past closing time, we waited at the door for the manager to arrive.
3. During our conversation with the client, we determined we would need additional resources to complete her project.
4. A blast of heat came through the vent and blew our proofs throughout the room.
5. The carpenter as well as the file clerks had to help push the bookcase back into place.

The Tech Editor's Cubicle

Why must a technical editor be able to identify prepositions and prepositional phrases? Read our example below from the *U.S. Army Field Manual 27-10*. Suppose you were asked to copyedit it; could you locate subject-verb agreement problems? By first locating the prepositional phrases and knowing that the subject of the sentence is never located with the prepositional phrase, your task becomes fairly easy. Remember, first find the verb and then state "who" or "what." Then ignore all prepositional phrases when looking for the subject.

The protection of civilian persons is governed by both GC and HR, the former supplementing the latter insofar as both relate to occupied territory.

Here the state-of-being verb *is* agrees with the single subject *protection*. *Persons* is the object of the preposition *of* and cannot be the subject.

EXERCISE 6.15

Unfortunately, you will likely encounter poorly written documents similar to this army manual. Strings of prepositional phrases make the writing particularly convoluted. Richard Lanham's book *Revising Prose* suggests using what he calls the Paramedic Method to revise and clarify the prose. Essentially, the Paramedic Method "resuscitates" weak prose by reducing the number of prepositional phrases and, when possible, incorporating action verbs.

Copyedit the following document by selecting the correct verb where indicated. Then try your hand at revising the document using the Paramedic Method by excluding and rewriting as many prepositional phrases as you can. For example, *The treaties pertaining to land warfare and ratified by the United States are listed in the abbreviations section of this manual* improves the first line of the foreword in the document below.

Foreword

A list of the treaties relating to the conduct of land warfare which have been ratified by the United States, with the abbreviated titles used in this Manual, **(are, is)** set forth in the abbreviations section of this manual. The official English texts or a translation of the principal treaty provisions **(are, is)** quoted *verbatim* in bold type in the relevant paragraphs throughout the Manual. It should be noted, however, that the official text of the Hague Conventions of 18 October 1907 **(are, is)** the French text which must be accepted as controlling in the event of a dispute as to the meaning of any provision of these particular conventions. (See TM 27-251.)

   The 1949 Geneva Conventions for the Protection of War Victims **(has been, have been)** ratified by the United States and came into force for this country on 2 February 1956. The **(effect, affect)** of these four conventions upon previous treaties to which the United States is a party **(are, is)** discussed in detail in paragraph 5 of the text. Each of the Hague Conventions of 1899 and 1907 and each of the Geneva Conventions of 1864, 1906, and 1929 will, of course, **(continues, continue)** in force as between the United States and such of the other parties to the respective conventions as have not yet ratified or adhered to the later, superseding convention(s) governing the same subject matter. Moreover, even though States may not be parties to, or strictly bound by, the 1907 Hague Conventions and the 1929 Geneva Convention relative to the Treatment of Prisoners of War, the general principles of these conventions **(has been, have been)** held declaratory of the customary law of war to which all States are subject. For this reason, the United States **(have, has)** adopted the policy of observing and enforcing the terms of these conventions in so far as they have not been superseded by the 1949 Geneva Conventions which necessarily govern the relations between the parties to the latter (see pars. 6 and 7 of the text).

   The essential provisions of each of the earlier conventions mentioned above **(has been, have been)** substantially incorporated into the more recent and more comprehensive conventions on the same subject matter, so that observance of the latter will usually **(includes, include)** observance of the former. For this reason, only the more recent 1949 Geneva Conventions and the relevant provisions of the 1907 Hague Conventions are quoted in this Manual.

Source: The United States Army.

INTERJECTIONS: WORDS THAT SHOW EMOTION

Functions of the Interjection

   Although you may have few occasions to use interjections as a technical editor, you need to be able to recognize them as a part of speech. Generally, interjections are used only in speech or in written dialogue, so we doubt you will encounter them very often. Interjections express extreme emotion or surprise.

   "Ahhh!" screamed the supervisor.
   "Quick! Get the fire extinguisher!" the reporter yelled.

Quick Reference Glossary

Adjectives.....................words that join elements of a sentence together
Prepositions................ words that communicate relationships in time and space
Interjections.................words that show emotion

EXPLETIVES: WORDS THAT ARE OFTEN UNNECESSARY

Perhaps you have heard the term "expletive deleted." This simply means that an obscene word or expression, an expletive, was cut from the text because it is offensive. Expletives also denote a word or phrase that otherwise has no specific meaning but merely fills in space. Common types of expletives include *There is* and *It is* constructions. For example, *There is a fee of $150 for this course.* A more direct statement would read, *The fee for this course is $150.* The sentence *It is the anticipation of the IT team to have the problem corrected before noon* reads better as *The IT team expects to correct the problem before noon.*

SUMMARY

Anyone nit-picking enough to write a letter of correction to an editor doubtless deserves the error that provoked it.
—*Alvin Toffler*

After reviewing the previous sections on the parts of speech, do you remember more than you originally thought? If you believe you don't have a good grammatical background, understand that your daily speech conforms to many of the conventions of grammar. And writing generally reflects those patterns of speech.

Before going further, be certain you comprehend the functions of each part of speech. As you know from looking up words in the dictionary, words can assume different roles in a sentence. You can *book* a flight or read a *book*. Basic editing requires an editor to distinguish between words that follow the conventions of grammar and those that do not. Remember, too, that your knowledge and command of grammar will in many ways determine your success as a technical editor.

CHAPTER 7

The Sentence: Giving Meaning to Words

At the conclusion of this chapter, you will be able to

- identify complete sentences;
- differentiate between phrases and clauses;
- categorize phrases and clauses according to their functions;
- identify the four major types of sentences;
- recognize sentence-level errors in a written contract;
- apply your knowledge of sentences to editing technical documents.

Perhaps you remember reading *The Old Man and the Sea* in high school. Maybe you appreciated Hemingway's style because it seemed straightforward, simple, and easy to comprehend. Seemingly, he could have written his short novel in a couple of afternoons poised behind his favorite table at Sloppy Joe's, his beloved bar in Key West. However, a closer analysis of the novella's outwardly effortless style reveals a splendidly inspired treatment of language that controls the setting, characters, and theme. When asked how he devised such a masterfully written work, Hemingway admitted he revised *The Old Man and the Sea* over 100 times before he felt satisfied.

> When something can be read without effort, great effort has gone into its writing.
> —*Enrique Poncela*

Hemingway's point is clear: good writing can only be achieved through revision and good editing. Certainly, no one expects you to revise either your own or someone else's manuscript 100 times! However, good writing can only be as good as its last revision. The final manuscript, no matter how many drafts it has taken to get there, becomes the only opportunity to communicate to the reader.

> If you reread your work, you will find on rereading that a great deal of repetition can be avoided by rereading and editing.
> —*William Safire*

Technical editors and editor-writers help give language meaning. They are the artisans who apply the final "sanding, staining, and varnishing" to create a professional document. From our experiences working with editors, all of them agree on one thing: editors must know the relationship between words and how they combine to create meaning. The conventions of grammar and usage, then, become the editor's tools for understanding these relationships.

The previous chapters have reviewed the eight grammatical classifications of speech (nouns, pronouns, verbs, adjectives, adverbs, conjunctions, prepositions, and interjections). The following sections are intended to help you understand how these classifications form meaningful sentences. If you already have a good background in grammar, use these sections as a quick review. However, if your background is lacking, use them to become familiar with these crucial concepts. Upon completion of each section, you should have a good grasp of the conventions of grammar and how to edit for correctness. Before going further, acquaint yourself with the "big picture." Learning grammar and usage is a systematic process requiring a firm foundation. The next four sections are designed to guide you through this building process.

The Big Picture: A Process Approach

What Is a Sentence?

- First, we will examine the parts of the sentence and their relationships (subject, predicate, complements).

Phrases and Clauses

- Second, we will see how groups of related words called phrases and clauses relate to other elements in a sentence.

Sentence Types

- Third, we examine the four different types of sentences and how to construct them.

Problem-Solving

- Finally, we present common problems and help you diagnose them.

WHAT IS A SENTENCE?

"I am." "Go!" These are the shortest sentences in the English language, but despite their brevity, they contain all the standard conventions that comprise a sentence. The first sentence possesses a subject, in this case the pronoun *I*. It contains a verb *am,* and it completes a thought or idea. This is all that is necessary. The second sentence contains the understood subject *you,* the verb *go,* and completes a thought by commanding you to take action.

Obviously, though, sentences are capable of containing much more meaning, depending on the author's intentions and use of language. For example, adding a single adjective to *I am* to read *I am cold* provides new information and completely changes the meaning.

Subject and Predicate

All sentences in the English language contain a *subject* that identifies what the sentence is about, and a *predicate* that either asks something about the subject or tells what the subject is doing.

Subject	Predicate
I	am
The Mets	lost the game
We	shall overcome
The old woman	swam the English Channel

The *simple subject* is a noun or pronoun; the *complete subject* is the simple subject plus all of its modifiers. A *compound subject* has two subjects usually joined by a conjunction. Similarly, the *simple predicate* is the verb (and any auxiliaries), while the *complete predicate* consists of the simple predicate plus complements or objects and their modifiers. A *compound predicate* has one subject but two verbs usually joined by a conjunction.

$$\underset{\text{subject}}{\text{}} \qquad \underset{\text{predicate}}{\text{}}$$

subject　　　　　　　　　*predicate*
Usually, the simple subject comes first, followed by the verb and its complements.

Basic Sentence Patterns

1. *The <u>laptop</u> <u>arrived</u> on the last shipment.*　　(simple subject + simple predicate)
2. *<u>The laptop</u> <u>was mailed to Sonja</u>.*　　(complete subject + complete predicate)
3. *<u>Yvette and Steve</u> <u>edited the journal</u>.*　　(compound subject + simple predicate)
4. *<u>Arin</u> <u>purchased a computer and borrowed a desk</u>.*　　(simple subject + compound predicate)

Subject Complement

When a subject is followed by a linking verb (*look, seem, appear, become*) or a "to be" verb (*is, are, was, were, am, have, been*) that identifies or describes the subject, it is called the *subject complement*. This word or group of words functions as a noun or adjective, depending on the writer's intention. If the complement acts as a noun, it is called a *predicate nominative* or *predicate noun*; if it acts as an adjective, it is called a *predicate adjective*.

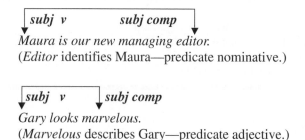

subj v   subj comp
Maura is our new managing editor.
(*Editor* identifies Maura—predicate nominative.)

subj   v   subj comp
Gary looks marvelous.
(*Marvelous* describes Gary—predicate adjective.)

Transitive and Intransitive Verbs

What happens, though, when the subject is not followed by a linking verb? In this case, the verb is said to be either *transitive* or *intransitive*. Transitive verbs require one type of complement (called a *direct object*) to complete the sentence, whereas intransitive verbs do not. To determine if a sentence requires a direct object, first locate the verb, then ask "who" or "what."

transitive verb

          subj   *v*   *dir obj*

Standardized exams require <u>a number two pencil</u>.
(Exams require what? Note: The sentence would not be complete without the direct object *a number two pencil*.)

intransitive verb

    subj   *v*

The flight due from Albany unexpectedly <u>crashed</u>.
(Note: The verb *crashed* does not require a direct object. "Who" or "what" isn't an issue in this sentence.)

EXERCISE 7.1

Locate the simple subject, the complete subject, the simple predicate, and the complete predicate in the following sentences. Determine if the sentence contains a linking verb and whether the verb is transitive or intransitive. Can you explain how each of these independent clauses constitutes a sentence? Why will this knowledge help you as an editor?

1. Michelle still edits without marking her changes.
2. Traditionally, writers are responsible for the contents of their edited documents.
3. Wilson prefers to use Microsoft Word software for editing, but Miller prefers hard-copy markup.
4. Copymarking instructions appear to be confusing at times.
5. We have finished with the project.

Quick Reference Glossary

Sentence.........................a group of words that contains a subject plus a predicate and completes a thought
Subject..........................what the sentence is about
Predicate......................tells about the subject or asks what the subject is doing
Complement.................a word or group of words that provides additional information about the subject
Transitive verb.............a verb that requires a direct object
Intransitive verb...........a verb that does not require a direct object

---

PHRASES AND CLAUSES

The previous sections have introduced you to the parts of speech and the basic components of the sentence. This section provides an overview of two more components, *phrases* and *clauses*. Similar to complements, these specialized word groups contribute to the sentence by shaping and adding supplemental information. However, one primary difference exists between them. Phrases do not contain a subject and a predicate, whereas clauses contain both. Some clauses can stand alone as complete sentences (*independent clauses*), whereas others cannot (*dependent clauses*) because they do not fulfill the third requirement of a sentence: completion of a thought.

Arguments over grammar and style are often as fierce as those over IBM versus Mac, and as fruitless as Coke versus Pepsi and boxers versus briefs.
—*Jack Lynch*

The engineer decided he couldn't complete the project.
(*independent clause*—can stand alone as a complete sentence)

Although she wanted to travel to Europe
(*subordinate clause*—contains a subject, verb but does not complete a thought)

Afraid to let go,

on the way,

determined by empirical reasoning,

a white building
(*phrases*—word groups that modify but do not contain either a subject or a predicate)

Knowing when to incorporate or delete phrases and clauses is an important editorial task. Below, we have concentrated on only the types of phrases and clauses that give novice writers, editors, and writer-editors the most trouble.

Prepositional Phrases

These are the easiest phrases to recognize in a sentence, especially if you had to memorize all the prepositions during high school. More than sixty prepositions exist in the English language, but you don't really need to memorize them. You may recall a mnemonic device that your teacher taught you: "A preposition is anything a squirrel can do to a stump." We agree that this catchphrase sounds bizarre, but it does work in most cases. A squirrel can go *among, between, over, through, inside, upon, around* (plus over thirty more prepositions) a stump; however, some prepositions such as *during, without, until, since, for,* and *except* won't fit this mnemonic.

Prepositional phrases begin with a preposition (*among, across, between, down, during, inside, through, over, up, upon,* etc.) and conclude with a noun or pronoun, the *object* of the preposition. For example, *Over the river and through the woods toward Grandmother's house we go* contains three prepositional phrases: <u>over</u> *the river,* <u>through</u> *the woods,* and <u>toward</u> *Grandmother's house.* Each prepositional phrase begins with a preposition and ends with a noun. Prepositional phrases can function as adverbs or adjectives depending on their context and what they modify.

*We climbed **up** the stairs, **over** the boxes, and **onto** the platform.*
(Here the prepositional phrases act as adverbs to answer the question *Where did we climb?*)

<pre>   prep   adj noun   prep   adv   adj</pre>
*We climbed **up** the spiral staircase, **over** the randomly scattered*

<pre>   adj   noun   prep   adj   noun</pre>
*cardboard boxes, and **onto** the dilapidated platform.*
(This sentence is similar to the previous example; however, modifiers in the form of adjectives and adverbs have been added within the phrases.)

<pre>   prep noun   prep   noun</pre>
*Our house **on** Fish River is **near** the Gulf of Mexico.*

(The prepositional phrase *on Fish River* acts as an adjective because it modifies the noun *house,* and answers the question *What house*? The prepositional phrase *near the Gulf of Mexico* functions as an adverbial by modifying the prepositional phrase *on Fish River* and answering the question *Where* on Fish River?)

One of the most common writing errors occurs when the subject of the sentence doesn't agree with the verb in number. This happens because writers embed prepositional phrases between the subject and verb and lose track of their original subject. When editing, ask whether the subject of the sentence agrees in number with the verb. First, locate the subject of the sentence. (To locate the actual subject, find the verb and ask "who" or "what" and then state the verb.) The subject will not be in a prepositional phrase. For example, in the sentence <u>One</u> *of the new employees seems* ~~seem~~ *lost,* the subject is *one,* not *employees.* The noun *employees* is the object of the preposition *of*; the subject *one* requires the singular form of the verb *seems.*

Another common problem occurs when writers string too many prepositional phrases together. You will know immediately when this happens because you will think, "Huh?" In other words, the sentence will not make much sense and will need revising, probably into two separate sentences. Both of these issues are addressed in detail in the "Problem-Solving" section later in this chapter.

EXERCISE 7.2

Locate the prepositional phrases in the following sentences. Determine if the phrases function as adverbs or adjectives.

1. Her acceptance speech, given from the hotel balcony, was filled with emotion.
2. Bo tossed his copy of *Technical Communication* over the desk and into the drawer.
3. Deciding on the most appropriate graphic for a particular audience can be a challenge.
4. During the storm, we crouched beneath our desks, except for Michael, who hid behind the curtains.
5. Copymarking symbols are used on hard-copy documents to illustrate where changes need to be made in a document.

Verbal Phrases

Verbal phrases are a bit more complex than prepositional phrases, but with a little practice, they are not difficult to locate. Although verbal phrases contain verbs, the entire phrase does not function as a verb; instead, verbal phrases function as adjectives, nouns, or adverbs. These specialized phrases are called participles, gerunds, and infinitives. It's important to remember that verbal phrases cannot stand alone in a sentence. Participles need an auxiliary verb to function as a verb; otherwise, participles are adjectives or nouns. For example, *The manager multitasking* is not a sentence. *The manager was multitasking* makes a complete sentence because the participle *multitasking* gets help from *was*; *was multitasking* is the verb. However, *The manager's multitasking drove us nuts* shows *multitasking* as a gerund (noun), which is also the subject of the sentence. *The manager, multitasking for sixteen hours straight, collapsed from exhaustion* shows *multitasking* as an adjective modifying the noun *manager*. These three sentences alone demonstrate why it is crucial for editors to know how words function in a sentence.

Verb	Infinitive	Present Participle	Past Participle
row	(to) row	rowing	rowed
go	(to) go	going	gone
dance	(to) dance	dancing	danced
walk	(to) walk	walking	walked
lend	(to) lend	lending	lent
leave	(to) leave	leaving	left

Three Types of Verbal Phrases

Type	Structure	Function
Infinitive phrases	**an infinitive (to + verb) + modifiers**	**adjective, adverb, or noun**

Ex. You need *to know* what you want in life. (infinitive phrase acting as an adverb)

Participial phrases	**participle (past or present) + modifiers**	**adjective**

Ex. *Seeing misplaced modifiers in his document,* I finally gave up. (participial phrase acting as an adjective, modifying *I*)

Gerund phrases	**present participle + modifiers**	**noun**

Ex. *Giving our time and energy* is how we got the new computer for free. (gerund phrase acting as a noun and as the subject of the sentence)

Infinitive Phrases

Infinitive phrases consist of a verb preceded by *to* (*to walk, to stop, to format,* etc.) and may include modifiers. These phrases function as adjectives, adverbs, or nouns. However, do not get them confused with prepositional phrases such as *Samantha walked to school.*

Nathan is the typist to beat. (infinitive phrase acting as an adjective)

TIP FOR TECHNICAL EDITORS

Some editors don't think splitting an infinitive with an adverb is a serious problem. However, for better readability, we recommend that you try not to interrupt infinitives with adverbs. For example, write *I want to install Quicken now* instead of *I want to now install Quicken*. Nevertheless, some adverbs sound very "natural" interrupting the infinitive. For instance, the title sequence of the famous *Star Trek* series featured the phrase "To boldly go where no man has gone before." Might this phrase have had the same impact if it had been "To go boldly" or "Boldly to go"?

When I split an infinitive, I split it so it stays split.
—*Raymond Chandler*

Participial Phrases

All participial phrases function as adjectives. Present participles end in *ing,* whereas past participles end in *ed* or *d* (except for irregular verbs). For example, *Striking the computer with his hand, Max got it to work again* consists of a present participial phrase that functions as an adjective modifying *Max.*

> *Preparing for the meeting,* the engineer reviewed her PowerPoint presentation. (present participial phrase functioning as an adjective modifying *engineer*)

> The new teacher, *stunned by the test scores,* began weeping with joy. (past participial phrase functioning as an adjective modifying *teacher*)

TIP FOR TECHNICAL EDITORS

Another error in writing occurs when a participial phrase does not clearly modify its noun. (Remember, participial phrases function as adjectives, which modify nouns or pronouns.) For example, *Running to catch the bus, my briefcase got caught in the door.* This sentence suggests that the briefcase was running! Here the participle has been left "dangling" because it's not clear what the participial phrase modifies. Revised, this sentence should read, *Running to catch the bus, I caught my briefcase in the door.*

Gerund Phrases

Gerund phrases are present participial phrases functioning as nouns. Because they are nouns, they function as the subject of a sentence, a direct object, or an object of a prepositional phrase.

> The young executive was anxious about *conducting his first meeting.* (object of the preposition)

> *Running* is an excellent way to lose weight. (subject)

> The technical editor's job was *ensuring quality control* among the team. (direct object)

Appositive Phrases

Using appositives is an effective way to add information to a subject or an object by renaming it. Appositive phrases are nouns or pronouns with modifiers that further identify another noun or pronoun. For example, *The intern Larry began work yesterday* incorporates the intern's name, Larry, into the sentence. This is much more concise than writing *The intern began work yesterday. His name is Larry.*

> Many technical document readers, *particularly engineers,* are already oriented to the task discussed in the document.

> She just finished reading The Chicago Manual of Style, *a detailed citation guide.*

Appositive phrases that are essential to the meaning of the sentence require commas, whereas nonessential appositives do not; please see Chapter 8 ("Punctuation") for a discussion of commas with restrictive and nonrestrictive elements.

Unlike phrases, however, clauses differ because they contain a subject and verb. Some clauses stand on their own as independent sentences (*main clauses*), whereas *subordinate clauses* (*dependent clauses*) do not because they fail to complete a thought. Clauses, when used effectively, create internal coherence by establishing logical connections between ideas. This principle is explained further in the "Four Types of Sentences" section below.

Editors need to be familiar with three types of clauses.
(This is an independent clause; the prepositional phrase *with three types of clauses* is a modifier and therefore part of the clause.)

When saving the file
(This dependent clause contains a subject and verb but does not complete a thought.)

TIP FOR TECHNICAL EDITORS

Although phrases and clauses add important details to a sentence, *main clauses* supply the most important information and should receive the most emphasis. While revising a document, technical editors need to remember that excessive use of *subordinate clauses* detracts from the main point and makes reading and comprehension difficult. Before addressing this issue, though, be familiar with the three types of subordinate clauses and their functions.

Noun Clauses

A noun clause is a type of subordinate clause that functions as a subject, an appositive, a complement, a direct object, or an object of the preposition in a sentence. They are easy to recognize because the clause usually begins with *that, how, what, when, whether, where, who, whom, whoever,* or *whomever.* Noun clauses answer the question "who" or "what."

<u>*What you have created*</u> *is a real mystery to the whole production team.*
(The noun clause *What you have created* functions as the subject of the sentence.)

The NASA manual was left with <u>*whoever was at the office*</u>.
(*Whoever was at the office* is the object of the preposition *with*.)

Even if you're not finished with the revision, give me <u>*whatever you have*</u> *completed.*
(This noun clause functions as a direct object.)

Adjectival Clauses

Adjectival clauses, another type of subordinate clause, serve as adjectives to modify nouns or pronouns and answer the question "what kind" or "which one." *Who, which,* or *that* introduce adjectival clauses. Notice that adjectival clauses differ in function from noun clauses that begin with the same relative pronoun.

They were the technicians <u>who networked the office</u>.
(*Who networked the office* modifies the noun *technicians*.)

I plan to seek a career in technical editing, <u>which has interested me for a long time</u>.
(This adjectival clause modifies the noun *editing*.)

Here is the document <u>(that) we sent</u>.
(Sometimes the understood introductory word *which* or *that* is missing.)

Adverbial Clauses

The third type of subordinate clause, the adverbial, modifies an adjective, verb, or adverb and answers the question "how," "when," "where," or "why."

<u>Because the network was down,</u> we could not finish the project on time.
(This adverbial clause modifies the verbs *could finish* and answers the question "why?")

Technical editors consult with the client <u>before they choose a specific style manual</u>.
(Here, the verb *consult* is modified by the adverbial clause answering the question "when?")

Adverbial clauses are introduced by *subordinating conjunctions*. The most common ones are listed below; however, consult a good grammar handbook for a complete list. Remember, words in the English language often have different functions depending on their role within a sentence. For example, the word might introduce an adverbial clause, adjectival clause, or noun clause. First determine what the clause modifies, and then what question it answers. Resolving these issues initially will indicate which type of clause is being used.

Some Common Conjunctions That Introduce Subordinate Clauses

as	if	when
although	since	why
after	that	while
before	until	whenever
because	unless	wherever

EXERCISE 7.3

Identify the clauses in the following sentences. State the kind of clause and describe how the clause functions in the sentence.

1. Whoever plans to attend the copymarking workshop needs to make reservations by tomorrow.
2. A good editor is one who provides writers with options.
3. Before beginning an editing project, the editor should write a short memo to the writer.
4. She is the one who brought us the style guides.
5. While you finish editing the document for correctness, the rest of us will go out to lunch.

Quick Reference Glossary

Phrase......................................a sequence of grammatically related words that provides additional information but does not contain a subject or predicate

Prepositional phrase.............a phrase that begins with a preposition and acts as an adjective or adverb

Verbal................................a verb that cannot stand by itself in a sentence; they include infinitives, participles, and gerunds

Verbal phrase.......................a verbal plus modifiers

Infinitive phrase..................a verbal phrase with the infinitive form of the verb plus modifiers and acts as noun, adjective, or adverb

Participial phrase.................a phrase that contains a participle and acts as an adjective

Gerund phrase.....................a noun phrase containing a present participle

Clause.....................................a group of words containing a subject and a verb

Main clause.........................a clause that can stand alone as a complete sentence (independent clause)

Subordinate clause..............a clause that cannot stand alone as a complete sentence (dependent clause)

Noun clause.........................a clause that can act as the subject of a sentence, an object, or a complement

Adjectival clause.................a clause that functions as an adjective by modifying a noun or a pronoun

Adverbial clause.................a clause that functions as an adverb and modifies a verb, adjective, or another adverb

---

THE FOUR TYPES OF SENTENCES

Before discussing style and asking you to apply what you have learned regarding the conventions of grammar, one more major issue needs to be addressed: types of sentences. Whereas the previous sections examined the fundamentals of grammar that give sentences meaning, this section explores how those sentences can be shaped to provide a coherent, logical flow of ideas. Four types of sentences exist in the English language: *simple, compound, complex,* and *compound-complex.* Each type delivers the writer's intended meaning in a different manner.

Nostalgia is like a grammar lesson: you find the present tense, but the past perfect!
—*Owens Lee Pomerov*

The Simple Sentence

Simple sentences are usually but not always relatively short. Generally, they contain a subject, verb, complement, and maybe an additional phrase. Stylistically, why would anyone want to use a short, simple sentence? First, simple sentences typically do not provide as much information as do more elaborate sentences. Because they contain fewer phrases and clauses, the subject and verb are normally clear. These types of sentences are especially effective when a document contains difficult subject matter that is best explained through detailed, sequential steps. Simple sentences slow the pace of information, which allows the writer to be more deliberate. In addition, they add variety to the writing when they follow longer, more complex sentences. Simple sentences can contain compound subjects and predicates.

Format the A: drive. (simple sentence)

Natasha and _Pedro_ *have already applied for the new opening in sales.* (simple sentence with compound subject)

The Compound Sentence

Compound sentences are actually two complete sentences (independent clauses) joined by a coordinate conjunction (*and, or, but,* or *for*) or a conjunctive adverb (*however, nevertheless, furthermore, therefore, consequently,* etc.). Don't confuse compound sentences with simple sentences that contain a compound subject or predicate. Compound sentences are easy to construct; simply unite two simple sentences with a coordinate conjunction or a conjunctive adverb.

Compound sentences have several functions. First, they move the reader along quickly by combining more information. Second, by incorporating conjunctions, compound sentences create consistency between ideas and allow the writer to coordinate closely related ideas. For example, when writers use the conjunction *and* (which means *plus*), they are combining two ideas that should be related. However, if they use *but* (which means *other than),* they are saying that an exception to the initial idea exists. Finally, the conjunction *or* (which means *otherwise*) asks the reader to acknowledge something else.

Examine the compound sentences below and pay close attention to the punctuation. Notice that (1) independent clauses joined by a coordinate conjunction require a comma, and (2) independent clauses joined by a conjunctive adverb require a semicolon and a comma.

Myra was late, so we had to start the meeting without her.
(A comma is necessary.)

Most companies already have editing procedures to follow, but freelance editors should establish a checklist for preparing materials for publication.
(A comma is necessary.)

Editors need to have excellent organizational skills; furthermore, they need to be able to analyze for details.
(A semicolon is necessary with a conjunctive adverb, followed by a comma.)

The Complex Sentence

Complex sentences share several attributes with compound sentences. Both sentence types help create internal coherence and consistency. In addition, both types establish relationships between ideas. Complex sentences, though, consist of only one main clause and at least one subordinate clause. These subordinate clauses can function as adjectives, nouns, or adverbs depending on the writer's intended meaning.

The skillful use of complex sentences not only adds to a document's readability but also to the inherent logic that connects information. For example, in the complex sentence _Because_

<u>the writer did not clarify the level of editing he desired,</u> *time and money were wasted,* the subordinate clause establishes a cause-and-effect relationship with the main clause. In short, *x* caused *y* to occur. Now, examine this sentence: *When you finish testing for usability, return the document to the SME.* Notice how the opening subordinate clause establishes a chronological sequence with the main clause—complete *x* first; afterward, complete *y*.

EXERCISE 7.4

Locate the subordinate and main clauses in the sentences below. Then write at least ten complex sentences of your own design. Be able to explain the relationship between the independent and dependent clauses in your sentences.

1. Whenever the rest of the team arrives, we can get started on the Murphy account.
2. If you improve your communication skills, you will become a better editor and writer.
3. We won't be able to go home until the brochure is completed.
4. Although editing can be a full-time job, many editors are also technical writers.
5. Editors often help writers get started on technical articles unless the writers prefer initially to work alone.

The Compound-Complex Sentence

Its name may sound daunting, but the compound-complex sentence is nothing more than two independent clauses plus at least one subordinate clause. That is, the compound-complex sentence consists of one complex sentence, *plus* a conjunction, *plus* one independent clause. When the complex sentence is joined to the independent clause by a conjunction, it becomes compounded, thus the name compound-complex. For example, *Because editors often work with subject matter experts, they should acquaint themselves with any pertinent technology, and they should recommend useful ideas such as photographs* is a compound-complex sentence.

Complex sentence + conjunction + independent clause = compound-complex sentence

Because editors often work with subject matter experts,	(subordinate clause)
they should acquaint themselves with any pertinent technology,	(independent clause)
and	(conjunction)
they should recommend useful ideas such as photographs.	(second independent clause)

This compound-complex sentence begins with a subordinate clause following an independent clause. A coordinating conjunction connects the second independent clause.

When should you use a compound-complex sentence? For obvious reasons, these sentences are used less frequently in technical writing than the other three. Not only are they the most complex structurally, but they provide a wealth of information that, if not carefully presented, may overwhelm the reader. Remember, a sentence comprises a unit of thought.

Technical editors must carefully judge how much and at what rate they want authors to present information to their readers. Compound-complex sentences should be avoided when the subject matter is highly technical or when addressing a global audience.

Examine the compound-complex sentences below. Locate the subordinate clauses, the independent clauses, and the conjunction.

If your supervisor sets limits on the extent of copyediting you are allowed to perform, you will have to be content to stop there, and you are ethically bound to report something that may harm the reader.

Onscreen markup systems are available for editors, but unless they improve substantially, I don't plan to purchase any for our company.

Because production schedules are important, editors and team leaders should work together, and they must establish realistic schedules if possible.

TIP FOR TECHNICAL EDITORS

When editing a document, make certain the writer uses a variety of sentence types. Sentence variety eliminates the choppy or the singsong effect that results from sentences that are near the same length. Simple sentences usually present the material clearly; however, when repeatedly strung together, they can become monotonous. Also, simple sentences that tag on too many prepositional phrases lose their readability. Used judiciously, compound and complex sentences that express logical relationships between ideas help to unify paragraphs. Too many compound-complex sentences will overload the reader with information, but they may be used occasionally when the information is straightforward. Strive for a balance, and always keep the level of difficulty and intended audience in mind.

EXERCISE 7.5

Locate a representative piece of your own writing that contains at least three paragraphs. First, test it for sentence variety. Label each sentence as one of the four sentence types. Does your writing contain a mixture of the four sentence types, especially simple, compound, and complex? Do the thoughts seem to flow smoothly throughout each paragraph? If your writing sounds choppy because you create too many short, simple sentences, combine a few into compound or complex sentences.

EXERCISE 7.6

Select another representative piece of your writing and examine the length of your sentences. Make a line chart for each paragraph like the example in Figure 7.1 to provide a visual illustration. What conclusion can you draw about your own writing? Do you write using a variety of sentence types and sentence lengths to avoid a "choppy" style?

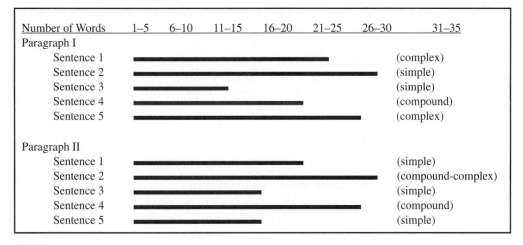

FIGURE 7.1
Line Chart of Sentence Type and Length.

PROBLEM-SOLVING: AVOIDING SENTENCE-LEVEL ERRORS

Although this section is not intended to replace a good grammar handbook, it does provide a helpful overview of the many types of sentence-level problems encountered by editors. As you work through each topic, diagnose your comprehension by completing each editing exercise. If you feel confident that you already have a solid foundation in grammar and sentence usage, complete the editing exercises first and return to any areas that may be problematic.

What is written without effort is in general read without pleasure.
—*Samuel Johnson*

This section is designed to help you:

- Eliminate fragments
- Revise fused sentences and comma splices
- Position modifiers correctly
- Select the proper verb
- Make subjects agree with their verbs
- Make antecedents agree with their pronouns

Sentence Fragments

Ah, the sentence fragment. Sentence fragments appear almost everywhere (like in the previous sentence), including novels, advertising, and conversation. They often are used for rhetorical or stylistic emphasis, and in these cases may be considered acceptable, even though they are grammatically incorrect. Sentence fragments consist of a group of words that are punctuated as a complete sentence but are missing a subject or a main verb. They will not stand alone as an independent clause.

Most frequently, fragments occur when the writer mistakenly confuses a subordinate clause for a complete sentence. For example, *Kelsey is an excellent writer and editor. Although he has problems with typing.* Notice that the second clause is actually a subordinate clause, not an independent clause. Revised, the sentence could read *Although he has problems with typing, Kelsey is an excellent writer-editor.* Usually, fragments can be revised either by adding an independent clause or changing the fragment into an independent clause by adding a subject or verb.

The computer program running slowly. (fragment)

The computer program was running slowly. (independent clause; verb added)

Unable to complete the manuscript on time. (fragment)

Unable to complete the manuscript on time, the young executive asked the team to pitch in. (independent clause added)

EXERCISE 7.7

As an editor, you must be able to recognize sentence fragments. Revise these fragments and make them complete sentences.

1. Since we have plenty of time to finish the specs.
2. To understand the importance of learning grammar.
3. His desire to have everyone in the office promoted.
4. Documents prepared for a global workplace.
5. Copyright laws involving the Internet.

Fused Sentences and Comma Splices

Similar to sentence fragments, fused sentences (also known as *run-on sentences*) and comma splices (also known as *comma faults*) can be found in modern novels, advertisements, and journalism. For example, in Faulkner's experimental novel *The Sound and the Fury,* some sentences continue for over a page, attempting to portray the characters' free-flowing thoughts. In other novels, writers use comma splices to imitate speech or dialogue for cultural realism or dramatic effect. Twain's *The Adventures of Huckleberry Finn* is such an example.

A *fused sentence* occurs when two sentences (*independent clauses*) run together without using terminal punctuation such as a period or a conjunction to separate the independent clauses.

A *comma splice* occurs when two independent clauses are joined only by a comma. Instead, the writer should have used a period or kept the comma but added a coordinate conjunction.

Imagine, though, if technical writers failed to observe the traditional conventions of usage and wrote in Faulkner's *stream-of-consciousness* style or attempted to imitate the speech patterns of the semiliterate Huck Finn? Although as an editor you might occasionally encounter writing that appears to be stream-of-consciousness or semiliterate, your task is to make editorial decisions that turn documents into clear, readable text. Because comma splices and fused sentences "rush" sentences together, they usually place an unnecessary burden on the reader's ability to comprehend.

Most style books shun the use of the generic pronoun "he" instead they recommend following guidelines that avoid sexist language. (fused sentences)

Most style books shun the use of the generic pronoun "he," instead they recommend following guidelines that avoid sexist language. (comma splice)

Most style books shun the use of the generic pronoun "he." Instead, they recommend following guidelines that avoid sexist language. (revised)

Most style books shun the use of the generic pronoun "he"; instead, they recommend following guidelines that avoid sexist language. (revised)

Editors use different types of illustrations to convey information you will need to know which type of illustration works best and in what circumstances. (fused sentences)

Editors use different types of illustrations to convey information, you will need to know which type of illustration works best and in what circumstances. (comma splice)

Editors use different types of illustrations to convey information, and you will need to know which type of illustration works best and in what circumstances. (revised)

Editors use different types of illustrations to convey information. You will need to know which type of illustration works best and in what circumstances. (revised)

Fused sentences and comma splices can be corrected several different ways, by dividing them into simple sentences or making them compound sentences. Before deciding which type of correction to use, ask yourself which revision works best in the context of the document you are editing. For example, technical instructions often contain fragments because itemized instructions are not in complete sentences. However, a feasibility proposal would allow for longer sentences.

> Writing is easy. All you do is stare at a blank sheet of paper until drops of blood form on your forehead.
> —*Gene Fowler*

Selecting the Appropriate Verb

Good writing incorporates well-chosen adjectives and adverbs, while strong, precise verbs contribute to a sentence's overall effectiveness. In other words, verbs hold authority over the subject of the sentence. Weak verbs, however, provide little authority and fail to capture the reader's attention. The revisions that follow indicate how sentences can be enlivened by selecting stronger, more effective verbs.

Continually working long hours at a computer <u>may be bad</u> for a person's eyes. (weak verb)

Continually working long hours at a computer <u>weakens</u> and <u>damages</u> a person's eyes. (revised)

The young apprentice <u>ate</u> his sandwich quickly at his desk and <u>went</u> back to work. (weak verb)

The young apprentice <u>gobbled</u> his sandwich down quickly and <u>resumed</u> working. (revised)

The waves <u>hit</u> the beach with incredible force. (weak verb)

The waves repeatedly <u>thundered</u> onto the beach with incredible force. (revised)

The most common verbs in the English language—forms of *to be* and *to do*—can easily be overused. If company policy allows it, even during the editing for correctness stage, the editor can direct the writer's attention to weak verbs and recommend a few changes, especially if variations of the verb *to be* (*am, is, was, were, being,* etc.) or the verb *to do* (*do, does, did, doing*, etc.) appear too often. Just a few strong verbs substituted for a few weak verbs can substantially improve a document.

Active and Passive Voice

Incorporating more action verbs into a document is only one method an editor uses to increase the readability of a document. Constructing sentences in the active voice is another. Active voice simply means that the sentence emphasizes who or what *is doing the action.* The passive voice, however, emphasizes who or what *is being acted upon.*

For example, in this case of the **active voice**, *Our spam blocker software redirects unwanted emails into a separate file,* the subject of the sentence, *software,* performs the action of the verb, *redirects.* If the sentence had been written in the **passive voice** it would read, *Unwanted emails are redirected into a separate file by our spam blocker software.* First, notice the change in the subject of the sentence from *software* to *emails.* Second, notice the change in the verb form, and third, notice how the subject now receives the action of the verb.

> *Dr. Butera gave the keynote speech at the convention.* (active)
>
> *The keynote speech at the convention was given by Dr. Butera.* (passive)
>
> *The computer science students designed a new type of editing software.* (active)
>
> *A new type of editing software was designed by the computer science students.* (passive)

Occasionally, writers deliberately use the passive voice, especially when they do not want to assign responsibility for action or in cases where "who" or "what" carried out the action is unknown or irrelevant. For example, *The editing of the manual was poorly executed* uses the passive voice to avoid assigning responsibility. In the sentence *The graphics are sketched in the margins*, "who" or "what" carried out the action is unknown and probably unimportant. Thus the passive voice has its place in technical writing. For example, lab and scientific reports often are written in the passive because the report seems more objective in tone. In this book, we use passive voice when the agent is unknown or when we want to stress the action over the doer of the action. In general, though, sentences continually furnish new information, and readers often prefer the sentence to present the subject first, followed by a verb that shows how that subject is acting.

TIP FOR TECHNICAL EDITORS

Many seasoned editors will admit that they had trouble with active and passive voice when they first began editing. Also, many new technical writers fail to appreciate the difference between active and passive voice; their writing often becomes infused with awkward passive sentences. When editing, always consider your readers' expectations. Above all, readers want information to be presented clearly and simply, which usually (but not always) means the active voice.

EXERCISE 7.8

Pick a page in this textbook. Find all the sentences that use passive voice. Revise the sentences using active voice. Which version sounds better? Why?

Subject-Verb Agreement

When two individuals agree with one another, they are said to be *of the same mind* or *in accord* with each other. Their agreement creates harmony, at least for the moment. The same is true for sentences. They, too, must be in accord, but unlike people, who may agree only occasionally, sentences must always agree, especially within their internal constructs.

Two agreement problems in particular arise within sentences: agreement between subject and verb, and agreement between a pronoun and its antecedent. These two problems are similar in nature; however, to eliminate confusion, we will address each issue separately.

Agreement between the subject and its verb remains one of the most common grammatical errors. Ironically, this problem becomes increasingly acute as individuals seek more sophistication by incorporating additional phrases and clauses within their sentences. As more phrases and clauses are added to provide supplemental information within the sentence, the subject frequently becomes separated from its verb. When this occurs, it's easy to confuse the actual subject for a noun or pronoun in a phrase or clause with what may seem to be the subject. For example, what is the subject in this sentence? *Any one of the company's new engineers who will meet with us tomorrow (is, are) capable of signing off on the project.* The subject is *one,* which must agree in number with the singular verb *is. Engineers* may seem to be the subject, but remember, the subject of a sentence is never the object of a prepositional phrase (*of the company's new engineers*).

Compound Subjects with And

Compound subjects are two or more subjects in the same sentence that use the same predicate. If the two subjects are connected by *and,* generally the sentence requires a plural form of the verb. *And* is similar to a plus sign: one and one *are* two.

> A usability <u>tester</u> and an <u>editor</u> validate ~~validates~~ that a manual is accurate. (compound subject)

> <u>Peter</u> and <u>Sally</u> support ~~supports~~ Mary's promotion to Editor in Chief. (compound subject)

Compound Subjects with Or or Nor

When using *or* or *nor* (or the correlative conjunctions *either/or* or *neither/nor*), the subject closer to the verb determines whether the verb is singular or plural.

> Either the chief <u>copy editor</u> or the workplace <u>editors</u> in section four review ~~reviews~~ our material.
> (*Editors* is plural, closer to the verb than *copy editor,* and therefore requires the plural form of the verb, *review.*)

Neither usability <u>experts</u> nor a company <u>editor</u> works ~~work~~ without consulting each other.
(*Editor* is singular, closer to the verb than *experts,* and therefore requires the singular form of the verb, *works.*)

Subjects with Intervening Phrases and Clauses

Often writers include one or more phrases or clauses between the subject and its verb. The verb must agree with the subject of the sentence and not a noun or pronoun within the intervening phrase.

<u>   *intervening participial phrase*   </u>
Normand's <u>photograph</u> hanging among photos of the company's baseball team <u>is</u> ~~are~~ very flattering.
(*Photograph* takes the singular form of the verb to be, *is.*)

<u>   *adjectival clause*   </u>
The <u>professor</u> who teaches technical editing <u>writes</u> ~~write~~ too fast on the board.
(*professor* takes the singular form of the verb, *to write*)

Collective Nouns

Collective nouns usually refer to a group. The group may be people (*army, audience, class, family, team, troop, jury*) or things (*herd, flock, committee*). In most cases, collective nouns require a singular verb because the people or things are considered as a single unit. Sometimes, though, collective nouns may need a plural verb depending on whether the writer intends for the reader to consider individual members of the group.

The jury <u>is waiting</u> for the judge to allow them to recess. (*jury* as one unit)

The jury <u>were arguing</u> among themselves. (*jury* members as individuals)

Ten hours <u>is</u> all it should take to finish the editing project.
(Ten hours is considered a single unit of time here.)

Measles <u>is</u> one disease for which you can receive a vaccination.
(Words such as *measles, news, mathematics, economics, physics*, and *civics* look plural but take a singular verb.)

Indefinite Pronouns

We have said that *collective nouns* usually require singular verbs depending on the context of the sentence. Pronouns, though, are a bit more straightforward. For example, the number of a *personal* pronoun—*I, he, she, it, we, you,* or *they*—is easy to determine. However, unlike personal pronouns, *indefinite pronouns—either, neither, one, everyone, everybody,* etc.—do not refer to anyone in particular and usually are considered to be singular. Although, similar to collective nouns, some indefinite pronouns can be plural depending on the context. Study the list below of common indefinite pronouns. Note that many of the pronouns end in *one* or *body,* which makes them singular.

anybody	either	neither	somebody
anyone	everybody	nobody	someone
each	everyone	no one	

Tricky Indefinite Pronouns

A few indefinite pronouns—*some, most, all,* and *none*—can be a bit tricky because, depending on the context of the sentence, they can be singular or plural. Often you can look inside an adjoining prepositional phrase to determine which form, singular or plural, is needed. Frequently, the best recourse for an editor is to simply revise sentences and avoid this problem altogether.

Singular	**Plural**
Most of the invoice *is* ruined	*Most* of the invoices *are* ruined
All of the computer *was* fried.	*All* of the switches *were* burned.
None of the food *was* wasted.	*None* of the biscuits *were* discarded.

EXERCISE 7.9

Failure of the subject and verb to agree in number is one of the most common errors that editors encounter. You should be able to determine the correct verb form in the following sentences with a little practice. (Remember that the subject of the sentence will not be in a prepositional phrase.)

1. All of the editors (like, likes) to relax at McGuire's after work.
2. A committee (was, were) formed to solve the technical problems.
3. The scout troop (travel, travels) on interesting camping trips.
4. Nobody (thinks, think) that we have enough time to complete the project.
5. Neither of the technicians (understand, understands) how to fix our computer.
6. None of the applicants (knows, know) grammar very well.
7. Some of his information (appear, appears) false.
8. Most of the punch (has, have) been consumed by the older faculty.
9. Either the graphics or the text (has, have) to be eliminated from the manual.
10. The Rolling Stones (was, were) one the of the best rock groups of the 1970s.

Pronoun-Antecedent Agreement

Although it is usually obvious, sometimes writers and editors face delicate problems making pronouns agree with their antecedents. Problems arise because pronouns stand for something else, and writers do not always make clear what that something else is: its antecedent. An *antecedent* is nothing more than a fancy term for what noun the pronoun refers back to.

For example, in the sentence *The television must have **its** volume turned off*, the pronoun *its* refers to its antecedent *television*. Or, in the example *Billings, Safwan, and Mavis tried to resolve **their** differences,* the pronoun *their* (functioning in this sentence as a possessive adjective modifying *differences*) refers back to *Billings, Safwan, and Mavis*. So far, agreement between the pronouns and their antecedents has been uncomplicated, but what happens in the following case? *One of the engineers left **his** briefcase in the office.* The pronoun *his* does agree with its antecedent, *one*; however, should we assume that all the engineers are male? Some would argue that this is a sexist statement.

Or suppose we change the sentence to read, *One of the engineers left **their** briefcase in the office. Their* can refer to both males and females, but it doesn't agree with its antecedent, *one*. In order to make this sentence gender neutral, we pluralized *their* but violated conventions of agreement. In such cases, one solution is to entirely reconstruct the sentence: *One of the engineers left **a** briefcase in the office.* Of course, some teachers suggest you can write *his or her* in place of *their—One of the engineers left **his or her** briefcase in the office.*

TIP FOR TECHNICAL EDITORS

Watch out for vague pronoun references as vague pronouns means weak writing. Editors should make sure that the language they are editing is concise and precise.

> *First, take the disk out of the hard drive to fix **it.***
> (*It* could be referring to *disk* or to the *hard drive*.)
>
> *To fix the disk, take **it** out of the hard drive first.* (revised for clarity)
>
> *I began to reassure myself that **this** could only be a phase.*
> (The pronoun *this* has no antecedent.)
>
> *I began to reassure myself that **this recession** could only be a phase.*
> (revised)

Relative Pronouns

The relative pronouns *who, whoever, whom, whomever,* and *whose* refer to people, whereas *that* and *which* refer to objects and animals.

The guy who dropped off the computer did not call back.

A writer who does not know the conventions of grammar must have written this brochure.

This is the dog that attacked the infant.

TIP FOR TECHNICAL EDITORS

It's easy to remember that the relative pronouns *who, whom,* and *whose* refer to people; however, deciding whether to use *who* or *whom,* or whether to use *whoever* or *whomever,* causes a lot of problems. Notice that people seldom make the *who/whom* distinction in speech, but you should always distinguish the correct relative pronoun in writing. Use *who* or *whoever* as the subject of sentences and *whom* and *whomever* as the direct object or object of the preposition.

> *Who is the author of this document?*
>
> *Whoever wants to go should meet in the lobby.*
>
> *The writer to whom I gave the copy never returned it.*
>
> *You can give it to whomever you choose.*
>
> *This is the engineer who completed the surveys.*
>
> *This is the technician whom I called to fix the computer.*
> (Notice that *whom* is used when it is the object in the noun clause.)
>
> *I am looking for whoever edited this document.*
> (Here *who* is used because it is the subject of the noun clause.)

Some editors are failed writers, but so are most writers.

—*T. S. Eliot*

TOP ERRORS IN CONVENTIONAL USAGE

Familiarize yourself with the following list of errors in usage and grammar, and determine areas where you perceive you may have weaknesses. Review previous selections in this text, or consult a grammar handbook for additional clarification.

1. Misuse of commas
   - Joining two independent sentences with a comma (comma splice)
   - Missing a comma after an introductory element
   - Missing commas in items in a series
   - Using unnecessary commas (restrictive clauses)
   - Placing a comma between a subject and its verb

2. Subject disagreement with its verb
3. Missing possessive-case apostrophe
4. Pronoun disagreement with its subject
5. Unclear pronoun reference to its subject
6. Misuse of a word (diction)
7. Fusing two sentences together
8. Dangling modifier
9. Sentence fragment
10. Confusing *it's* for *its*

SUMMARY

Ignorant people think it is the noise which fighting cats make that is so aggravating, but it ain't so; it is the sickening grammar that they use.

—*Mark Twain*

This chapter has introduced you to the most basic unit of information, the sentence, and how it functions to provide meaning. We approached the sentence by building on fundamental concepts such as the phrase and the clause. We then demonstrated how phrases and clauses can be combined to make any of four types of sentences. Finally, we provided some of the most common problems facing editors on the sentence level. If you have a good understanding of this section, you are ready to move forward to more substantive editing issues.

CHAPTER 8

Punctuation: Adding Readability to Sentences

At the conclusion of this chapter, you will be able to

- summarize the importance of punctuation;
- identify the major types of punctuation in the English language;
- incorporate punctuation accurately into your writing;
- apply your knowledge of punctuation to editing technical documents.

Supposeyouhadtoreadeverythinginthistextwithoutthebenefitofspac esbetweenthewords. Although possible, reading these sentences without the benefit of spacing undoubtedly would be a mentally taxing and time-consuming undertaking. Yet for centuries the earliest writers did not space between symbols, and even after writing evolved, they didn't insert spacing between their words. Egyptian hieroglyphics and Mesopotamian cuneiform are primary examples of early writing, but despite the unprecedented advances of both cultures, neither envisioned the importance of spacing.

Although linguists cannot say when spacing was "invented," most likely it began in Greece several centuries before the common era. Aristophanes, a Greek librarian, recognized the need for written signals and developed the earliest systematic approach for punctuating. His punctuation system called for the *periodos* (period), *kolon* (colon), and *komma* (comma) to be integrated within written text, primarily to alert the reader when to pause. However, Aristophanes's system was generally ignored until the ninth century AD when cultures that depended on writing began to emerge.

Johann Gutenberg's invention of the printing press in Germany around 1440 and the inevitable expansion of education to the average citizen created a further need for a logical, efficient approach to writing. Presently we employ numerous punctuation marks, including commas, periods, dashes, ellipses, colons, semicolons, brackets, and question marks, to provide clarity and meaning to our written expression. Without punctuation, confusion and misreading are a certainty. For example, punctuate the "Dear Jean-Luc" passage below. How you perceive Nicolette's intentions will determine your punctuation.

> English usage is sometimes more than mere taste, judgement, and education— sometimes it's sheer luck, like getting across the street.
>
> —*E. B. White*

> Dear Jean-Luc I want a man who knows what love is all about you are generous kind thoughtful people who are not like you admit to being useless and inferior you have ruined me for other men I yearn for you I have no feelings whatsoever when we're apart I can be forever happy will you let me be yours Nicolette

Now compare your understanding of Nicolette's letter to Jean-Luc with two possible ways of punctuating this letter. Notice the difference in meaning.

> Dear Jean-Luc:
>
> I want a man who knows what love is. All about you are generous, kind, thoughtful people who are not like you. Admit to being useless and inferior. You have ruined me. For other men, I yearn. For you, I have no feelings whatsoever. When we're apart, I can be forever happy. Will you let me be?
>
> Yours, Nicolette

> Dear Jean-Luc:
>
> I want a man who knows what love is all about. You are generous, kind, thoughtful. People who are not like you admit to being useless and inferior. You have ruined me for other men. I yearn for you; I have no feelings whatsoever when we're apart. I can be forever happy. Will you let me be yours?
>
> Nicolette

THE IMPORTANCE OF PUNCTUATION

Previous chapters have discussed the parts of speech, phrases, clauses, modifiers, types of sentences, and some problems that occur in sentence construction. To become a successful editor, you need to know these terms, their functions, and how to rectify surface-level problems.

Readers expect to read a document without having to pause to reread a sentence for content or understanding. The most essential task for the editor, then, is to address the readers' expectations for clarity. An editor's ability to create clear prose for readers comes from knowing the structure of language and how linguistic elements interact to complete ideas. Punctuation serves these ends as well.

Similar to a conductor who leads a symphony, punctuation controls the rhythm, emphasis, and flow of sentences. For example, commas can slow a sentence's tempo to make the ideas more deliberate, or a semicolon can be used to hasten two closely related ideas together. Learn the conventions of punctuation usage if you are not already familiar with them. Use them as guidelines for creating clear, forceful prose.

Aristophanes's punctuation system began with only three marks. Today, we use no fewer than eleven punctuation marks. Undoubtedly, you have been using punctuation for many years, and you are familiar with its functions. However, from experience we know that

From now on, ending a sentence with a preposition is something up with which I will not put.
—*Winston Churchill*

some punctuation remains tricky even for accomplished writers and editors. The following sections, therefore, have been selected because they cover the most troublesome areas and will provide you with a good overview.

First, examine the structure of the four types of sentences and how they are punctuated. Figure 8.1 illustrates the basic structure of these four types of sentences. Remember, though,

Simple sentence
Grammar rules are fun to learn.

Independent clause

Compound sentence using a coordinate conjunction
*The editors were late for the meeting, **so** we decided to begin without them.*

Independent clause	, so	*Independent clause*

Compound sentence using a conjunctive adverb
*The editors were late for the meeting; **however,** we couldn't begin without them.*

Independent clause	; however,	*Independent clause*

Complex sentence
Because we had not finished the graphics, we had to wait on the final edit.
or
We had to wait on the final edit because we had not finished the graphics.

Dependent clause	,	*Independent clause*
Independent clause	no comma	*Dependent clause*

Compound-complex sentence
Although we do allow our editors to use laptops, a hard-copy manuscript is easier on the eyes, and editors generally feel more confident editing hard-copy documents.

or

A hard-copy manuscript is easier on the eyes, and editors generally feel more confident editing hard-copy documents, although we do allow our editors to use laptops.

Dependent clause	,	*Independent clause*	, + conj	*Independent clause*
Independent clause	, + conj	*Independent clause*	, +	*Dependent clause*

FIGURE 8.1
Sentence Structure.

that these structures are flexible. For example, dependent clauses can be interchanged with independent clauses for complex and compound-complex sentences. Notice, too, that punctuation may change. For example, when complex sentences begin with a subordinate clause, a comma is added afterward to separate it from the main, independent clause. However, when the subordinate clause follows the main clause, a comma is not necessary.

EXERCISE 8.1

Research Aristophanes and the first three punctuation marks. In what ways has punctuation evolved?

The Comma

The comma is the most widely used (and misused) form of punctuation in the English language. When Aristophanes first conceived of the comma, early cultures, especially the Greeks, communicated their ideas orally from memorized speeches. Once orators began to write out their speeches, the comma emerged as a marker, reminding the orator to pause and briefly take a breath before continuing. Even today, many English teachers tell their students to insert commas into text where a person would naturally pause. Although well-intentioned, this advice doesn't work well, as evidenced in the preceding "Dear Jean-Luc" letter. Furthermore, non-native speakers who learn English as a second language find the natural pause rule problematic.

We believe that you should follow one directive when using the comma: *If a comma clarifies an expression, use it; if it doesn't, lose it!* Another common phrase among editors regarding comma use is *When in doubt, leave it out!* Of course, we are not advocating abandoning the conventions of usage. On the contrary, conventions governing the use of commas and other punctuation evolved because they clarify meaning. For example, using a comma after an introductory element separates that element into a unit of meaning psychologists call *chunking*, making the meaning easier for the reader to retain.

We do suggest a commonsense approach to using commas. Follow the standard conventions first; however, if a comma is needed to clarify a point, use it. Generally, though, avoid the excessive use of commas if they impede the rhythm of the sentence. Also, be prepared to encounter editors or in-house style guides that abide rigorously by the standard conventions of usage.

Use commas to

- *Join two independent clauses (compound sentence) linked by a coordinate conjunction*

    *Roberta's proofreading is marginal at best, **but** she is an excellent graphics designer.*

    *Gururaj, you should strive to reduce the document's wordiness, **and** Jo, you need to look at the punctuation.*

    *Let's finish tagging the soft copy, **or** we can begin the graphics project.*

- *Join an introductory clause (complex sentence) to an independent clause*

   Although our car broke down, *we weren't late for work.*

   Because hard-copy editing has ergonomic advantages, *Chico prefers it to onscreen editing.*

   When you get finished with the textbook, *please let me use it.*

Note: When the subordinate clause follows the independent clause, commas are not necessary: *She left work early **because she had to run an errand.***

- *Set off items in a series*

   *Students taking this course come from **Germany, Spain,** and **the U.S.*** (series of nouns)

   *Sarah **walked to her car, drove to the office supply store,** and **bought herself Adobe PageMaker**.* (series of verb phrases)

   The final document was printed on red, yellow, and white paper. (series of adjectives)

Note: Sometimes, items that appear to be in a series are considered a single unit. For example, in the sentence *I'll have grits, coffee, ham and eggs, and toast and jelly for breakfast,* notice that *ham and eggs* and *toast and jelly* are considered as single items. In addition, some style guides (e.g., APA) do not recommend a comma after the item immediately preceding *and*. Also, each item in a series should be parallel, that is, should begin with the same part of speech, as in a bullet list of job duties on a résumé.

- *Set off an introductory phrase*

   At the end of the meeting with the CEO, *we decided to break for coffee.*

Note: Sometimes, when the phrase is relatively short, such as ***In general*** *we like to edit,* commas are not necessarily needed. Check with your company's in-house style to confirm.

- *Set off nonrestrictive clauses and phrases*

   *Mr. Bartolotti, **the senior editor for the magazine,** will stop by here this afternoon.*

Note: The distinction between restrictive and nonrestrictive clauses is essential. Restrictive clauses contain essential information, whereas nonrestrictive clauses do not provide essential information in the sentence. Ask yourself if the information within the phrase or clause can be omitted without changing the writer's intended meaning. If the phrase or clause can be omitted, it requires commas to offset it from the rest of the sentence. If the information is essential to avoid a misreading or confusion, then commas are not used. For example, in the sentence *The documents **that need to be heavily proofread** have been left for the new editor,* the clause ***that need to be heavily proofread*** is essential because it identifies a subset of all the documents to be proofread. Without this information, the new editor would not know which documents need proofreading.

Technically, clauses beginning with *that* are restrictive, whereas clauses beginning with *which* are nonrestrictive. Be certain, though, not to use *that* or *which* to refer to people; always use *who* or *whom*.

- *Set off a noun of direct address*

   Mr. President, *can you come over and check these statistics?*

*Please come over and check these statistics, **Mr. President**.*

*Can you come over, **Mr. President,** and check these statistics?*

- ***Set off certain introductory words***

    For example, The Chicago Manual of Style *is one of the most commonly used style guides.*

    Moreover, *companies may have their own style guide for editors to follow.*

    In addition, *we may have to employ several additional copy editors before June.*

- ***Set off items in an address or date***

    *Deliver this correspondence to **1238 Hyland Ave., New York, NY, USA**.*

    *Our next conference won't be until **July 14, 2009,** in **Orlando, Florida**.*

TIP FOR TECHNICAL EDITORS

One way to help you remember whether or not to use commas with nonrestrictive (also called nonessential) clauses is to think of the two commas as a pair of scissors. If a clause is not needed in a sentence, you place commas before and after the clause because you can "cut" the clause out of the sentence without affecting meaning.

EXERCISE 8.2

Punctuate the following sentences by adding commas where needed. Be prepared to justify your answers.

1. Hiring the new editor it turns out was an excellent idea.
2. Before you begin to edit a manuscript you need to know your audience's level of expertise.
3. *The Chicago Manual of Style* considered the most widely accepted style guide describes substantive editing and mechanical editing.
4. When producing a technical document it's not unusual for the graphics to be sent to the design team and the manuscript to be sent separately to a copy editor.
5. Many companies incorporate their editors as an integral part of the design process and editors usually develop a close working relationship with the design team.
6. If you expect me to finish on time Ms. Nguyen you will have to get me some more help from Mandy the head software developer.
7. The duties of technical editors include meeting deadlines educating others working as team members and continuing to expand their knowledge.
8. After completing a markup copy editors should write their initials in the upper right-hand corner of the document.
9. If you receive a document written or typed in all capital letters don't mark all the letters that should be lowercase but instead mark only the letters that should be capitalized.
10. I want you to edit the document yourself because it's your responsibility.

EXERCISE 8.3

Supply the missing punctuation in the letter in Figure 8.2, and then check your answers using the corrected version in Figure 8.3 as a guide.

Dear Mr. Morris

The Loan Department at our bank recently received a request from your business PhotosAreUs for an additional $10000 in start-up funds. We understand that $5,000 of those funds is needed for two color printers furthermore you desire to retain the remaining $5000 to ensure adequate cash flow. Unfortunately we cannot honor your request at this time.

First our bank was assured that the contract we signed with PhotosAreUs for $100,000 would be sufficient for all start-up costs. Although we understand that your costs may have exceeded the original amount you anticipated we feel that two color printers and an additional $5000 are not warranted.

Second if PhotosAreUs needs additional monies we recommend you contact the Better Business Bureau and seek a small business loan through them. The BBB frequently provides assistance to new businesses and they have many contacts in the lending industry.

We wish you well with you new enterprise and look forward to working with you in the future.

Sincerely

Federika Smith
Chief Loan Officer

FIGURE 8.2
Incorrect Letter.

Dear Mr. Morris,

The Loan Department at our bank recently received a request from your business, PhotosAreUs, for an additional $10,000 in start-up funds. We understand that $5,000 of those funds is needed for two color printers; furthermore, you desire to retain the remaining $5,000 to ensure adequate cash flow. Unfortunately, we cannot honor your request at this time.

First, our bank was assured that the contract we signed with PhotosAreUs for $100,000 would be sufficient for all start-up costs. Although we understand that your costs may have exceeded the original amount you anticipated, we feel that two color printers and an additional $5,000 are not warranted.

Second, if PhotosAreUs needs additional monies, we recommend you contact the Better Business Bureau, and seek a small business loan through them. The BBB frequently provides assistance to new businesses, and they have many contacts in the lending industry.

We wish you well with you new enterprise and look forward to working with you in the future.

Sincerely,

Federika Smith
Chief Loan Officer

FIGURE 8.3
Corrected Letter.

The Semicolon

Semicolons separate two independent clauses that are closely related. With few exceptions, you can use a period to replace a semicolon. Avoid using semicolons too often; use them primarily to separate closely related ideas and to connect independent clauses with a conjunctive adverb.

> *The doctor did not keep me waiting long; she was right on time.*
>
> *Many technical editors first assume that onscreen markup is automatically tracked; however, depending on the software, track changes may not be saved in a source file.*
>
> *Elect Jacobs; he'll give you the business.*

Semicolons can also separate items in a series. Normally, commas are used; however, when the items in a series contain internal punctuation, semicolons separate the items more clearly for the reader.

> *Computers have three central features that include a hard drive, a disk that stores information; a microprocessor, a chip made of silicon that makes calculations; and an operating system, a software program that creates an environment for all other programs to work.*

The Colon

Similar to the semicolon, the colon is preceded by an independent clause. However, colons are followed by items in a series or an explanation. For example, in the sentence *Technical editors must acquire many competencies: with language, with technology, and with people,* the independent clause is followed by a colon and then a series of competencies that explain which skills. Colons have other uses that include separating elements of time, Bible chapters and verses, and titles and subtitles; and introducing a bulleted or numbered list.

> *The skills needed to become a technical writer are: language, technology, and people.* (incorrect example with misused colon)
>
> *The skills needed to become a technical writer are language, technology, and people.* (revised)
>
> *The English language consists of eight parts of speech: noun, pronoun, adjective, verb, adverb, conjunction, interjection, and prepositions.* (colon used correctly)

TIP FOR TECHNICAL EDITORS

In text, a colon means *namely*. Therefore, if you can substitute the word *namely* for the colon and the sentence still makes sense, the colon is probably used correctly. Colons are also used after introductory elements in letters (*To Whom It May Concern:*), with scriptures, such as *Exodus 1:8,* and with time, *12:30 a.m.* A common error is to use a colon to separate an object from its verb or the object of a preposition from the preposition itself.

Parentheses

Writers use parentheses (often called parenthetical expressions) to enclose information that comments on, clarifies, or supplements more important information within the main clause or a previous sentence. Also, parentheses are sometimes used around numbers, such as *(1)*.

Hyphens and Dashes

Although word processors have almost eliminated the need to hyphenate words at the end of a line, should you need to divide a word, do so between syllables (e.g., *abili-ty,* not *ab-ility*). You may have to consult a dictionary to determine the separation of a word into syllables. Often, the syllables appear with dots or hyphens between them. However, you should not believe that hyphens can be used to divide any syllable at the end of a line.

Dashes are constructed by typing two adjacent hyphens; if turned on, the AutoFormat selection (in Microsoft Word) will automatically change two typed hyphens in a row into a dash. Don't confuse hyphens with dashes, though. Dashes are generally used to denote a break in thought, such as *Mary said she was taking us to the theater—but wait, doesn't she have to go home first?* Also, dashes may be substituted for commas to emphasize parenthetical elements. *Joe Ferguson—definitely an overachiever—writes a novel a year.* The *em dash* (typed as two hyphens, or —) does not require spacing but functions to demonstrate a brief digression from the main idea (*The report—recently published—is on my desk*). The *en dash* (which is shorter than the em dash but longer than a hyphen and is used in quantitative material as a minus sign) is frequently used to indicate a range, length, or duration, such as *Maine–Miami Amtrak,* or *4–5 feet.*

Hyphens and dashes must be typed differently. For example, a hyphen does not have a space before or after it (e.g., mother-in-law). Hyphens usually join compound adjectives, such as *long-term care* and *part-time worker,* as well as compound proper nouns such as Smith-Barney.

Dashes and hyphens can be a bit daunting. If you encounter difficulty using either, we recommend you consult a good reference guide such as *The Chicago Manual of Style* or one selected by your instructor.

The Period

Periods mark the end of declarative sentences such as this one. Also, periods are used in abbreviations: *Ms., Mr., Jr., Ph.D., etc., U.S.A.*

The Question Mark

When a direct statement is asked, a question mark is used as terminal punctuation.

Does anyone know when the copy is due back to layout?

What day is best for you?

Quotation Marks

Quotation marks are not as difficult to use as some people believe. With a little practice, you can easily master them. Remember, quotation marks are used in pairs: one set to open the quotation and another to close it. Also, quotation marks are used for what is actually said, not what is said indirectly.

For example, *"I am sick and tired of reality television shows," said John.* Here, John's words can be attributed directly to him. Also, notice that the quotation marks only surround what is actually said, not the expression *said John.* You would not, however, use quotes if you wrote, *John said he was sick and tired of reality television shows.*

Quotation marks are also necessary when a portion of someone's statement becomes an integral part of a sentence. For example, *According to Claudia, winning the lottery was "something worth taking a chance on."*

Be certain to consult your company's in-house style guide for the following uses of quotation marks, because style guides vary. Generally, though, use the following guidelines:

- Use quotation marks around the titles of short stories, short poems, songs, essays, articles in books, and subdivisions of books. (Don't confuse quotation marks with italics. Italics substitute for what might be underlined, such as titles of books, magazines, newspapers, plays, movies, television programs, long poems, software, and music CDs. Also, italics can be used occasionally to provide emphasis.)
- Use quotation marks around idiomatic, colloquial, slang, or other nonstandard words and expressions, or those used in an ironic sense. Such expressions normally would not be used in technical documents. *Stella's "limo" was a broken-down 1956 Chevy.*

Finally, remember these general guidelines for punctuating quotation marks: place periods and commas inside and place colons and semicolons outside. Other punctuation such as question marks and exclamations go inside when they are part of the quotation, outside when they are not.

"What books are common to both the Bible and the Koran?" Nakita asked.

What is the point of editing the manuscript "halfway"?

John asked, "What is the point of editing the manuscript 'halfway'?"

(This sentence requires quotes within quotes. Use single quotation marks within the double quotes and notice the placement of the question mark.)

The Apostrophe

For many individuals, the apostrophe is a troubling punctuation mark, but with a little review it isn't difficult to master. The apostrophe originated over 700 years ago when Middle English (Chaucer's English) was spoken. Middle English nouns in particular were problematic. A noun used as a subject was written and spoken one way; however, when it was used as a direct object, it often acquired a different ending. The solution, in short, was the apostrophe to show ownership. The castle of Lear became written as *Leares his castle*. Its final variant form, which we use today, *Lear's castle,* the apostrophe takes the place of the *e* in *Leares,* and *his,* which is redundant, was dropped.

Use apostrophes to

- *Indicate the possessive case in singular nouns and indefinite pronouns*

    Omar's *camera seems to be lost.*

    Heidi's *function in this office is to provide computer support.*

    It is **anyone's** *guess how much longer we will need to finish this assignment.*

- *Indicate the possessive case of plural nouns*

    Please wait a minute until I return from the **men's** *room.*

    I injured my foot when I stepped on the **children's** *Legos.*

    We are waiting to see if the **engineers'** *manual is ready for print.* (This sentence refers to more than one engineer. If the manual belonged to only one engineer, it would read *engineer's.*)

    Desiree deserves a **week's** *vacation.*

    When plural nouns do not end in *s,* add an apostrophe and then the *s.* However, when the noun ends in *s,* simply add an apostrophe, unless the pronunciation requires an additional *s,* such as *Miles's train* (see the "Tip for Technical Editors" below).

TIP FOR TECHNICAL EDITORS

Here is where the English language can be tricky. How do you show possession for an individual's name, such as *Jules* or *Lars*? Should you write *Jules's book* or *Jules' book*? Should you write *Lars's pen* or *Lars' pen*? Before you answer, try to say *Jules' book* without pronouncing a second *s*. Notice that you must pronounce the second *s*. The best way to understand this problem is to ask if the second *s* is actually pronounced. Because the second *s* for both names is indeed pronounced, the correct possessive form becomes *Jules's* and *Lars's*. However, be prepared for individuals who argue that *Jules'* and *Lars'* are grammatically correct. (Just ask them to pronounce both names without actually saying the second *s,* and see what happens.)

- *Indicate the possessive case for compound words*

    The hotel's porter lost his own **mother-in-law's** *suitcase.*

    The **secretary of state's** *speech compelled the commander to take action.*

- *Indicate a contraction*

    Didn't *you know that* **they'd** *gone to the meeting?*

    They're *finished with the markup, but they* **don't** *want to tell the developmental editor yet.*

    It's *up to you when you complete the usability testing.*

TIP FOR TECHNICAL EDITORS

It's = It is   **Its = possessive case**

Be careful not to confuse *it's* with *its*—a common mistake made by many writers because these forms do not follow the standard conventions of usage. *It's* is not possessive as you might assume but rather is a contraction for *it is*. On the other hand, *its* without the apostrophe is the possessive case. For example **It's** *going to rain tonight* can also be written as **It is** *going to rain tonight*. In the following sentence, however, *its* indicates possession but does not require the apostrophe: *The Ruby Corporation gave* **its** *approval to hire four more technical editors for our project*. Whenever you are in doubt whether to use *its* or *it's*, read the sentence aloud and substitute *it is*. Your ear will tell you the difference.

EXERCISE 8.4

Insert punctuation and revise words as needed in the following sentences.

1. Ms. Jones arrival time for the conference is scheduled for two o'clock however, she informed us that she would not arrive until sometime after four.
2. Heres what you need to do before you leave the office unplug the computers, turn off the lights, and lock the doors.
3. Its time for the accounting department to run the numbers.
4. When dealing with customers, one of a technical editors main responsibilities is to know the requirements for the particular document to be revised.
5. Quick grab the boss PDA and tell her when Thursday's meeting is.

EXERCISE 8.5

Before continuing, we recommend that you work through the next two *editing for correctness* exercises as a copy editor. You do not have to be concerned with sentence structure, spelling, and stylistic changes at this point; we will address those points in the following sections and chapters. For this exercise, simply concentrate on grammar and punctuation. You may want to review before attempting this exercise. Be sure to consult the "Copymarking Symbols for Hard-Copy Markup" table from Chapter 3 (Figure 3.1) and insert the proper caret (insertion symbol) where appropriate to make the necessary corrections. Insert the necessary punctuation into Figure 8.4. Compare your answers to Figure 8.5.

Frederick S. Caldwell CEO
Macrohard Software

MEMO
To: R & D engineers
Re: Security threats
Date: December 12 2005

As all of you know protecting computers from cyberattacks is becoming increasingly difficult. Todays practice of applying security patches after a virus or worm has been discovered will simply not be adequate in the future. No doubt you are familiar with some of the new super fast spreading worms and viruses that has already surfaced. Johnny Johnson CEO at Symantron says that computers will soon face the "Warhol" effect that will infect worldwide systems in less than 15 minutes. He predicts that within a few years these 15 minutes will become "flash" threats that occur within seconds. Our current technologies cannot stop these threats furthermore businesses will need to have multiple layers of security that not only alert them to cyberattacks but eliminate the virus or worm on the "fly."

After carefully considering whats at stake Im directing our engineering department to begin immediately to assess the situation conduct a needs analysis and begin developing software that can deal with these threats. Tomorrow you will receive specific instructions regarding issues of personnel time for development and implementation and budget at our weekly 900 am meeting.

Next week I will publicly announce to our stockholders this new but important direction our company will be taking. Its going to be challenging I know all of you are up to the task.

FIGURE 8.4
Memo Missing Punctuation.

Frederick S. Caldwell, CEO
Macrohard Software

MEMO
To: R & D engineers
Re: Security threats
Date: December 12, 2005

As all of you know, protecting computers from cyberattacks is becoming increasingly difficult. Today's practice of applying security patches after a virus or worm has been discovered will simply not be adequate in the future. No doubt you are familiar with some of the new super-fast-spreading worms and viruses that have already surfaced. Johnny Johnson, CEO at Symantron, says that computers will soon face the "Warhol" effect that will infect worldwide systems in less than 15 minutes. He predicts that within a few years, these 15 minutes will become "flash" threats that occur within seconds. Our current technologies cannot stop these threats; furthermore, businesses will need to have multiple layers of security that not only alert them to cyberattacks but eliminate the virus or worm on the "fly."

After carefully considering what's at stake, I'm directing our engineering department to begin immediately to assess the situation, conduct a needs analysis, and begin developing software that can deal with these threats. Tomorrow you will receive specific instructions regarding issues of personnel time for development and implementation and budget at our weekly 9:00 a.m. meeting.

Next week, I will publicly announce to our stockholders this new but important direction our company will be taking. It's going to be challenging; I know all of you are up to the task.

FIGURE 8.5
Corrected Version.

EXERCISE 8.6

Insert the necessary punctuation into Figure 8.6. Compare your answers with Figure 8.7.

Widgets and Gadgets, Inc.
664 Sprocket Blvd.
Spokane, WA, 22333

MEMO

To: Members of the Management Team

From: Donald Spades, CFO

Date: September 16th, 2006

Re: Homeland Security

Last week representatives from the Department of Homeland Security visited our facilities during their five hour inspection they made numerous recommendations how our company can recover in the event of a terrorist attack. Mr. Johnson the Director of Homeland Recovery said The only sure way to get back into business is to plan and conduct emergency exercises. Based on the Departments recommendations Im directing all members of the Management Team to develop and implement measures that will insure our recovery if we experience a terrorist attack.

First develop plans that insure all co workers well being. Involve everyone from all levels in devising emergency plans. Use newsletters intranets staff meetings and word of mouth to communicate plans. You will probably want to set up phone trees email alerts and emergency telephone numbers. Also, Im directing the IT Department to develop a password protected web page on our companys website with instructions telephone numbers and individuals to contact during an emergency. Be certain your division establishes an out of town phone number where employees can leave an I'm Okay message.

Second we will need to plan and practice what we intend to do if we are faced with a disaster. Begin by making certain all employees know the evacuation procedures for this building. Soon Im going to ask each division to conduct training seminars for current and new employees to maintain emergency preparedness.

Plan to meet next Wednesday at 2:45 pm in the Board Room to discuss how we can best implement these recommendations from the Department of Homeland Security. Ill look forward to hearing your ideas.

FIGURE 8.6
Memo Missing Punctuation.

Widgets and Gadgets, Inc.
664 Sprocket Blvd.
Spokane, WA, 22333

MEMO

To: Members of the Management Team

From: Donald Spades, CFO

Date: September 16th, 2006

Re: Homeland Security

Last week, representatives from the Department of Homeland Security visited our facilities; during their five-hour inspection, they made numerous recommendations how our company can recover in the event of a terrorist attack. Mr. Johnson, the Director of Recovery, said, "The only sure way to get back into business is to plan and conduct emergency exercises." Based on the Department's recommendations, I'm directing all members of the Management Team to develop and implement measures that will insure our recovery if we experience a terrorist attack.

First, develop plans that insure all co-workers' well-being. Involve everyone from all levels in devising emergency plans. Use newsletters, intranets, staff meetings, and word of mouth to communicate plans. You will probably want to set up phone trees, email alerts, and emergency telephone numbers. Also, I'm directing the IT Department to develop a password-protected web page on our company's website with instructions, telephone numbers, and individuals to contact during an emergency. Be certain your division establishes an out-of-town phone number where employees can leave an "I'm Okay" message.

Second, we will need to plan and practice what we intend to do if we are faced with a disaster. Begin by making certain all employees know the evacuation procedures for this building. Soon, I'm going to ask each division to conduct training seminars for current and new employees to maintain emergency preparedness.

Plan to meet next Wednesday at 2:45 p.m. in the Board Room to discuss how we can best implement these recommendations from the Department of Homeland Security. I'll look forward to hearing your ideas.

FIGURE 8.7
Corrected Version.

SUMMARY

Punctuation usage evolved out of the necessity to add readability and meaning to writing. Punctuation helps ideas flow smoothly and often prevents readers from misinterpreting those ideas. Experienced technical editors know how and when to use punctuation to avoid confusion, promote coherence, and create effective documents.

Mechanics: Odds and Ends That You Need to Know

At the conclusion of this chapter, you will be able to

- determine when to use capitalization;
- recognize how to use abbreviations;
- distinguish between italics and underlining;
- explain the importance of correct spelling;
- evaluate the overall correctness of technical documents;
- diagnose your ability to edit grammatical and mechanical errors.

Most style guides discuss mechanics independently from grammar and punctuation; a few don't make a distinction. However, we believe you should at least be aware that a difference exists and that mechanics usually refers to capitalization, abbreviations, numbers, italics versus underlining, spelling, and hyphenation. Just as correct grammar and punctuation are essential to the readability of any document, proper use of mechanics adds clarity to writing and demonstrates accuracy and professionalism.

CAPITALIZATION

Conventional usage requires that proper nouns be capitalized but not common nouns. Proper nouns refer to the name of a specific person, place, or thing. For example, capitalize words that name religions (e.g., *Islam*), religious followers (*Christians*), nationalities (*French*), races (*African-American*), governmental agencies (*Environmental Protection Agency*), historical and literary movements (*the Romantic Age*), documents (*the Declaration of Independence*), specific places (*Juneau, Alaska*), organizations (*Parent-Teacher Association*), months (*May*), holidays (*Easter*), days of the week (*Wednesday*), languages (*English*), etc. Other items to capitalize include

- ***Nouns as part of a person's title***

    Before we could contact him, Dr. Vertigo had left the office.
    She sent the law brief to Francis Jones, J.D.

> *Mayor Smith lost the election when the newspaper revealed a past criminal record.*

- **Nouns as titles and subtitles, including books, articles, and songs**

    For Whom the Bell Tolls
    The Rime of the Ancient Mariner
    "It's Been a Hard Day's Night"

Note: Articles, coordinate conjunctions, and prepositions are not capitalized unless they begin the title. When words are hyphenated, capitalize the second word if it is important, such as *F-Prot Antivirus*. Also, depending on the discipline, style guides vary in their rules for capitalization of titles. For example, the MLA style guide in humanities does not capitalize prepositions in a title; the APA style guide for the social sciences only capitalizes the first word and proper names when they appear in an article or book citation. Again, you will want to check the particular style guide recommended for a specific field or company.

- **Nouns naming specific persons, places, holidays, religions, days of the week, organizations, races and nationalities, and historical events**

    Michael Phelps
    London, England
    Yom Kippur
    Monday
    United Nations
    B'nai B'rith
    Caucasian
    German
    the Renaissance

ABBREVIATIONS

Abbreviations can be used to shorten titles before and after an individual's name; with specific dates, times, numbers, and amounts; and with the names of organizations, corporations, or countries.

Ms. Frances Smith
Rev. Peter Stone
Prof. Sara Caldwell
William Yancey, M.D.
Ophelia Sexton, D.D.S.
567 BC
7:00 p.m.
no. 43

Note: Avoid redundant abbreviations in titles. Don't write *Dr. Tristan Sterling, Ph.D.* Use one title or the other, but not both. Also, many familiar organizations, countries, or corporations can be abbreviated without periods as acronyms, such as *IBM, YMCA, USSR,*

USA, NAACP, etc. (An acronym is a word formed from initials, such as *NATO,* for *North Atlantic Treaty Organization.*)

However, when using an unfamiliar abbreviation, write out the full title first and then immediately following the title, write the abbreviation in parentheses. After you have established the abbreviation, you can use the abbreviated form in the rest of the document. For example, *The Society for Technical Editors (STE) will hold its annual meeting in March. All STE members are encouraged to register early for the conference.*

Also, avoid inappropriate abbreviations. In formal writing, do not abbreviate personal names, units of measurement, states, months, days of the week, and holidays. For example, a memo to employees about upcoming holiday time probably should not read *All RNs should alert Dr. M about scheduling for X-mas and Jan. 1, which both fall on Sun. Thanx, AY.* Instead, the memo might read, *All nurses should alert Dr. Mazey for scheduling Christmas and New Year's holiday as both of these holidays fall on a Sunday this year. Thank you, Anita Yount.*

TIP FOR TECHNICAL EDITORS

Be familiar with the following Latin abbreviations:

Abbreviation	Latin	Meaning
cf.	*confer*	compare
e.g.	*exempli gratia*	for example
et al.	*et alii*	and others
etc.	*et cetera*	and so forth
i.e.	*id est*	that is
N.B.	*nota bene*	note well

Many communities maintain outdated ordinances; e.g., in Mobile, Alabama, it's against the law to row a boat in the downtown area.
Campbell, Wang, Sioux, et al. are the authors of the newly published text on anatomy.

NUMBERS AND SYMBOLS

When using numbers and symbols (such as %) in documents, you will want to check closely with your client's preferred style guide or your company's in-house style guide. For example, some guides specify spelling out percent, while others prefer the symbol %. Traditionally, numbers between one and ten are written out, whereas numbers above ten may be written as numerals (11, 12, 13, etc.). However, some style guides will tell you to spell out numbers that are only one or two words and use numerals for numbers of more than two words. When the dates or numbers appear at the beginning of a sentence, however, they should be written out. *Forty people attended the workshop, but 15 did not attend the advanced session after lunch.*

I knocked over a total of eight (not 8) *pins the first time I bowled.*
In the year 1776 (not seventeen seventy-six), *the colonies signed the Declaration of Independence.*

One hundred twenty (not 120) *children paraded through the school halls this afternoon.*

June 22, 2009

153 Hyland Avenue

35%

TIP FOR TECHNICAL EDITORS

In technical and business writing, numerals are the preferred style because they stand out as more readable in a sea of words and tend to be more concise. However, sometimes you will need to use a combination of numerals and words, such as *45 million*. We recommend using numerals according to the standard practice of your field.

ITALICS AND UNDERLINING

Before word processors, writers used underlining to indicate when words were to be typeset in *italics*. Word processing has made this process obsolete; anyone can create italics with a couple of keystrokes or one mouseclick.

Use italics to indicate the title of books, newspapers, long poems, films, television programs, musical compositions, works of art, magazines, pamphlets, and plays.

The Sound and the Fury (book)

Newsweek (magazine)

The New York Times (newspaper)

A Streetcar Named Desire (play)

The Passion of the Christ (movie)

The Scream (work of art)

The Four Seasons (musical composition)

Spacecraft, ships, trains, foreign words, and aircraft are sometimes underlined.

USS Constitution (ship)

Amtrak (train)

siesta (foreign word)

Again, be sure to check with your company's or the client's style guide; often underlined words such as the ones given above are italicized instead.

SPELLING

Nothing will stop a reader faster and besmirch the credibility of a document more than misspelled words. If you are editing in soft copy, it's easy to turn on the spelling checker and locate misspellings. Don't assume that the writer (or even the software!) has caught all the spelling errors. Also, because the spelling of words differs throughout the English-speaking countries, be sure to adjust your spelling checker for the appropriate country if the document is intended for an audience outside the United States such as the United Kingdom, Australia, India, or Trinidad. Be especially sensitive to an audience whose second language is English. Many may have studied English under the British system of education.

  If you are editing hard copy and do not have an electronic text that you can submit to a spelling checker, you will need to recognize misspellings yourself. Challenge any word that you are not absolutely sure of and look it up in a good dictionary; there are now some very good online dictionaries, such as www.websters.com. Although modern spelling follows conventions of usage for the most part, many exceptions to these conventions still exist. Below is a small sample of differences that occur in spelling. (We provide a more comprehensive list of spelling differences across English-speaking countries in Chapter 21.)

American spelling	*British spelling*
color	colour
judgment	judgement
check	cheque
fetus	foetus
defense	defence
utilize	utilise

It's a damn poor mind that can think of only one way to spell a word.
 —*Andrew Jackson*

  Which spelling is correct—*all right* or *alright*? *receive* or *recieve*? *a lot* or *alot*? *omitted* or *omited*? *mispell* or *misspell*? *suprise* or *surprise*? *Artic* or *Arctic*? *embarass* or *embarrass*? These are only a few of the most commonly misspelled words. Dictionaries, grammar handbooks, and the Web provide lists of literally thousands of everyday words that people have trouble spelling correctly.

  If you have spelling difficulties, you have plenty of company. Even many of us who studied phonetics in school have trouble spelling words correctly, first because people pronounce words differently, and second because many words do not follow "the rules." Even a word that follows the rule, such as *receive*—*i* before *e* except after *c*—looks okay when misspelled as *recieve*. When you hear someone say the word *surprise,* does the individual pronounce the first *r*? Probably not. If you don't hear the *r* sound, it's reasonable to assume you will spell the word incorrectly as *suprise.* One of the occupational hazards facing editors is seeing the same words misspelled so often that they forget the correct spelling.

My spelling is Wobbly. It's good spelling but it Wobbles, and the letters get in the wrong places.
 —*A.A. Milne*

  Take the opportunity to examine a few of the many websites that list misspelled words. In addition to the standard dictionaries and handbook websites, we recommend Paul Brian's *Common Errors in English Usage,* which can easily be found on the Web. Not only does this online book provide a long list of common misspellings, but each word links to an explanation of why it is commonly misspelled.

  Just for fun, but also to give you a sense how diction and misspellings can severely alter the meaning of a sentence, we have included a few excerpts from Richard Lederer's "History of the World," a compilation of student bloopers that Lederer collected from eighth graders through college-level students. The bloopers will give you pause before you totally rely on your spelling checker again.

1. The inhabitants of ancient Egypt were called mummies. They lived in the Sarah Dessert and traveled by Camelot. The climate of the Sarah is such that the inhabitants have to live elsewhere.

2. The Bible is full on interesting caricatures. In the first book of the Bible, Guinesses, Adam and Eve were created from an apple tree.

3. Moses went up Mount Cyanide to get the ten commandments.

4. The Greeks invented three kinds of columns—Corinthian, Doric, and Ironic.

5. Socrates was a famous Greek teacher who went around giving people advice. They killed him. Socrates died from an overdose of wedlock.

6. Sir Walter Raleigh is a historical figure because he invented cigarettes. Another important invention was the circulation of blood. Sir Francis Drake circumcised the world with a 100-foot clipper.

7. Christopher Columbus was a great navigator who discovered America while cursing about the Atlantic.

8. One of the causes of the Revolutionary Wars was that the English put tacks in their tea. Finally, the colonists won the War and no longer had to pay for taxis.

9. Abraham Lincoln became America's greatest Precedent. Lincoln's mother died in infancy, and he was born in a log cabin which he built with his own hands. He said, "In onion there is strength."

10. Louis Pasteur discovered a cure for rabbis. Charles Darwin was a naturalist who wrote the "Organ of the Species." The First World War, caused by the assignation of the Arch-Duck by a surf, ushered in a new error in the anals of human history.

Although it is by no means complete, in the following list we have compiled some words we feel are particularly abused when it comes to spelling.

absence	convenience	intellectual	representative
acceptable	correlate	irrelevant	roommate
accommodate	definitely	irresistible	sheriff
accuracy	description	library	shining
achievement	develop	lightning	skiing
across	disapprove	loneliness	sophomore
advice	discussion	manual	studying
advised	ecstatic	mathematics	succeed
affected	effect	mischievous	summary
all right	eligible	mysterious	surprise
annual	environment	nickel	technique
arguing	escape	nuisance	thorough
athlete	exaggerate	occasionally	tragedy
athletics	exceed	occur	twelfth
basically	exhaust	occurred	unnecessary
bulletin	February	occurrence	vacuum
bureaucracy	forfeit	omission	vengeance
candidate	forty	omitted	venomous
chief	government	pastime	waive
commitment	guaranteed	permissible	weather
compelled	humor	physician	weird
conceivable	hypocrisy	precede	wield
conscientious	illogical	preparation	writing

EXERCISE 9.1

Figure 9.1 is a short technical document similar to what you might encounter in your editing career. Using the appropriate editing marks, edit the document based on what you have previously learned. You do not need to revise sentences yet; we will discuss sentence revision thoroughly in Unit Four.

Nicobar industries is pleased to introduce an inexpensive lighter to our regular line of personal lighters now the user can chose between disposing the lighter or refilling it. Our engineering department has added a removable screw to the base making it possible for the user to access the butane compartment and refill it.

The D-Lite lighters overall height is 2 inches with a thickness of one fourth of an inch. The body is primarily cylindrical in shape made of high impact plastic and easy to hold. The D-Lite consists of a ribbed rotating wheel that when struck by the user creates a spark from the flint insert. As long as the user depresses the access lever butane gas jets upwards in a controlled flame. The butane chamber holds approximately one ounce of butane that lasts the normal user around three hundred lights. If the user decides not to dispose of the lighter a screw on the base is easily removed with a coin.

Through extensive testing our engineering department has determined that a typical D-Lite lighter can be reused six times before the flint wears completely down. Additional manufacturing costs are negligible we estimate an additional 3 cents per unit. Marketing estimates that overall sales will increase substantially because customers are provided an option but still paying the same price for a disposal lighter.

Furthermore Nicobar industries plans to introduce on March 15th 2005 a combination package with a D-Lite lighter and a 3 ounce can of butane. The combination package will retail for $3.95 and appear in most retail outlets. Circle X convenience stores has consented to exclusively sell D-Lite this alone will gross our company over one million dollars the first year according to early estimates.

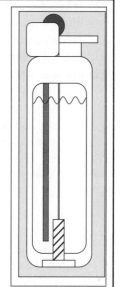

FIGURE 9.1
The D-Lite Refillable/Disposable Lighter.

SELF-DIAGNOSTIC: HOW MUCH HAVE YOU LEARNED?

Directions: Below are fifteen sentences that contain at least one grammatical or usage error. Determine the error, and then check your answers from the answer key that follows. Some sentences may have numerous errors.

1. Each of the engineers were anxious to return to their work.
2. The woman that you said would review the graphics has not shown up.
3. Neither the editor or her assistant are planning to stay late for the seminar.
4. Mr. Sanders feels real well about being able to outsource the job to India.
5. The data is complete, please download it to your computer.
6. Its time for the employees to stay after work finish the project and begin the Chan account.

7. When a technical writer submits a draft to be edited its important to know the intended audience.

8. Technical editing a skill that takes considerable time and energy to master is a rewarding career for many people.

9. Margaret please return this document to the copy editor and when you see him ask him if Mr. Isselbaches account will be completed by next Tuesday.

10. Often editors are part of the initial planning for a project and consult with the writer during the documents production.

11. During the planning stage an editor needs to assess their needs and estimate production schedules and costs.

12. When documents contain classified information editors frequently complete control forms to ensure the documents security.

13. Technical editing students should remember five main objectives when completing the final edit purpose accuracy clearly developed ideas supporting details and audience.

14. My boss who is next in line to become VP of this company has gone on a weeks vacation to Key West Florida.

15. Ms. Wentworth offered to loan Ryland a staff member for the afternoon, however, because he had not recieved the reports on time an alot of time already had been wasted he declined her offer.

Answers: Compare your answers with ours. If you missed any items, review before going on.

1. Each of the engineers were anxious to return to their work.
   *Each of the engineers **was** anxious to return to his or her work.* (subject-verb agreement; pronoun-antecedent agreement)
   ***Better:** The engineers were anxious to return to their work.* (third person and nonsexist)

2. The woman that you said would review the graphics has not shown up.
   *The woman **who** you said would review the graphics has not shown up.*

3. Neither the editor or her assistant are planning to stay late for the seminar.
   *Neither the editor **nor** her assistant **is** planning to stay late for the seminar.*
   (Use *nor* with *neither* and *or* with *either*. Use *is*, not *are*, because *assistant* is the subject closest to the verb and is singular.)

4. Mr. Sanders feels real well about being able to outsource the job to India.
   Mr. Sanders feels great about being able to outsource the job to India.

5. The data is complete, please download it to your computer.
   *The data **are** complete; please download **them** to your computer.*
   (Don't take off if you missed this one; it is very tricky. Most scientists and engineers use *data* as the plural form of *datum*. *Datum* is a single piece of *data*. However, some style guides—especially those used in the computer industry—refer to *data* as a collective noun that requires a singular verb. If you go with data being plural, then *it* should be *them*. Check your style guide when using *datum* and *data*.)

6. Its time for the employees to stay after work finish the project and begin the Chan account.
   ***It's** time for the employees to stay after work, finish the project, and begin the Chan account.*

(*It's* means *it is* and requires the apostrophe here. Also, items in a series need commas to set them off. The comma after *project* is optional in some cases. Be sure that items in a series are parallel: the first word of each item should begin with the same part of speech.)

7. When a technical writer submits a draft to be edited its important to know the intended audience.
   When a technical writer submits a draft to be edited, **it's** *important to know the intended audience.*

   (*When a technical writer submits a draft to be edited* is an introductory subordinate clause and is separated from the independent clause by a comma. *It's* requires the apostrophe as in the previous example.)

8. Technical editing a skill that takes considerable time and energy to master is a rewarding career for many people.
   Technical editing, a skill that takes considerable time and energy to master, is a rewarding career for many people.

   (The phrase *a skill that takes considerable time and energy to master* is an appositive that identifies the subject and requires commas because it is nonessential.)

9. Margaret please return this document to the copy editor and when you see him ask him if Mr. Isselbaches account will be completed by next Tuesday.
   Margaret, *please return this document to the copy editor,* *and when you see him,* *ask him if Mr.* **Isselbaches's** *account will be completed by next Tuesday.*

   (First, nouns of direct address are set off by commas. Second, did you determine that this was a compound-complex sentence? A comma is necessary after *copy editor* because it ends an independent clause that is joined to another independent clause with the coordinate conjunction *and*. A comma follows *him* because *when you see him* is an introductory subordinate clause. Finally, *Isselbaches's* needs an apostrophe and an *s* added.)

10. Often editors are part of the initial planning for a project and consult with the writer during the documents production.
    Often, *editors are part of the initial planning for a project and consult with the writer during the* **document's** *production.*

    (Introductory words such as *often* are frequently set off with commas; *documents* is possessive and requires an apostrophe.)

11. During the planning stage an editor needs to assess their needs and estimate production schedules and costs.
    During the planning stage, **editors** *need to assess their needs and estimate production schedules and costs.*

    (Introductory phrases are set off with commas. Pluralize subject for subject-pronoun agreement.)

12. When documents contain classified information editors frequently complete control forms to ensure the documents security.
    When documents contain classified information, editors frequently complete control forms to ensure the **document's** *security.*

    (The introductory subordinate clause needs a comma, and possessive case needs an apostrophe.)

13. Technical editing students should remember five main objectives when completing the final edit purpose accuracy clearly developed ideas supporting details and audience.

Technical editing students should remember five main objectives when completing the final edit: purpose, accuracy, clearly developed ideas, supporting details, and audience.

(A colon means *namely,* and commas are needed to set off the items.)

14. My boss who is next in line to become VP of this company has gone on a weeks vacation to Key West Florida.
My boss, who is next in line to become vice president of this company, has gone on a week's vacation to Key West, Florida.

(This sentence contains an appositive that requires commas. The word *week's* has an apostrophe because time takes the possessive form when used as an adjective and therefore must have an apostrophe before the *s*. Finally, the name of cities, towns, counties, countries, etc. are set off by commas.)

15. Ms. Wentworth offered to loan Ryland a staff member for the afternoon, however, because he had not recieved the reports on time and alot of time already had been wasted he declined her offer.
*Ms. Wentworth offered to **lend** Ryland a staff member for the afternoon; however, because he had not **received** the reports on time and **a lot** of time already had been wasted, he declined her offer.*

(We salute you if you got everything correct in this sentence. Notice that *loan* is a noun and *lend* is the verb. The bank will provide you with a *loan,* but I want you to *lend* me ten dollars. *Receive* and *a lot* are two of the most common spelling errors. A semicolon, not a comma, is necessary after *afternoon* to avoid a comma splice.)

SUMMARY

We hope you have done well on the previous exercises; however, if you had difficulty, consider reviewing the previous sections or consulting a good grammar handbook. Bookstores and libraries contain many grammar books that are beneficial. Exam review books (GRE, LSAT) are good sources. Remember, though, that like any learned skill, the more you work at grammar, the better at it you will become.

Editing for Visual Readability

Not surprisingly, *editing for visual readability* means exactly what it says: editing a document so that the target audience finds it more visually readable. By *readable* we mean how clearly and how quickly the reader is able to absorb the necessary information from the technical document. A text that scores high on a readability scale is one that is easily digested; the reader does not have to struggle to find information, follow instructions, or understand authorial intention.

You most certainly have encountered plenty of texts that were barely readable in terms of visual presentation: small typeface in a legal contract, letters obscured by a dark background in a PowerPoint presentation, or margins so narrow in a novel that your head ached as you struggled to read what seemed like a thousand words crammed into each of the pages. Would these documents be more effective by increasing the font size, using better contrast of text to background, and including more white space?

This unit is devoted to making a document more readable in the visual sense. Just as it matters what you wear to a business interview, how a document is "dressed" and designed affects its readability. The following two chapters on document design and graphics provide the fundamentals of editing technical documents for visual readability. These skills are crucial because readers and end users need to be able to digest information in a technical document easily and rapidly. If they can't, they won't continue reading.

Editing for visual readability normally occurs at the page layout stage. Page layout for most technical documents should be done after the content (text and graphics) has been edited, not at the same time. Formatting during the writing and editing stages distracts you and the writer from the document's content. If the final layout of the document will be done using a different program, then any layout done in the word processor will be lost anyway. Of course, during the writing and editing stage the editor should make suggestions on topics such as chunking, parallel headings, "would a graphic be useful here?" and similar content issues, but details of fonts, margins, and other visual readability concerns come later.

CHAPTER 10

Document Design: Improving Visual Readability

At the conclusion of this chapter, you will be able to

- explain the Gestalt theory of document design;
- edit a document for visual readability;
- employ a critical eye for white space, margins, borders, and text alignment;
- choose font attributes (style, size, emphasis) appropriately;
- format text for cohesion and parallelism;
- use headings, subheadings, separators, and indicators effectively.

The concept of formatting a document to be readable and attractive for a target audience has a long history, but the term *document design* came to the forefront in professional and technical communication fields in 1996 after Karen Schriver published *Dynamics in Document Design: Creating Text for Readers*. According to Schriver, document design is "the field concerned with creating texts (broadly defined) that integrate words and pictures in ways that help people to achieve their specific goals for using texts at home, school, or work." Effective document design, then, is how well the text, white space, and graphics are arranged on the page or screen.

Some technical communicators may disagree with this definition of document design and formatting, but these terms share the same essential principle. How well a document is formatted in terms of text in relation to white space, graphics, width of margin, location and size of headings, etc., determines how readable it will be for the target audience. Effective document design means readers will be happy with the document and desire to read or examine it further. This is particularly important with technical manuals or other technical documents that are read by users, such as instructions on how to assemble a crib for a proud new parent. What if the font (also called typeface) is too small to read? What if the instructions are jammed into one corner? What if the crib warnings or cautionary instructions are not set apart with emphasized text, such as boldface, larger font, or a border?

You can imagine the problems. At the very least, the customer will be unhappy and unable to assemble the crib, which means money lost for the crib company (via the customer returning the crib or calling technical support to request assistance or services); at the very worst, the parent assembles the crib without the help of the instructions and misses a crucial warning in the assembly, which causes the crib to collapse after just six months of use. Such a disaster could have been avoided by employing just a few basic document design principles. Please understand that *editing for document design is part of an editor's responsibility*. Document design does not mean the document's decoration but rather is another mode of language that needs an editor's critical attention.

Examine Figures 10.1 and 10.2. The first is a résumé by a job-seeker who does not know how to use effectively the elements of document design. The second figure shows the work of a job-seeker who uses white space, font sizes, emphases, margins, lines, and bullets effectively. Think about the differences between these two documents and what you can do to make sure that your documents, your clients' documents, and your company's documents employ the proper fundamentals of document design and formatting.

Ivan T. Ayjob
505 West Seventh Street
Charlotte, NC 28202

ivantayjob@email.com
(704) 968-2701

I have senior experience in a business development environment establishing strategic partnerships, identifying and pursuing revenue generating relationships, prospecting and courting sales leads, and networking within the venture capital community and high-technology sectors. My primary professional aptitudes encompass business development, and consultative sales in the areas related to Brand Strategy development and positioning, naming, corporate identity design, strategic marketing planning and implementation, and PR. I am interested in leveraging both my experience in business development and management consulting with my thorough knowledge of brand management and strategic marketing.

VentureWorx—Venture Management Consulting Firm, April 2000–Present
A subsidiary of Addison Whitney, one of the leading strategic branding firms in the world. VentureWorx was established to manage growth strategies for emerging technology companies. VentureWorx specialized in the areas of brand management, marketing strategies, PR, and additional operational disciplines. Recruited by the CEO and founder of Addison Whitney to drive the vision and profitability of the company.

Managing Director, Charlotte, NC.
Identified and developed strategic partnerships with top-tier Venture Capital firms, positioning VentureWorx to be the premier service provider to their portfolio companies. The above strategy yielded over $2 million in interactive client projects within the first year of operation. Responsible for the negotiation and delivery of over 60% of the firm's revenue. Partnered with senior executives from client companies to help develop new branding and naming strategies that translate their solutions into Internet experience. Responsible for the strategic oversight of interactive client engagements, managing client relationships and directing over 20 staff members. Managed multi-disciplinary teams throughout project lifecycle—including business development, project scoping, proposal writing, solution design, presentation, implementation, progress and success review—ensuring that both the client and VentureWorx objectives were met. Led specialized initiatives specifically geared to achieve critical milestones for emerging technology companies—attracting venture financing, pioneering new markets with the latest infrastructure solutions, strategic build-out of executive management— operations emphasized market research, internal branding, advertising, PR and IR.

First Union Corporation, with headquarters in Charlotte, North Carolina, is the nation's sixth largest banking company based on assets of $253 billion at March 31, 2001. First Union serves 15 million customers.

FIGURE 10.1
Résumé with Poor Document Design.

<u>Online Marketing Strategist,</u> Charlotte, NC.
Developed innovative online brand strategies and e-business implementation plans that helped to extend brands to new digital offerings. Produced discovery audits for each division of the consumer bank. Generated competitive analysis reports, product positioning and development plans, and pricing suggestions. In collaboration with senior management, evaluated and developed brand guidelines for consumer products and services to ensure continuity across online and offline channels. Expanded online presence through partnerships, which included co-branded ventures with Internet service providers, online loyalty programs, and wireless device services. Developed and wrote content for new product introduction, including web content, advertising copy, positioning statements, marketing materials and training guides to ensure cohesive and consistent communication practices. Organized and conducted internal marketing and branding campaigns promoting online services, these efforts included representing the department on internal television segments. Initiated and performed due diligence on inactive online customers, methods of research included focus groups and customer surveys. Subsequently drafted and presented an activation-marketing plan to the management of First Union's Consumer Bank division, the project received additional funding based on the problems identified in the research.

<u>Associate,</u> Charlotte, NC.
Member of the brand strategy group, advising Fortune 500 on brand initiatives from the conceptual to the implementation stages. Project managed a diverse set of client relationships-including financial services, biotechnology, telecommunications, and high technology. Developed new business opportunities with existing clients resulting in additional client projects, as well as successfully landing new prospective business through direct sales to Fortune 500 companies. Led expansion in verticals, including financial services and e-business solutions. Interfaced with each of the firm's disciplines—including creative, linguistic, digital branding/web design, and legal—to create "world-class" brands for some of the most profitable public companies. Wrote new business proposals to procure new projects and clients.

Duke University, Durham, NC, Major: Psychology, cum laude, Minor: Marketing
Member of the American Management Association's (AMA) Sales and Marketing Council.
Serve on the advisory board of Vynamic, a PKI security software company venture backed by Cabletron Systems.

FIGURE 10.1 (*continued*) Résumé with Poor Document Design.

DOCUMENT DESIGN THEORIES

Several movements in the field of art and more recently in visual rhetoric studies have helped us to understand why we see the way we do. A brief background on some of these areas will help you better understand document design principles when editing. As readers, we tend to read text but see visuals. As technical editors, you also need to learn to see the text as an image or a group of images and to read visuals for their properties beyond aesthetics. Unfortunately, many students still view visuals as ornamentation only, not as valuable vehicles for transferring information to the user or reader.

Think about how you perceive police officers: very rarely would you see police officers as people separate from their uniforms, unless of course you knew them personally and had seen them in other contexts. More than likely, you see the uniform as part of the police officer, not as a separate ornamentation or dress. We tend to have the same perception of soldiers and other uniformed professions: the clothes, essentially, are performative in that they make the individual (that male hospital employee) seem transformed (to a nurse tech in his scrubs).

The same goes for text and visuals: we tend to see sentences grouped together as paragraphs or lists or poems, depending on how they are grouped. We see the elements of the text (the words, the font style and emphasis, the white space between letters and words) all at the

Stashia Pundleby
slpundle@bcc.cba.ua.edu

Present Address: Home Address:
P.O. Box 871600   267 Meadowbrook
Tuscaloosa, AL 35487   Hoover, AL 35242
(205) 445-9808   (205) 810-4007

CAREER OBJECTIVE	A position in Assurance and Advisory Services with Ernst & Young
EDUCATION	*Candidate for Bachelor of Science in Business Administration* **The University of Alabama** Major: Accounting Expected Date of Graduation: May 2007
WORK EXPERIENCE May 2004–Present	**Assistance Aide,** University Testing Services • Register students for graduate-level tests • Schedule dates for tests to be given • Issue test scores to students
May 2003–August 2003	**Imaging Clerk,** Amsouth Bank, Birmingham, AL • Organized records and files using a computerized filing system • Assisted in researching customer records
August 2002–August 2003	**Secretary,** AC Systems, Birmingham, AL • Processed customer orders for computer sales • Created and balanced accounting records related to computer sales
May 1999–July 2002	**Cashier,** Champs Sports, Birmingham, AL • Worked cash register for merchandise sales • Organized and maintained athletic merchandise • Assisted customers with purchase decisions through positive communication
HONORS and ACTIVITIES	American Society for Women Accountants CCSO Student Recruitment Committee Dean's List, 1996–1999
REFERENCES	Available upon request

FIGURE 10.2
Résumé with Improved Document Design.

same time, but we still process this information in order to help us understand the meaning of the text (e.g., a capitalized letter helps us to know that a new sentence is starting or that the word is a proper noun). With visuals, it is much easier for us to see the image as a whole than to process the elements of the graphic individually, and yet these elements or parts are necessary to understanding how to read the visual. Although we still live in a word-heavy world, our dependency on visuals over text for meaning is increasing, which is all the more reason why technical editors need to be able to edit visuals—including text as a visual—effectively. (See Chapter 11 for a more in-depth discussion of editing graphics.)

Design is not just what it looks like and feels like. Design is how it works.
—*Steve Jobs*

Gestalt Theory

Seeing something as a whole is precisely the premise behind Gestalt theory. Largely attributed to psychologist Rudolf Arnheim, Gestalt theory means more than the analytic "a whole is the sum of its parts"; it means "a physical, biological, psychological, or symbolic configuration or pattern of elements so unified as a whole that its properties cannot be derived from a simple summation of its parts." The critical point of Gestalt theory is that the mind doing the perceiving contributes something extra to whatever is perceived. More is seen than what is physically present in the stimulus. Take the following paragraph as a textual example of Gestalt theory. You may have seen this paragraph on the Internet or received an email about it:

> Aoccdring to rscheearch at an Elingsh uinervtisy, it deosn't mttaer in waht oredr the ltteers in a wrod are, olny taht the frist and lsat ltteres are at the rghit pcleas. The rset can be a toatl mses and you can sitll raed it wouthit a porbelm. Tihs is bcuseae we do not raed ervey lteter by ilstef, but the wrod as a wlohe.

Please note that this letter-scramble trick only works with the longer words (here the nouns, verbs, and adjectives). Shorter words like prepositions have to remain largely untouched, or the passage is not interpretable. In other words, the mind contributes much to perception, but certain things have to be physically present in the perception (the first and last letters of lexical items in the right places, and the short grammatical signposts undisturbed) or else the mind cannot build a coherent whole or "gestalt."

A word is more than the sum of its letters. Clearly "Elingsh" is not a word in our vocabulary, but we recognize it as "English" because all the letters are there. The word starts with the same first and last letters, it is a capitalized adjective describing "university" ("uinervtisy"), and it is long but is surrounded by short grammatical signposts that are undisturbed (*at, an, it*). We are able to determine meaning in large part by context.

In terms of visuals, we see an octagon as more than eight lines; we actually see a geometric shape, a meaning we are able to get because the arrangement of eight equal lines in such a shape transforms our thinking into "octagon," much in the same way we label someone a doctor, not just a woman in a white lab coat. This is the beauty and fascination of psychology and language!

We can take this cognitive theory one step further by stating that in certain cultures, an octagon on a pole becomes a stop sign, even if it is not red and lacks the letters "STOP." (See Figures 10.3 and 10.4.)

Eight lines   **Octagon**   **Stop Sign**

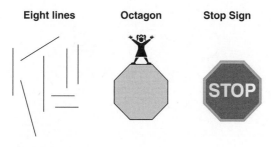

FIGURE 10.3
How We Perceive Meaning in Visuals.

FIGURE 10.4
Letters and Colors Are Not Always
Required for Meaning.

Visual Perception: Gestalt Laws

More than seeing a whole as more than the sum of its parts, Gestalt theory describes why we perceive things the way we do and why what we see as a visual does not necessarily mirror reality (in fact, it usually doesn't). We will discuss just a couple of areas that are relevant to technical editing.

Natural Perception

Gestalt theory alerts us that we do not always see things in their natural state. For example, the circles in the middle of each image in Figure 10.5 are the same size, even though the center circle on the right appears larger. This "misperception" is caused by the size and proximity of the circles around the center one.

In terms of editing for natural perception, you might have a situation where your client wants two objects in a visual that are identical in size also to *appear* identical to the reader. You as editor would have to make sure, for instance, that these objects' sizes appear identical by not surrounding them with differently sized shapes.

Grouping

The example with the circles in Figure 10.5 also demonstrates the gestalt *law of grouping*: circles in the example appear to be two (and only two) separate groups of circles because of their *proximity* (or lack thereof) to one another. This is why we often label two separate

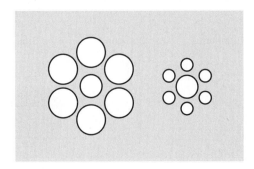

FIGURE 10.5
Gestalt Example of Grouping and Natural
Perception.

individuals sitting next to each other as a couple or a pair or related in some way (business colleagues, friends). They are never one body (unless they are Siamese twins!), and yet they are a unit because of how close they are to one another (proximity).

In addition, if objects move together in the same direction, we perceive them as a group or unit. This is called the *law of common fate*. We can't show object movement in this book very well, but we know you have seen the law of common fate in your own lives, such as individual leaves of grass swaying together, runners beginning the Boston Marathon, or moving bulleted list items during a PowerPoint presentation.

Finally, with regard to visual rhetoric theory and its effect on your role as a document design editor, there is the *law of good continuation,* which simply states that items that follow a curved or a straight line are perceived as a unit. Because minds try to make order out of visuals (such as grouping), we sometimes create the order. According to Paul Lester, "the law of continuation rests on the principle, again assumed by Gestalt psychologists, that the brain does not prefer sudden or unusual changes in the movement of a line. In other words, the brain seeks as much as possible a smooth continuation of a line." In the image in Figure 10.6, our minds follow the lines from *a* to *c* and *b* to *d* but not *a* to *b* and *c* to *d* because we tend to follow a curved or a straight line all the way through, from end to end.

FIGURE 10.6
The Gestalt Law of Good Continuation.

WHITE SPACE

One of the central elements of effective document design is something we rarely think about: white space. White space, of course, is the space on a document or web page that does not contain text or graphics or other illustrations and borders. Of course, white space isn't always white; it is the document's background. For simplicity's sake, though, we will refer to all document background color as *white space*.

White space is a key element in visual readability. Too little makes the document appear cluttered, but too much makes the text look like it is suspended or floating above the page. Not only does effective use of white space make a document easier to digest, but white space also aids interpretation by indicating relationships between information. For example, the white space at the beginning of the first line of this paragraph lets you know that this block of text is indeed a paragraph. The difference in spacing between letters versus words helps us distinguish where one word ends and another begins, even when we are looking at a text in a language we don't understand.

White space is also an important component of document design because of eye fatigue. We are inundated with information in a variety of media: emails, letters, magazines, newspapers, billboards, signs, web pages, blogs, advertisements, books, instant messages, cell phone text messages, etc. It is difficult for our eyes to process much information at one time when there is not enough space or an effective contrast between the words and images and the background of the message. Figures 10.1 and 10.2 demonstrate the difference made by the use of white space. Another example is given in the documents in Figures 10.7

> To design is to communicate by whatever means you can control and master.
> *—Milton Glaser*

> Design is directed toward human beings. To design is to solve human problems by identifying them and executing the best solution.
> *—Ivan Chermayeff*

Make sure to describe the study in enough detail to permit another investigator to replicate it. The Method section is often divided into three subsections, but this division is optional. The three subsections are subjects, apparatus, and procedures. The subjects are what are being investigated or examined. The apparatus is the materials or equipment needed to perform the investigation. The procedures are the step-by-step instructions explaining how the investigation was performed.

FIGURE 10.7
Ineffective Use of White Space in Document Design.

Method

The Method section is often divided into three subsections, but this division is optional.

Subjects: what is being investigated or examined
Apparatus: the materials or equipment needed to perform the investigation
Procedures: the step-by-step instructions explaining how the investigation was performed.

Make sure to describe the study in enough detail to permit another investigator to replicate it.

FIGURE 10.8
More Effective Use of White Space in Document Design.

and 10.8. We use the same words in both to demonstrate that a lack of white space (Figure 10.7) leads to us seeing the text as an image, not as text, which makes reading it difficult. Effective white space, alignment, and chunking vastly improve its readability.

Note that the actual content of the document doesn't matter. When you first looked at Figure 10.7, you might have groaned either silently or out loud. The document is difficult to read because there is not enough white space to help your eyes rest as you read the text. Figure 10.8, however, splits up the text by inserting white space between short paragraphs. The use of short paragraphs and lists is one example of text *chunking*. Figure 10.8 provides a good example of text that has been chunked into short paragraphs for better visual readability. Other types of text chunking include lists and headings. Chunking is an effective method because readers can process short chunks of information quickly and easily into their short-term memory. In addition, Figure 10.8 uses font emphases to highlight parts of the text that are important to the document's organization. Note also how the emphasized words are surrounded by white space to accentuate the text and provide extra eye rest for the reader.

TIP FOR TECHNICAL EDITORS

Be careful not to use too much white space, as this has the same effect on readability as too little: the document is difficult to read, with text and images appearing to "float" on the page, like islands in a sea of white (or whatever color the background may be). When a text has too much white space, the eye longs to rest in the white areas. Thus readers have to exert extra effort to force their eyes away and onto the words or graphics that contain the information. This continual effort is tiring, both mentally and physically (eyestrain).

MARGINS AND BORDERS

When editing for visual readability, you must spend time editing the white space surrounding the text as well as the text itself. Many technical documents such as engineering reports and specifications are printed on full-size paper. These documents typically have a one-inch margin of white space surrounding the entire page. Other documents such as books are more likely to be printed on a different paper size and have different, often narrower margins.

Some books have wide left margins on the main text, often with other elements (headings, illustrations, icons associated with notes or tips) extending into the margin.

Text Alignment

White space is also crucial between words. It is controlled by a feature called *justification* or *alignment*. Left-aligned text (also called *flush left* or *left-justified*) is aligned vertically on the left margin, but the right margin is ragged (see Figure 10.9). Right-aligned (*flush right* or *right-justified*) text is just the opposite: the right side is aligned vertically but the left side is ragged. Centered text has the midpoint of each line of text at the center between the left and right margins. Finally, there is fully justified text, where both the left and the right sides of the text are aligned vertically with the margins, leaving the white space to fall between the words. Most software programs control the amount of white space in each line of left-aligned, right-aligned, and centered text by providing equal space between words. However, in fully justified text, this line-by-line spacing between words does not equally distribute the white space in all lines. The resulting uneven distribution of space may cause some words to look crowded and others to appear isolated, aspects that affect the readability of the document.

Left-aligned text

The Method section is often divided into three subsections, but this division is optional. The three subsections are subjects, apparatus, and procedures. The subjects are what are being investigated or examined. The apparatus is the materials or equipment needed to perform the investigation. The procedures are the step-by-step instructions explaining how the investigation was performed. Make sure to describe the study in enough detail to permit another investigator to replicate it.

Right-aligned text

The Method section is often divided into three subsections, but this division is optional. The three subsections are subjects, apparatus, and procedures. The subjects are what are being investigated or examined. The apparatus is the materials or equipment needed to perform the investigation. The procedures are the step-by-step instructions explaining how the investigation was performed. Make sure to describe the study in enough detail to permit another investigator to replicate it.

Center-aligned text

The Method section is often divided into three subsections, but this division is optional. The three subsections are subjects, apparatus, and procedures. The subjects are what are being investigated or examined. The apparatus is the materials or equipment needed to perform the investigation. The procedures are the step-by-step instructions explaining how the investigation was performed. Make sure to describe the study in enough detail to permit another investigator to replicate it.

Fully justified text

The Method section is often divided into three subsections, but this division is optional. The three subsections are subjects, apparatus, and procedures. The subjects are what are being investigated or examined. The apparatus is the materials or equipment needed to perform the investigation. The procedures are the step-by-step instructions explaining how the investigation was performed. Make sure to describe the study in enough detail to permit another investigator to replicate it.

FIGURE 10.9
Examples of Text Alignment to Margins.

TIP FOR TECHNICAL EDITORS

Ragged right is generally thought to be more readable than fully justified text, just as text with mixed upper- and lowercase letters is more readable than text in all caps. However, right-aligned text is often used for headings such as chapter titles, as well as for some information in headers or footers. For example, in MLA format the header on each page contains the author's last name and the page number. These two items are right-aligned.

Borders and Lines

As an editor, you also need to pay special attention to borders around text and blocks of text. Like bullets and font emphases, borders draw attention to text by outlining it and setting it apart from the rest of the text. Decorative lines, often found above or below headings and other elements, are a type of border. They differ from underlining; line borders are often thick and extend from margin to margin even when the text does not. Bordered paragraphs are often of a different width from the main text or they may extend into a wide left margin, further drawing attention to these elements. Again, watch out for writers who go wild with border types and settings. As usual in technical writing, the KISS principle applies: "Keep It Simple, Stupid." When in doubt, choose something plainer and more professional-looking for a technical document, unless the document's content or intended audience calls for snazzier design.

Some Types of Borders for Text or Graphics

Empty border: white space (or padding) surrounding a component.
Line border: a colored border of a single thickness.
Bevel border: a 3-D raised or lowered beveled border.
Etched border: a 3-D etched-in border.
Matte border: a colored or tiled-image border.
Titled border: a border containing a title with a specified position and alignment.
Compound border: the nesting of an outer and an inner border.

TYPOGRAPHY

Typography is the art and technique of designing, choosing, and using type. When text was composed on a typewriter, the *typewritten* result was *typeset* by a specialist into the final copy that was then printed onto paper. That specialist typesetter chose the font, size, and other attributes of the typefaces to be used. Today, the vast majority of the documents you will encounter as a technical editor will be composed on a computer. Because we now rely predominantly on computers instead of typewriters, the terms commonly used for the preparation of final copy have changed. The terms *typewritten, typescript,* and *typeset*—although still used—are less common. Now that anyone can design and format a document, we use the typographers' terms *typeface* or *font*.

When editing for visual readability, you will want to edit the selected typeface for *font style, size,* and *emphasis.* Editing for these three areas is becoming increasingly complex as there are approximately 120,000 different fonts! One way to edit for font style is to decide what Jo Mackiewicz and Rachel Moeller call a typeface's "personality." In a study of fifteen fonts, Mackiewicz and Moeller discovered that readers find certain fonts (e.g., Times New Roman

Design can be art. Design can be aesthetics. Design is so simple, that's why it is so complicated.
—*Paul Rand*

and Helvetica) to look more "professional," whereas others, like Comic Sans and *Bradley Hand*, were considered "friendlier." Mackiewicz and Moeller studied different attributes of the font: the height of the letters "g" and "a" and whether the crossbar in the "e" is horizontal, slanted, or slightly slanted. They found that, for example, readers thought a horizontal cross-bar (as in Times New Roman) looked professional, a slanted crossbar made the font appear very friendly, and a slightly slanted crossbar seemed between friendly and professional.

EXERCISE 10.1

Go to www.fonts.com and closely examine at least five different fonts. Try to categorize them based on Mackiewicz and Moeller's typeface personality criteria. Make a graphic to demonstrate how your font choices are friendly, professional, or in between.

EXERCISE 10.2

Analyze a piece of handwritten text and compare it to typeface. What are the characteristics of the handwriting in comparison to the computer-generated font? What positive and negative inferences can you make about the perception of documents that are typed as opposed to handwritten?

Font Styles

Appropriate font style is also crucial to an effective document. Fonts can be divided into two basic styles: *serif,* which means "with feet," and *sans serif,* which means "without feet." The feet are merely cross-strokes in the font. Fonts can be further distinguished by whether or not the edges of the individual letters are rounded or straight-edged, whether or not the letters fall below the text baseline, and whether or not the letters are a certain height. Not surprisingly, letters that look more angular and pointed are perceived as more profes-sional, whereas rounded, loopy letters are considered more casual and, in Mackiewicz and Moeller's terms, "friendly." Figure 10.10 shows a variety of fonts.

Sans serif font is usually thought to be most appropriate for headings and titles because the absence of cross-strokes makes the font style clearer and more readable. Arial is a popu-lar sans serif font, whereas Times New Roman is perhaps the most commonly used serif font for editors and writers with PCs (as opposed to Macs). Serif fonts are ideal for body text in printed material, as they are considered easier to read. Many people find that body text in

Arial	**Arial Black**	Broadway	Calligrapher
Garamond	**Haettenschweiler**	**Impact**	Lucida Sans
New Zurica	Old Century	Tahoma	Times New Roman

FIGURE 10.10
Some Font Styles.

online material is easier to read if the font is sans serif, but serif fonts are being designed for onscreen reading.

As you can see, there is more to fonts than size and feet. Font *type* is also important. Choose the font type that is appropriate for the document you are editing. For example, a résumé would look strange in ALGERIAN type, which is a decorative font designed for stationery, invitations, programs, etc. Likewise, **Broadway** would not work because its style and round letters are associated with entertainment, signs, and banners, not a professional résumé. Times New Roman is considered more standard for a résumé, although there are several professional font types (visit www.fonts.com). Some fonts, such as Georgia (serif) and Verdana (sans serif), are designed specifically for online use.

Font Sizes

It is important that the font be readable. Many documents use 12- or 11-point font sizes, but technical instructions, fliers, brochures, posters, etc. may require larger or smaller font

f	8
f	9
f	10
f	11
f	12
f	14
f	16
f	18
f	20
f	22
f	24
f	26
f	28
f	36
f	48
f	72

FIGURE 10.11
Some Font Sizes.

sizes. Some commonly used font sizes are shown in Figure 10.11. The height of the letters varies from one font to another, so 12-point type in one font may look considerably smaller than 12-point type in another font. It is important for you as an editor to determine whether or not the font size is appropriate for the document and, more importantly, the audience. Fonts should be clear and readable but not overbearing. Using different font sizes is also important, but only when appropriate, not just because the author wants to be decorative. A document is unlikely to be as effective if the writer is using several different font sizes in one paragraph. Likewise, lack of font size variety between headings and body text can impede visual readability and effectiveness.

Font Emphasis

Font emphasis, or the way typeface is accentuated in a text, can be divided into six basic methods: boldface, italics, color, underlining, capitalization, and switching font type or size. There are other non-typeface ways to emphasize text, such as indentation, bullets, and borders. All of these methods draw attention to an item by making it distinct from the surrounding text and graphics.

- **Boldface** makes the letters of a word thicker.
- *Italic* puts the font at a 15° angle.
- Color involves changing the color of the font.
- <u>Underline</u> places a line underneath text.
- Capitalization takes place in the form of ALL CAPS, where all of the letters of a word are capitalized, or in SMALL ALL CAPS, where all the letters are capitalized, but the size of the first letter of each word is still bigger than the others (provided that letter is capitalized).
- Changing font simply means shifting the font style (or size). Here we have changed the font from Bookman Old Style to Times New Roman.

Some industries and many organizations have developed conventions for using font emphasis for specific purposes, as a means of conveying extra information to the reader. For example, software user guides commonly use a monospaced font such as `Courier` to indicate text to be typed by the user, **boldface** to indicate buttons to click on the screen (such as OK or Cancel), and *italics* for keyboard keys to press (Enter, F1, etc.). Before inventing a scheme for font emphasis, be sure to check whether you should be following a particular convention or style guide.

The most common mistake of even good writers and editors is overuse of font emphases, even for computer manuals where the "emphasis" carries extra information for the reader. A text that contains several different font emphases often looks cluttered and unprofessional. Usually, one or two font emphases are sufficient.

EXERCISE 10.3

Critique the use of fonts in Figure 10.12. Why is the font in the right column a poor choice? Try to find characteristics of the font to support your answer.

Good	Bad
When in the Course of human events, it becomes necessary for one people to dissolve the political bands which have connected them with another, and to assume among the powers of the earth, the separate and equal station to which the Laws of Nature and of Nature's God entitle them, a decent respect to the opinions of mankind requires that they should declare the causes which impel them to the separation.	*When in the Course of human events, it becomes necessary for one people to dissolve the political bands which have connected them with another, and to assume among the powers of the earth, the separate and equal station to which the Laws of Nature and of Nature's God entitle them, a decent respect to the opinions of mankind requires that they should declare the causes which impel them to the separation.*

FIGURE 10.12
A Comparison of Good Versus Bad Font Use.

HEADINGS AND SUBHEADINGS

Headings and subheadings are important for identifying "chunks" of information in technical documents. When editing headings and subheadings, consistency is key. You have probably noticed in this chapter and elsewhere in this text that all of the chapter titles appear in the same font size, style, and emphasis. Similarly, all section headings appear with the same font size, style, and emphasis. The subheadings for each section are the same format as all other subheadings. As mentioned earlier, heading alignment may differ from the alignment of text in paragraphs. Headings are often centered, extended into the left margin, or right-aligned.

One of the best ways to make your document readable is consistency of titles and subheadings. Visual consistency allows your audience to find its way through the text using predictable markers. Consistency is easily achieved by using heading styles and tagging headings with those styles. Editors should check that heading styles have been correctly and consistently applied and that no manual formatting has been used.

> To dismiss front-end design as mere "icing" is to jeopardize the success of any site.
> —*Curt Cloninger*

Cohesion

When something coheres, it "sticks together." What is the purpose of cohesion in a technical document? For technical editors, revising a document so that the text and/or graphics stick together is another aspect of visual readability. In Unit Four, "Editing for Effectiveness," we demonstrate how content sticks together based on organizational logic. However, there is also visual cohesion, which can be accomplished through text alignment, parallelism, chunking, bulleted and numbered lists, the placement of graphics, and, of course, headings and subheadings. Using headings and subheadings consistently helps the text and graphics "stick" to one another in a logical, cohesive fashion.

Parallelism

Just as grammar calls for parallel syntax, document formatting has visual parallelism. If you decide to chunk your text—as most authors and editors of technical documents do—then the section headings and subheadings need to be parallel in terms of both wording and appearance (font style, size, emphasis, location on page, color, etc.). For example, the three levels of edit discussed in this textbook are all worded using parallelism: editing for correctness, editing for visual readability, and editing for effectiveness. We did not start each level of edit with "editing for" to bore you; instead, we used the same words to establish a parallel structure. Likewise, we made each of the chapter headings using the same font and the same graphic. This type of visual parallelism helps the audience understand how the text and graphics are organized. The more parallel the document design, the easier it is for the audience to find the necessary information or to perform the tasks or instructions at hand.

Forward arrow	⇒
Checkmark	√
Open box	□
Asterisk	*
Diamond	♦
Point/circle	•
Box	■

FIGURE 10.13
Some Bullet Types.

SEPARATORS AND INDICATORS

Linear, sequential exposition is not always the best approach to presenting technical information in a clear and understandable way for a range of readers. Many documents include examples, tips, notes, summaries, digressions into related information, and expanded or detailed discussion, as well as figures and tables. In online documents, some of these items are best provided on separate pages hyperlinked from the main text flow, or made to drop down additional information when the reader clicks a link. In printed documents, they may be separated from the main text by visual indicators such as boxes, marginal icons, shaded or colored backgrounds, or different-colored headings. For example, a software user guide might cover versions for Windows, Mac, and Linux operating systems. Where differences exist, marginal icons and colored backgrounds could indicate the operating system to which the information applies. No matter what strategy you use for indicators, keep your system consistent.

Luck is the residue of design.
—*Branch Rickey*

SUMMARY

Editing for visual readability goes beyond editing for correctness to address the visual aspects of a technical document. Grammar, punctuation, and mechanics are not the only types of obvious document errors. Readers will readily see mistakes in typography choice and emphasis, margins and borders, and section headings and subheadings. These mistakes can be avoided by editing for effective document design or formatting. Remember, document design is almost as important as the document's content. Editors must remember to always edit for consistency in typeface, font style, and font size. Technical editors who master the skills from this chapter will be more knowledgeable in the areas of graphic design and web page design.

11

Graphics: Intelligence Made Visible

At the conclusion of this chapter, you will be able to

- explain why graphics are visible intelligence;
- distinguish the different types of graphics and their functions;
- select appropriate graphics for technical documents;
- edit graphics for accuracy, appropriateness, and organization;
- categorize a table into its appropriate parts;
- state the purpose of each major type of graphic.

Many graphic and design artists agree that Charles Minard's classic illustration of Napoleon's Russian campaign of 1812 "may well be the best statistical graphic ever drawn" (Tufte). Published in 1861, this graphic captures what E. J. Marey calls the "brutal eloquence" of Napoleon's abysmal losses during his campaign to seize Moscow and his subsequent retreat back to the Niemen River. Edward Tufte, in his book *The Visual Display of Quantitative Information,* reveals how Minard's graphic subtly interweaves the multivariate dimensions of time, space, and environment within a narrative framework.

Study Figure 11.1, a recreation of Minard's 1861 original. The steadily diminishing width of the two bands represents the size of Napoleon's army at any given time during the campaign. The light band depicts Napoleon's advance on Moscow, whereas the dark band illustrates his retreat. Smaller bands that veer off indicate where Napoleon sent small contingents to protect his northern flank.

Notice that Napoleon began on the Polish-Russian border in the fall of 1812 with 422,000 men, reached Moscow with 100,000 men, and returned in December with fewer than 10,000. Most of his army died as a result of fatigue, disease, or exposure; however, a substantial number apparently deserted. Also, notice how Minard connects the dark band depicting Napoleon's retreat to the declining temperatures throughout November and December. This was one of the coldest winters in Russia. On December 6, Minard's graphic reports that Napoleonic Grand Army faced a staggering temperature of minus 38 degrees Celsius. Now locate where the retreating army crossed the Berezina River on November 28. Although Minard provides no explanatory text, can you surmise from the graph what happened when the army attempted to cross the river?

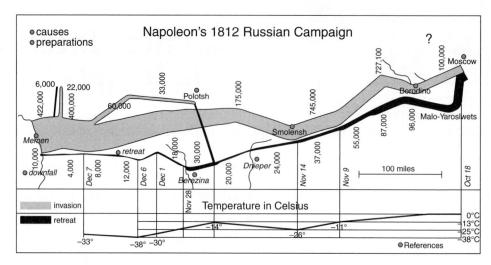

FIGURE 11.1
Recreation of Charles Minard's Graph.

Graphical elegance is often found in simplicity of design and complexity of data.

—*Edward Tufte*

Even if you know nothing about Napoleon's Russian campaign of 1812, could you now provide an historical narrative for that disastrous campaign? Suppose you were asked to translate this graphic into text. How many paragraphs do you think it would take you to convey all the same information that Minard infuses into his graph? Which do you think a reader would prefer: your textual account of Napoleon's campaign or Minard's graphic illustration? The primary function of technical graphics is to communicate information *efficiently*, *effectively*, and *accurately*.

EXERCISE 11.1

Based on what you have already learned about effective technical editing, identify the errors (including spelling) in Figure 11.1. Create a list of what you like and dislike about this graph. Do you find it easy to understand?

Edward Tufte, author of numerous books on visual design, uses a unique metaphor to define graphics: "intelligence made visible." This human passion, to make intelligence visible, can be traced first to the Paleolithic period (circa 30,000 BC), when early humans created pictorial icons by fashioning mixtures of soot and crushed berries onto cave walls. Millennia later, Egyptians recorded their daily lives as hieroglyphics on temples and pyramids, while the Greeks across the Mediterranean carved out symbols and images of their gods on temples and buildings. However, not until the Renaissance in the 15th century AD did any significant advancements take place in technical illustrations. Earlier artisans had not solved the problems of dimensions; their creations remained two-dimensional and flat. Even the drawings from the Middle Ages depicted objects and people without depth, dimension, or perspective. During the early Renaissance, though, masters such as Leonardo da Vinci, Raphael, and Jan van Eyck created the *illusionistic perspective* that added the spatial illusion of a third dimension to their paintings and drawings. Leonardo produced such masterful technical drawings of would-be inventions and the human anatomy that some of his scientific illustrations are still used today. The overlays of muscle tissue in your biology book most likely were adapted from Leonardo's original drawings.

The Scientific Revolution of the seventeenth century, followed by the Industrial Revolution of the eighteenth century, created a further need for technical graphics. Drawings and illustrations became essential to the advancement of science and industry. With the advent of mass production and specialization of parts, the need for standardization became critical. Ultimately, as the middle class developed an intense appetite for material goods, technical illustrations had to become simple enough for the layperson to understand and artisans to work from. In 1905, a person could actually receive an entire house from Sears, Roebuck & Co. in the mail, complete down to the last nail. Of course, instructions on how to assemble the house had to be included. Many families took the opportunity to purchase these relatively low-cost (albeit ordinary) homes and assemble them right out of the crates. Would you like to have written the construction manual for these homes?

During the second half of the twentieth century, the need to "make intelligence visible" increased exponentially from train schedules to statistical scatter plots, musical scores to digital maps, and data charts to engineering schematics. Termed by some as "cognitive art" or "information images," graphic illustrations that once extolled the human form or religious motifs shifted toward digital graphics such as CT scans, mathematical phase spaces, and GIS maps. In 1986, the National Science Foundation funded a project that analyzed the use of computer graphics for scientific purposes. Out of this report came a new term: *scientific visualization*. Scientific visualization means computer-generated graphics that allow scientists to "see" what is normally unseen, such as computations and simulations. Today, many disciplines have begun to use scientific visualization to produce graphic representations of everything from digital maps to structural music. Computers can now translate raw data from satellites into 3-D images of the terrain on Mars and Venus. Perhaps you have had an MRI or CT scan as a result of an injury, or if you are a woman who has had a child, a sonogram that transforms digitized sound into an image of your baby.

Graphic creations of technical information have become increasingly complex; however, we believe that there will always be a need for humans to determine how these sophisticated drawings will be used. This means that technical editors will also be needed to recommend, correct, and improve upon an ever-increasing area of technical communication: graphics that make intelligence visible.

> Although modern technology has become so complex that it is beyond the grasp of most consumers, human curiosity and the desire to "know" how things work will provide fertile ground for the technical illustrators of the future.
>
> —*Kevin Hulsey*

WHY ARE GRAPHICS IMPORTANT?

In Unit Four, we discuss how people learn through *data* and *process* schemata. According to this theory, the mind stores knowledge in bundles or patterns (data schemata) analogous to data files stored in your computer. Process schemata allow us to amplify upon previously perceived patterns and assimilate those patterns into a plan or script of action. For example, if you are going to lunch at a restaurant, your data schemata recalls the specifics associated with eating, such as fork, spoon, napkin, lunch booth, menu, and literally thousands of other pieces of information. Your process schemata tells you to walk over to the booth, sit down, wait for a server to come over, examine the menu, and so on. This theory seems straightforward enough. However, researchers want to know how we learn information and what can be done to make our learning more efficient and effective. One answer is to use graphics.

TIP FOR TECHNICAL EDITORS

The terms *image, visual,* and *graphic* are often used as synonyms, and some software refers to graphics as *pictures,* whether or not they are photographs. *Image processing* refers to turning digital data into photograph-like visuals.

Technical graphics refer to any type of visual illustration: diagrams, drawings, photographs, flowcharts, tables, graphs, and charts. Each type of graphic has a specific purpose and function and, if properly designed, reinforces meaning. Details that are insignificant to the intended meaning are omitted or deemphasized. Graphics frequently accompany text and create visual information for the reader; occasionally, they can completely replace text. For example, examine Figure 11.2 on the number of births and deaths during 2004. The graphic is self-explanatory and needs no textual elaboration.

Graphic representations of technical information are often the most efficient means to convey information to a reader. The quantitative information contained in Figure 11.2 might be difficult to understand if written out in text form. Graphics also allow information to be compared, to be contrasted, and to demonstrate relationships. For example, Figure 11.3 allows you to compare the number of men and women over age 20 who were employed full-time from 1994 to January 1, 2004. Notice that during those ten years, the number of men employed full-time continued to rise; however, the number of fully employed women leveled off from 2000 to 2004. Can you hypothesize why?

Furthermore, well-designed technical graphics communicate information efficiently by simplifying ideas and creating interest. A labeled line drawing of a plant cell clearly distinguishes its parts more readily than does text or even a photograph. A chart that represents data in the form of bars, lines, or pie wedges efficiently communicates quantitative information that would be lost by merely presenting raw numbers. How could anyone understand geometry without graphic representations of angles, lines, arcs, and circles that demonstrate relationships to mathematical principles? Also, graphics add interest by directing the reader's attention toward the subject. We all are drawn more to a textbook, magazine, instruction manual, or technical report that supplements the text with graphics. Today, readers expect graphics to be integrated into documents to support the text as well as keep their interest. As Marshall McLuhan said, "The medium is the message."

World Vital Events Per Time Unit: 2004

(Figures may not add to totals due to rounding.)

Time unit	Births	Deaths	Natural increase
Year	129,108,390	56,540,896	72,567,494
Month	10,759,033	4,711,741	6,047,291
Day	352,755	154,483	198,272
Hour	14,698	6,437	8,261
Minute	245	107	138
Second	4.1	1.8	2.3

FIGURE 11.2
Example of a Data Table.
Source: U.S. Census Bureau.

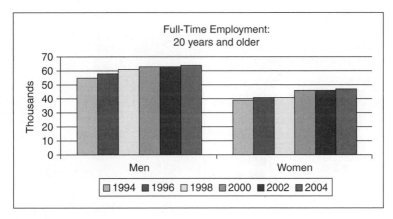

FIGURE 11.3

Example of a Comparative Graph.

Source: U.S. Department of Labor.

The charts in Figure 11.4 provide graphic representations of data. Each chart has a different function, but all attempt to present information in a pleasing visual manner.

Research conducted by Levie and Lentz identifies four important rationales for including graphic illustrations with text. Graphic illustrations (1) direct the reader's *attention* to the subject, (2) positively affect the reader's *attitude* toward the subject, (3) enhance the reader's *comprehension* of the subject, and (4) *compensate* for poor reading skills by providing pictorial clues. Levie and Lentz also provide research that shows a 36% gain in learning when graphic illustrations accompany text. Additional research conducted by Gatlin maintains that 83% of our learning is derived from what we see, whereas only 11% comes from what we hear. Morrison and Jimmerson's research concluded that we remember 43% more when graphics are integrated with word passages.

This last point is important: graphics with text may increase readers' long-term memory. Significant research conducted during the past two decades by Paivio, Freedman, Haber, Alesandrini, Buzan, and other scientists and psychologists has determined that our brains maintain an astonishing capacity for graphic recall; in fact, this capacity is limitless. (Graphics, especially in the form of pictures, create direct neuro-pathways through the brain to where long-term images are stored as chemical traces.) Moreover, their research has further underlined how the two hemispheres of our brain sustain different functions. Although both hemispheres become activated by the same sensory stimuli, the left hemisphere processes analytical, verbal, sequential, and symbolic information, whereas the right hemisphere triggers our imagination and our ability to synthesize and combine ideas.

The implications from this research are clear: readers exposed to text integrated with visual illustrations retain substantially more information than those who read only text. This is true for both paper-based and computer-based learning. However, we need to note that you can't simply throw a lot of graphics at a document and make the learning stick. According to the latest research, a person's long-term memory is affected the greatest when the graphics are

- *balanced* proportionately with the text, not overemphasized or garish;
- *congruous* with text-redundant information;
- *essential* to the text;
- *uncluttered* with irrelevant information;
- *as simple* as possible;
- *as large* as reasonably possible;
- *inoffensive* to users from other cultures.

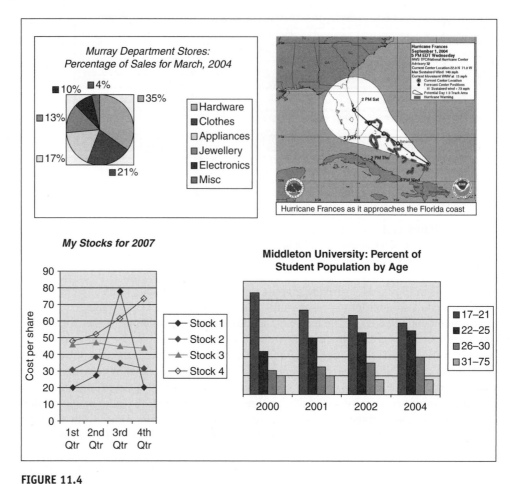

FIGURE 11.4

Graphics Simplify Textual Information.

Source: U.S. Commerce Department and National Oceanic and Atmospheric Administration.

When well designed, graphics are powerful tools to facilitate learning quantitative and qualitative information. Most commonly, they structure information by showing numerical, logical, or spatial relationships and by organizing large amounts of data that cannot easily be understood through words. And remember, information presented visually is coded by both hemispheres of our brain and is much more likely to be retained as schemata over long periods of time.

EDITING GRAPHICS

The responsibilities of a technical editor, whether editing text or graphics, depend principally upon company policy and the level of editing requested by the client. We suggest, though, that individuals considering a career in technical editing begin preparing themselves for what seems to be a dramatic shift in responsibilities. Although many companies continue to embrace the traditional, linear, lockstep approach to document design, others are beginning to use a team-approach model that involves collaboration of writers, graphic artists, editors, and production staff. As a result, responsibilities and duties are progressively shared as well.

This paradigm shift within the workplace, called the *network economy,* parallels the rapid change in technology.

Previously, an editor's graphic responsibilities (in addition to editing for correctness, visual readability, and effectiveness of the text) may have involved a simple review of tables, charts, diagrams, and illustrations for proper labeling, parallel form, appropriateness of titles, and placement within a document. As additional companies adopt a network economy, technical editors are finding themselves as equal contributors to the document design process. At some companies, editors now take part in the planning, designing, and writing of technical documents. We therefore suggest that you prepare for the additional editing responsibilities that will undoubtedly come your way. Take advantage of any opportunity to attend special courses to increase your knowledge or work with editors and writers to gain invaluable experience. Also, note that many companies have become global and appeal to international markets. For example, Mercedes and Hyundai cars are now built in the United States as well as in Germany and South Korea. In the near future, it would not be unusual for you to be team-editing documents for an international company.

When editing technical graphics, remember that the principal purpose behind graphics is to convey information effectively and efficiently to the reader. This becomes particularly important when quantitative or qualitative data can be understood only visually. Displaying information in the form of a chart, table, graph, drawing, illustration, etc. permits the reader to study the whole picture, make comparisons, observe relationships, and, in the case of illustrations and drawings, associate the components spatially. Think how much you would dread putting a bicycle together if the instructions were solely text-based with no accompanying illustrations.

Comprehensive editing requires attention to the *appropriateness, organization,* and *accuracy* of each graphic. Ideally, graphic design issues will be addressed during a collaborative planning phase with everyone involved; however, due to time or budgetary constraints, this collaboration may not always be possible. Decisions based on the three above principles will directly affect the choice of whether to use a graphic, and if so, which type. Generally, graphics or visuals, as they are also known, are chosen based on their function: whether they *represent* information or whether they *convey* information. Illustrations, in the form of line drawings, cutaways, or photographs, *represent* information by depicting an actual object. Informative visuals, on the other hand, *convey* information in the form of tables, diagrams, charts, and graphs.

> Where I grew up, learning was a collective activity. But when I got to school and tried to share learning with other students that was called cheating. The curriculum sent the clear message to me that learning was a highly individualistic, almost secretive, endeavor. My working class experience . . . was disparaged.
> — *Henry A. Giroux*

> Individuals, working together, construct shared understandings and knowledge.
> —*David Johnson*

Appropriateness

A traditional, comprehensive approach to visual editing begins with judging the appropriateness of each graphic. The graphic itself and its placement within the text must make sense to readers. In other words, can they clearly understand what they are seeing based on what they have previously read? Do the visuals complete the reader's need for understanding or for completing a task? What degree of complexity should the graphic represent? Does the reader need to see a detailed, exploded diagram, or would a simple photograph be sufficient? Some documents may be totally pictorial. Emergency instructions on airlines, for example, are almost totally graphic to avoid misinterpretation by speakers of different languages.

The type of illustration used should reflect the purpose of the accompanying text. Is the document's purpose to instruct, report, describe, propose, argue, compare, contrast, or inform? For example, if the text discusses the effects of a temperature change on a certain material over a period of time, a line graph would be the likely choice. However, if quantitative information needs to be visually *compared,* usually a bar/column graph would be more

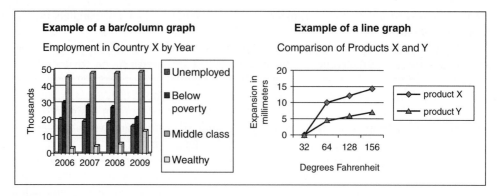

FIGURE 11.5
Two Types of Graphs.

suitable. (See Figure 11.5.) Tables are often better than graphs for simple data. Noncomparative data sets usually belong in tables, not charts.

Explaining a process may call for diagrams such as flowcharts, Gantt charts, schematics, or other appropriate types of drawings that further the writer's purpose. Schematics, line drawings, or cross-sectional drawings should be chosen to represent the function of the data. Schematics generally depict how something is structured or wired; flowcharts demonstrate steps in a process; line drawings outline how something looks without shadows, with shading, or other details; and cross-sectionals outline how something looks inside (see Figure 11.6).

When determining the appropriateness of a graphic for a particular situation, technical editors need to consider the *rhetorical situation,* just as they would for the text. These considerations include an analysis of the *audience, purpose,* and *context.* For example, if the intended audience consists of children or perhaps individuals who have little technical background, the visuals should be less complex, such as photographs or simple line drawings. Experts, however, likely would not benefit from simplistic illustrations or photographs. Editing graphics, just like text, requires good judgment based on the audience's needs. A fourteen-year-old assembling a model Corsair fighter plane for the first time probably is not interested in a composite chart depicting the airplane's capabilities of thrust, roll, and pitch. Instead, a representational line drawing illustrating the relationships between the model's parts would be appropriate. The sole purpose of the instructions is to focus on assembling the model; extraneous information is unwarranted and confusing.

Organization

Organizing graphics on a page follows many of the same principles as organizing text: both forms of communication should be clear, concise, and simple. Also, decisions regarding the text and graphics must consider the needs of the audience, the purpose of the document, and the context in which the document is created. For example, graphics can make lengthy and difficult documents more interesting and easier to understand.

To be effective, each graphic should be integrated within the document by placing it next to the text to which it refers—the closer, the better. The text itself should refer to a graphic as a *table* if it contains a set of data in ordered columns or words, or a *figure* if it is an illustration, drawing, or photograph. (Notice that we call Figure 11.2 and the other tables in this chapter *figures* rather than *tables* because they function as *illustrations* of tables.) Editors determine if the tables and figures have been properly numbered throughout the document. Tables and figures are numbered in separate sequences. Longer documents may be subdivided into sections

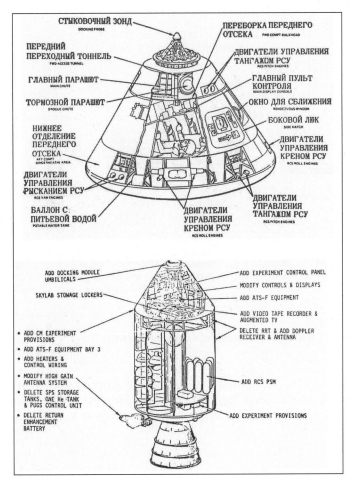

FIGURE 11.6

Comparative Illustrations.

Source: NASA.

or chapters. In such cases, the section number precedes the sequence number. For example, Figure 11.4 identifies the fourth figure in Chapter 11. Table captions are placed above the table, whereas figure captions are placed below. Remember that different style guides, such as APA, MLA, *Chicago,* or your company's own in-house guide, may differ in how to represent tables and figures. For example, many software user guides number figures sequentially through the entire book.

Editors must also determine if the graphics need additional labeling. As shown in Figure 11.7, line and bar graphs have x and y axes divided into units. They always require a label to be clear. Other types, such as multiple-bar, stacked-bar, and multiple-line graphs, require legends that identify which items are being graphed. Editors should also make certain that each graphic has a caption that succinctly describes the whole graphic, such as *Tax Table for Single Individuals* or *Percentage of Sales for 2004*. Sometimes editors suggest that the writer add a short, one- to three-sentence description to the caption to either explain the graphic or draw a conclusion. However, if the text is well written and clearly refers to the graphic, descriptions in captions usually are not necessary. When it is necessary to place graphics at some distance from the explanatory text (for example, in an appendix, as is often done with engineering drawings) explanatory captions are quite useful. Finally, editors

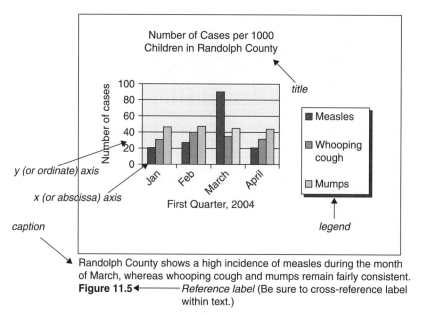

FIGURE 11.7
Example of a Bar/Column Graph with Complete Labeling.

should ensure that any data taken from another source is properly cited in the graphic. Using complete tables and graphics will require permission from the author unless they are considered fair use, are from the U.S. government, or are otherwise in the public domain.

Accuracy

Technical editors are not responsible for the actual data; this responsibility lies with the engineers, scientists, technicians, subject matter experts, or writers. However, if you are involved in a comprehensive edit, you will be responsible for determining if the data provide an *accurate reflection* of the information. Data, especially when represented in the form of a numerical table, chart, or graph, can be distorted and thus provide inaccurate information. For example, examine Figures 11.8 and 11.9. Both graphs consist of the same unemployment data. However, one presents the data more accurately. Which one is better and why? How would you edit the other one?

Two problems exist with Figure 11.8. First, the scaling between the x (abscissa) and y (ordinate) axes creates a false impression that unemployment has not changed much over a five-year period. The y scale is disproportionally high and flattens the data. Second, readers normally read from left to right and expect trends and chronologies to move in the same direction as they read. Figure 11.8 presents the latest year, 2004, first. Because we view trends from left to right, a cursory examination of this graph suggests that unemployment among women has slightly decreased. Actually, just the opposite is true. Beginning at the right of the graph with the year 2000 and reading backward, you will see that unemployment has actually increased. Also, because the y axis is scaled from zero to 100 in Figure 11.8, unemployment doesn't seem to be a problem. If you were an incumbent politician, which graph would you present to the public?

Figure 11.9's scale provides a more accurate picture of unemployment during the past five years. It demonstrates a realistic decline in unemployment.

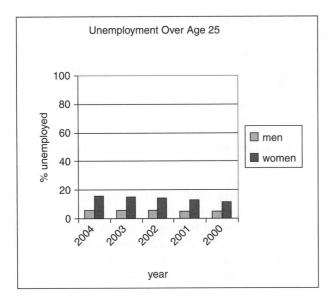

FIGURE 11.8
Bar Graph A.

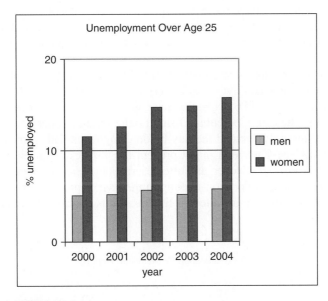

FIGURE 11.9
Bar Graph B.

When editing, remember that the *y* axis typically depicts values, whereas the *x* axis typically illustrates time. Look carefully at both axes and ask if the reader is likely to infer a misleading conclusion because one or both of the axes distort information. Are the values in the *y* axis divided into appropriate intervals? Are the values depicted too small or too large and consequently overexaggerate the effect? Is there sufficient accompanying text that explains the graphic? Finally, are all parts of the graph labeled properly?

Although Graph B is an improvement over Graph A, there are still some possible problems with Graph B. Why? Depending on the point you want to make with the graphic, the scaling could be construed as an unethical design choice. In other words, the rhetorical situation—the choice of audience would be the most important factor here—affects the kind of graph that would best and most accurately relay the necessary information to the target audience as truthfully as possible. Although we discuss ethics more completely in Unit Six, it is important that you understand how ethical choices underlie almost every move you make as an editor, because you are constantly dealing with language, and text and visuals can be misperceived. The editor is obligated to discuss any confusing graphics with the author before production. Of course, one way to clarify ambiguous graphics is to accompany them with explanatory text in your document. Another method is to simply revise the graphic. The second method is usually preferable, because many people will absorb the information in the graphic but not read the text closely.

Distortion of data occurs when graphs are displayed three-dimensionally. This is particularly true of pie charts. In Figure 11.10, compare Wednesday at 21 percent time allocation to Thursday at 7 percent. Although Wednesday's share is three times Thursday's, notice how it appears to be only twice as large. Similar to the previous examples of distorted graphics, the reader is not presented with an accurate reflection of the data.

A solution that we recommend is simply to avoid extreme 3-D perspectives altogether and present the data in two dimensions as in Figure 11.11, where each pie wedge conveys a precise percentage without a distorted perspective. Do not forget to check all graphs to make sure the numbers add up to the designated totals. If you are working with percentages, the numbers need to add up to 100 percent.

A pie chart should not have more than six wedges, as it becomes too difficult to separate the units. Instead, consider using a miscellaneous wedge that includes everything the other five wedges do not include; just do not allow the miscellaneous wedge to exceed 10 percent of the total.

We cannot overstate the consequences resulting from inaccurate, confusing, or misread data. Time, money, careers, and lives may be seriously affected. One of the best examples of this occurred on January 28, 1986, when the space shuttle *Challenger* exploded seventy-three seconds after liftoff. The cause, potential O-ring failure on the primary boosters during cold weather, had been intensely debated the day before. Engineers had warned that the below-freezing weather at Cape Canaveral might prevent the O-rings from sealing and allow hot gases to escape the booster. However, as Tufte describes in *Visual Explanations,* a combination of "groupthink, technical decision-making in the face of political pressure, and bureaucratic failures to communicate" resulted in the disastrous decision to launch anyway, and

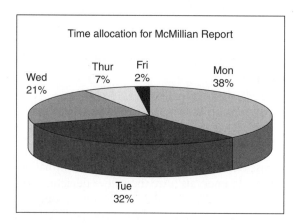

FIGURE 11.10
Sample Pie Chart.

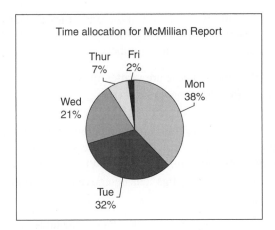

FIGURE 11.11
Revised Pie Chart.

seven astronauts lost their lives as a consequence. Although Tufte cites many variables that led to the fateful decision to launch, he points primarily to thirteen poorly designed charts and data tables as the principal reason the launch was not aborted. We recommend that you read Tufte's full account of how wrong decisions were made from poorly designed graphics.

EDITING TABLES

Anyone going into the technical editing field must be able to edit any form of graphic that presents raw data. Technical editors should not be surprised if called upon to actually create the graphics themselves by synthesizing raw data into a visual illustration such as a graph (bar, pie, linear, column, area, scatter, etc.) or a table (textual, statistical, numerical, etc.).

Whether editing or creating graphics that present data, technical editors must be able to organize, summarize, and shape the data in a form that complements the text. Graphs and charts are best to use when you want to show a trend or a relationship between variables. Tables, however, are best when you want to create a *tabular presentation*: specific data that draw comparisons between variables. Tables express these values as totals, subtotals, statistics, percentages, means, medians, modes, frequencies, etc. By providing a visual display of the data, tables relieve the writer from having to discuss every individual numerical value in the text. Instead, the author is free to use text to generalize and make conclusions about the specifics presented in a table.

Previously, we have shown how to label and edit a graph; we now turn your attention to tables, the most widely used graphic form for presenting data. See Figure 11.12.

Parts of a Table

1. *Caption.* This includes the title and number of the table, and is placed above the table.
   - The title provides a concise description of the table, but never more than two lines long. The first letter of each word should be capitalized following the rules for titles. No period is placed at the end of the title.

- Tables are sequentially numbered in the order they appears. The table number is followed by a period and space separating it from the title. Table numbers are in a separate sequence from figure numbers.

2. *Headings and subheadings.* These are written in singular form unless the data refer to a group, such as *children* or *women.*
   - Use title-style capitalization, unless your style guide specifies otherwise.
   - *Column headings* identify the data listed below. When the column head is located in the uppermost left, it is referred to as the *stub column* and its heading is the *stub heading.* Data following the stub column form the *stub.*
   - *Column spanners* are headings that lie above two or more columns and classify them into a group.

3. *Table body.* The variables organized as the data in the table.
4. *Dividers.* Lines that separate the table into various components.
5. *Table spanners.* Located across the entire top of a table, they are frequently used to combine two or more tables into a single table to avoid repetition.
6. *Notes.* These are used to explain what may not be self-evident in the table. Standard abbreviations are acceptable, such as "%" for percent, "N" for number, "SD" for standard deviation, etc. They also may contain the definitions of any technical terms not readily familiar.

Guidelines for Tables

Data Presentation

- Include only relevant data.
- Make certain the table "speaks for itself."
- Right-align numbers.
- If numbers include decimals, align them on the decimal point.
- Round numbers up or down if possible.
- Create tables only if you can make more than two columns.
- Provide the units of measurement.
- Keep the title as succinct as possible.
- Keep the amount of data reasonable.
- Be consistent throughout your tables.
- Provide totals for columns when appropriate.

TIP FOR TECHNICAL EDITORS

If the number of decimal places is not the same for all figures in a column, do not add trailing zeros to make them the same, as this may be misleading. The number of decimal places has mathematical significance; adding zeros indicates greater precision than may be true.

Caption (Table number and Title)

Stub head

Column spanner

Stub

Column head

Table body

Table spanner

Notes

Table 1   Percent of workers with access to retirement and health care benefits, by selected characteristics, private industry, National Compensation Survey, March 2004

Characteristics	Retirement benefits			Health care benefits		
	All plans[1]	Defined benefit	Defined contribution	Medical care	Dental care	Vision care
All workers	59	21	53	69	46	29
Worker characteristics:						
White-collar occupations	69	24	64	76	53	33
Blue-collar occupations	59	26	49	76	47	29
Service occupations	31	6	27	42	25	18
Full time	68	25	62	84	56	35
Part time	27	9	23	20	13	8
Union	84	70	48	89	73	56
Nonunion	56	16	53	67	43	26
Average wage less than $15 per hour	46	11	41	57	34	20
Average wage $15 per hour or higher	77	35	68	86	63	41
Establishment characteristics:						
Goods-producing	70	32	60	83	56	36
Service-producing	55	18	50	65	43	27
1-99 workers	44	9	40	58	31	18
100 workers or more	77	35	68	82	64	42
Geographic areas:						
Metropolitan areas	59	22	53	70	47	30
Nonmetropolitan areas	55	15	51	66	39	24
New England	56	22	50	68	49	25
Middle Atlantic	59	29	50	71	47	34
East North Central	65	24	58	70	46	27
West North Central	65	21	57	66	40	21
South Atlantic	57	17	54	69	46	25
East South Central	57	14	55	72	45	34
West South Central	56	18	52	68	40	24
Mountain	59	17	54	68	45	30
Pacific	52	22	46	70	51	38

[1]Includes defined benefit pension plans and defined contribution retirement plans. The total is less than the sum of the individual items because many employees have access to both types of plans.

NOTE: Because of rounding, sums of individual items may not equal totals. Where applicable, dash indicates no employees in this category or data do not meet publication criteria.

FIGURE 11.12
Parts of a Table.
Source: U.S. Bureau of Labor Statistics.

Data Appearance

- *Contrast.* Make the key elements easy to identify by using boldface headings, spanners, dividers, or titles. You can use different-size fonts and different font styles to distinguish or highlight specific areas; however, just remember to keep it as simple as possible.
- *Structure.* Group the data in a way that is not only pleasing to the eye but makes the information easy to follow.
- *Arrangement.* Keep the table neat and uncluttered. *Less is more* when creating or editing tables. Make certain all the elements are aligned, such as the numbers in columns under the correct headings, titles, spanners, etc.
- *Separation.* Use white space to your advantage by generously incorporating it among the elements of the table.

Well-Designed Elements

Examine the table in Figure 11.13. Notice how it combines contrast, structure, arrangement, and separation into an easy-to-read, attractive table. The data are uncluttered and generous use is made of white space. Key elements such as the title and stubs use italics and boldface type to set them off. Study the table carefully, and be able to apply these guidelines when editing tables.

Poorly Designed Elements

In contrast, the table in Figure 11.14 is difficult to read. Although the columns are aligned and generous use is made of white space, notice how the data are not separated clearly into defined stubs, and data totals are not clearly distinguished from the data summarized. For additional clarity, the column headings need an explanation. Compare the poorly designed elements of this table with an edited portion of the table in Figure 11.15. Pay particular attention to how the data on meat, pork, poultry, and fish have been emphasized by visual alignment.

TIP FOR TECHNICAL EDITORS

Although you won't generate the actual data, you do have an ethical responsibility to ask questions of those who are responsible for the data and graphics. Inquire if any information may have been omitted. Determine if the text supports the graphics. Are the tables and charts labeled properly? Do they provide a clear and accurate picture without distorting information? Do the graphics reflect who is responsible for the data? Do the conclusions that have been drawn seem logical? Will the intended audience be able to make similar conclusions based on the information and graphics provided? Of course, the extent of inquiry you conduct will be dictated by deadlines, the complexity of the project, your experience, and the level of editing required. Regardless, though, review the graphics thoughtfully. As more companies move toward network economies and as technical editors assume additional responsibilities, you will be well served to acquire as much knowledge and experience as you can.

Reporting Location	Sector of Institution			
On-Campus		**2001**	**2002**	**2003**
	Public, 4-year or above	1,246	1,288	1,333
	Private nonprofit, 4-year or above	815	914	1,046
	Private for-profit, 4-year or above	4	4	8
	Public-2-year	120	125	162
	Private nonprofit, 2-year	3	8	9
	Private for-profit, 2-year	12	8	15
	Public, less-than-2-year	5	2	3
	Private nonprofit, less-than-2-year	1	0	3
	Private for-profit, less-than-2-year	0	1	2
	Total	2,206	2,350	2,581
Residence Halls				
(included in on-campus)		**2001**	**2002**	**2003**
	Public, 4-year or above	885	904	942
	Private nonprofit, 4-year or above	643	719	800
	Private for-profit, 4-year or above	1	3	1
	Public-2-year	33	34	49
	Private nonprofit, 2-year	1	5	5
	Private for-profit, 2-year	8	5	9
	Public, less-than-2-year	1	1	1
	Private nonprofit, less-than-2-year	0	0	1
	Private for-profit, less-than-2-year	0	0	0
	Total	1,572	1,671	1,808
Non-Campus		**2001**	**2002**	**2003**
	Public, 4-year or above	239	231	283
	Private nonprofit, 4-year or above	118	70	68
	Private for-profit, 4-year or above	6	10	4
	Public-2-year	32	8	11
	Private nonprofit, 2-year	7	0	1
	Private for-profit, 2-year	46	0	0
	Public, less-than-2-year	2	0	0
	Private nonprofit, less-than-2-year	4	0	0
	Private for-profit, less-than-2-year	36	0	0
	Total	490	319	367

Note: Totals include crimes at administrative units of institutions not summarized elsewhere.

FIGURE 11.13
Forcible Sex Offenses on U.S. College Campuses.
Source: U.S. Department of Education

Item and group	U.S. city average	
	CPI-U	CPI-W
Expenditure category		
All items...	100.000	100.000
Food and beverages	15.291	17.024
Food	14.295	15.940
Food at home	8.183	9.540
Cereals and bakery products	1.185	1.342
Cereals and cereal products	.400	.467
Flour and prepared flour mixes	.050	.061
Breakfast cereal	.221	.249
Rice, pasta, cornmeal	.128	.157
Bakery products	.784	.875
Bread	.225	.260
Fresh biscuits, rolls, muffins	.109	.116
Cakes, cupcakes, and cookies	.220	.251
Other bakery products	.230	.249
Meats, poultry, fish, and eggs...........................	2.272	2.845
Meats, poultry, and fish	2.178	2.725
Meats	1.456	1.841
Beef and veal	.725	.916
Uncooked ground beef.............................	.268	.354
Uncooked beef roasts.............................	.131	.165
Uncooked beef steaks.............................	.269	.326
Uncooked other beef and veal	.057	.071
Pork	.449	.570
Bacon, breakfast sausage, and related products	.147	.177
Ham	.096	.118
Pork chops	.103	.141
Other pork including roasts and picnics...................	.102	.133
Other meats	.282	.354
Poultry	.413	.530
Chicken...	.332	.437
Other poultry including turkey.........................	.081	.093
Fish and seafood...................................	.309	.354
Fresh fish and seafood	.181	.210
Processed fish and seafood..............................	.127	.145

FIGURE 11.14

Relative Importance of Components in the Consumer Price Indexes: U.S. Average, December 2004.

Source: U.S. Bureau of Labor Statistics.

Item and Group	U.S. City Average by Percentage	
	CPI-U*	CPI-W*
All Items	*100.000%*	*100.000%*
All Food and Beverages Consumed	**15.291**	**17.024**
Foods consumed at home	**8.183**	**9.540**
Cereals and cereal products		
Flour and prepared flour mixes	.050	.061
Breakfast cereal	.221	.249
Rice, pasta, cornmeal	.128	.157
Total	**.400**	**.467**
Bakery products		
Bread	.225	.260
Fresh biscuits, rolls, muffins	.109	.116
Cakes, cupcakes, and cookies	.220	.251
Other bakery products	.230	.249
Total	**.784**	**.875**
Cereal and bakery products	**1.185**	**1.342**
Beef and veal		
Uncooked ground beef	.268	.354
Uncooked beef roasts	.131	.165
Uncooked beef steaks	.269	.326
Uncooked other beef and veal	.057	.071
Total	**.725**	**.916**
Pork		
Bacon, breakfast sausage, and related products	.147	.177
Ham	.096	.118
Pork chops	.103	.141
Other pork including roasts and picnics	.102	.133
Total	**.449**	**.570**
Poultry		
Chicken	.332	.437
Other poultry including turkey	.081	.093
Total	**.413**	**.530**
Fish and seafood		
Fresh fish and seafood	.181	.210
Processed fish and seafood	.127	.145
Total	**.309**	**.354**
Meats, poultry, and fish	**2.178**	**2.725**

*Consumer Price Index for All Urban Consumers (CPI-U) and the Consumer Price Index for Urban Wage Earners and Clerical Workers (CPI-W). These data are to be used in conjunction with the CPI-U and CPI-W released in that same year (December, 2004).

FIGURE 11.15
Relative Importance of Consumer Consumption in U.S. Cities.
Source: U.S. Bureau of Labor Statistics.

Appropriateness

- Do the graphics seem appropriate for the intended purpose of the document?
- Can the landscape-oriented graphs be changed to portrait-oriented graphs?
- Does the reader have to process complex data that could be clarified with the insertion of a graphic?
- Are the graphics appropriate for the level of complexity in the text?
- Are the graphics appropriate for the level of the reader's understanding?
- Has the "rhetorical situation" for the graphics been considered?
- Do the graphics contain *chartjunk*—unnecessary lines, drawings, icons, ink, etc. that detract from the intended purpose of the chart?
- Can vertical lines be eliminated from any tables to make them seem more open and appealing?
- Do the graphics present the desired message?
- Has the best chart been chosen to represent the desired comparison?

Organization

- Are the graphics clear, concise, and uncluttered with nonessentials?
- Does the accompanying text refer to the graphic?
- Are the graphics placed closely to the relevant text?
- Have proper credits been cited for the graphics?
- Are the graphics' titles short but complete enough to provide an overview of their function?
- Have all measurements, variables, and units been properly labeled?

Accuracy

- Do the graphic data accurately reflect the purpose of the document?
- Do the graphics distort the data?
- Are charts and graphs proportional?
- Do any conclusions seem illogical based on the data provided?
- Have labels, axes, and titles been included?

FIGURE 11.16
Checklist for Editing Technical Graphics.

CHECKLIST FOR EDITING GRAPHICS

Figure 11.16 provides a general checklist for editing graphics based on their appropriateness, organization, and accuracy. Above all, those guidelines remind you to consider your audience and the purpose of the graphics. Depending on the level of edit requested and your company's policies, though, you may have to make adjustments to our checklist. Use photographs, drawings, diagrams, or schematics for objects. Numerical data are best represented with bar charts, pie charts, line graphs, and tables. Concepts that show processes or hierarchical relationships are usually displayed using flowcharts.

Figure 11.17 is a quick reference guide for which type of graphic to use for different types of data. Figure 11.18 includes additional specific recommendations for you to consider when editing technical graphics.

Function	Type	Attributes

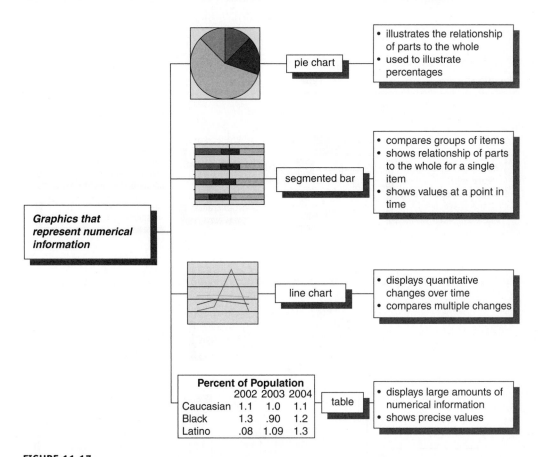

- pie chart
  - illustrates the relationship of parts to the whole
  - used to illustrate percentages

- segmented bar
  - compares groups of items
  - shows relationship of parts to the whole for a single item
  - shows values at a point in time

Graphics that represent numerical information

- line chart
  - displays quantitative changes over time
  - compares multiple changes

Percent of Population

	2002	2003	2004
Caucasian	1.1	1.0	1.1
Black	1.3	.90	1.2
Latino	.08	1.09	1.3

- table
  - displays large amounts of numerical information
  - shows precise values

FIGURE 11.17

Quick Reference Guide for Graphic Illustrations.

Sources: U.S. Department of Energy; NASA; National Oceanic and Atmospheric Administration; U.S. Equal Employment Opportunity Commission; Environmental Protection Agency; and Social Security Administration.

EDITING GRAPHICS FOR MULTICULTURAL READERS

In his article "The Almost Universal Language: Graphics for International Documents," William Horton reminds us that in today's global economy, graphics may have certain cultural implications of which we are unaware. What may seem like a harmless graphic of a hand gesture in one culture may be offensive in another. Technical editors need to be sensitive to multicultural issues when editing, especially when it comes to graphics. For example, some cultures place more emphasis on politeness when giving directions and will tell the reader to *please* do this or that and will thank them after they complete the task. A photograph of a woman wearing Western-style clothing may be offensive in Islamic cultures where some women wear burqas to cover most of their face. Individuals depicted using their left hand may be offensive to some Asian cultures where the left hand is considered to be unsanitary. Colors, hand gestures, and symbols also may be misinterpreted by different cultures. Before you begin your technical editing career, we recommend that you familiarize yourself with these multicultural issues, which we discuss in detail in Chapter 21, particularly as they relate to graphic representations.

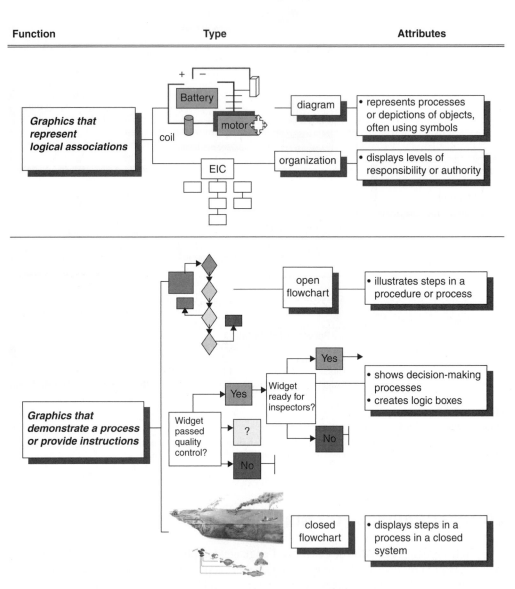

Function	Type	Attributes

FIGURE 11.17 (*continued*) Quick Reference Guide for Graphic Illustrations.

Sources: U.S. Department of Energy; NASA; National Oceanic and Atmospheric Administration; U.S. Equal Employment Opportunity Commission; Environmental Protection Agency; and Social Security Administration.

Function	Type	Attributes

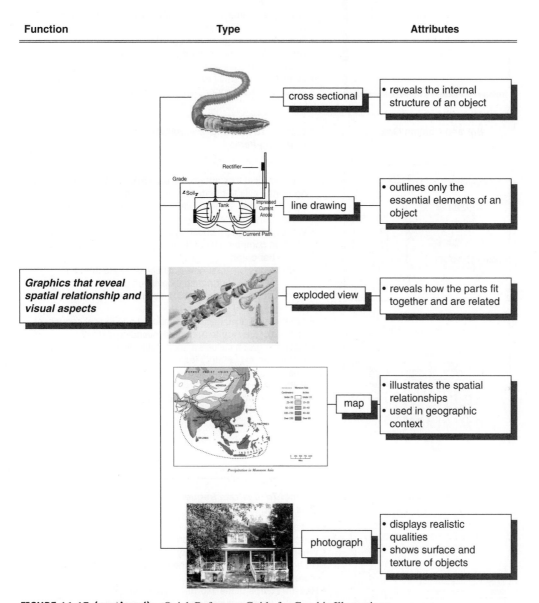

cross sectional — • reveals the internal structure of an object

line drawing — • outlines only the essential elements of an object

exploded view — • reveals how the parts fit together and are related

map — • illustrates the spatial relationships • used in geographic context

photograph — • displays realistic qualities • shows surface and texture of objects

Graphics that reveal spatial relationship and visual aspects

FIGURE 11.17 (*continued*)   Quick Reference Guide for Graphic Illustrations.

Sources: U.S. Department of Energy; NASA; National Oceanic and Atmospheric Administration; U.S. Equal Employment Opportunity Commission; Environmental Protection Agency; and Social Security Administration.

Tables

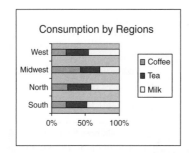

Purpose: *To illustrate large quantities of numerical data*

- Have the units of measurement been clearly indicated?
- Does the left-hand column (the sub) indicate the items being compared?
- Is the title as succinct as possible while clearly defining what the table illustrates?
- Has credit been given to all outside sources?
- Are the data columns lined up logically and clearly with decimal points falling directly below each other?
- Is the table as small as possible with its data uncluttered?
- Would an explanatory caption benefit the reader?

Bar and Column Graphs

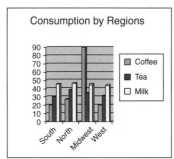

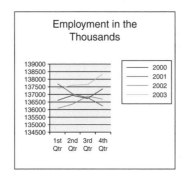

Purpose: *To illustrate comparative numerical values for ranking items*

- Is the vertical axis approximately 25 percent shorter than the horizontal axis to create fair proportions?
- When possible, do the quantity scales begin at zero?
- Are all parts labeled properly?
- Is the type of graph chose appropriate for the data?
    - (a) To compare two or three quantities per item, use a bar graph
    - (b) To demonstrate how quantities deviate from a standard norm, use a deviation graph
    - (c) To compare combined units, use segregated bar graphs
- Are the bars arranged in consecutive and logical sequences (chronological, top to bottom, descending order)?
- Does the graph include a short but well-crafted title?
- Is there an explanatory legend?
- Has the graph been listed as a *figure*?
- Is the graph close to the text?
- Has the source of the information been cited?

Line Graphs

Purpose: *To illustrate frequency distributions over a time period*

- Is the vertical axis approximately 25 percent shorter than the horizontal axis to create fair proportions?
- When possible, do the quantity scales begin at zero?
- Do horizontal grid lines help the reader analyze the graph more clearly?
- Has the source of the information been cited?
- Does the graph include a short but well crafted title? Are the *x* and *y* axes properly labeled?
- Is the graph an accurate representation of the text?
- Is the graph located near the relevant text and labeled with a *figure* number?

FIGURE 11.18

Detailed Guidelines for Editing Technical Graphics.

Source: U.S. Department of Energy; NASA; National Oceanic and Atmosphere Administration; U.S. Equal Employment Opportunity Commission; Environmental Protection Agency; Social Security Administration; U.S. Census Bureau; and U.S. Bureau of Labor Statistics.

Pie Charts

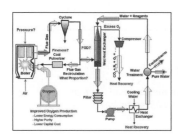

Types of Waste Deposited
into Municipal Waste System

Misc 22%

Paper 36%

Metal 8%

Plastic 11%

Food 11%

Glass 12%

Purpose: *To illustrate relative size of parts to the whole*

- Does the largest wedge appear towards the top of the pie?
- Have the number of wedges been restricted to six?
- If a 3D pie has been used, does it distort the relative size of the pieces?
- Do the wedges move around in a clockwise motion?
- Do all the wedges add up to 100%?
- If a wedge is emphasized, does it contain a brighter, contrasting color to the other wedges?
- If one wedge is over 10% miscellaneous, is it noted below the chart?
- Is there any unnecessary chartjunk?
- Do tick lines help or distract from the graph?

Note: Miscellaneous wastes includes rubber, leather, and textiles 7.1%; glass, 5.5%; wood, 5.7% and other materials, 3.4%

Process Diagram

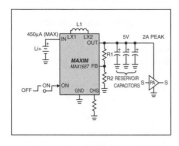

Purpose: *To provide a working representation of projects or processes using symbols*

- Is the diagram in the form of a blueprint, schematic, or writing diagram?
- Does the diagram clearly show logical relationships among the parts?
- Are all the parts clearly labeled?
- If the diagram demonstrates a process, is the flow or movement of the internal elements shown using arrows?
- Does a demonstrated process seem to portray a logical sequence of events?

Schematic Diagram

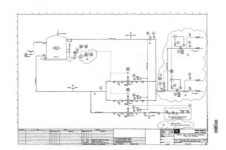

Blueprint Diagram

FIGURE 11.18 (*continued***)** Detailed Guidelines for Editing Technical Graphics.

Source: U.S. Department of Energy; NASA; National Oceanic and Atmosphere Administration; U.S. Equal Employment Opportunity Commission; Environmental Protection Agency; Social Security Administration; U.S. Census Bureau; and U.S. Bureau of Labor Statistics.

Drawings	Purpose: *To create simplified representations of objects*

Line Drawing

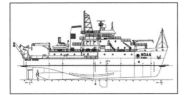

- Does the drawing specifically focus on the desired information intended for the reader?
- Would a photograph be a better graphic choice?
- Would an exploded or cutaway drawing be a better graphic choice?
- Are all the important aspects clearly labeled?
- Would thicker lines highlighting important areas of the object help the reader's focus?
- Is the drawing sufficiently detailed? Too detailed?

Exploded View

Cutaway View

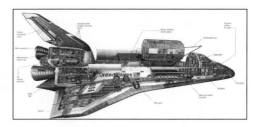

Flowcharts	Purpose: *To demonstrate hierarchical lines of authority or responsibility; closed-system flowcharts demonstrate a process*

Organizational Flowchart

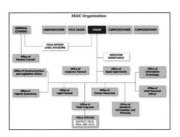

- Have any steps in the lines of authority or procedures been omitted?
- If the flowchart is open, does it require a start or finish label?
- If different geometric figures are used to represent a change in the process, does the flowchart need an explanatory legend?
- Does the chart need a citation identifying the original source?

FIGURE 11.18 (*continued*) Detailed Guidelines for Editing Technical Graphics.

Source: U.S. Department of Energy; NASA; National Oceanic and Atmosphere Administration; U.S. Equal Employment Opportunity Commission; Environmental Protection Agency; Social Security Administration; U.S. Census Bureau; and U.S. Bureau of Labor Statistics.

Closed-system Flowchart

Decision-making Flowchart

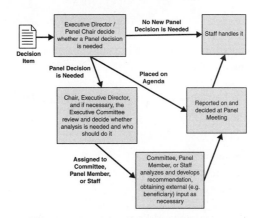

FIGURE 11.18 (*continued*)   Detailed Guidelines for Editing Technical Graphics.
Source: U.S. Department of Energy; NASA; National Oceanic and Atmosphere Administration; U.S. Equal Employment
Opportunity Commission; Environmental Protection Agency; Social Security Administration; U.S. Census Bureau; and
U.S. Bureau of Labor Statistics.

EXERCISE 11.2

Locate a graphic from a newspaper such as *USA Today* or a magazine such as *Newsweek* or
Time. Make separate copies of the article and the graphic. Exchange your article for a classmate's but withhold the graphic. After you have read your classmate's article, sketch what
you consider to be an appropriate graphic. Be sure to label all the parts. After you have completed your sketch, compare it to the graphic that accompanied the original article. Decide
if your graphic is appropriate for the type of information provided in the article. You may
want to repeat this exercise several times.

EXERCISE 11.3

Develop a paragraph that contains enough data to warrant a graphic. Decide in advance
what type of graphic is best suited to complement your paragraph. Exchange your paragraph with a classmate, and from your classmate's paragraph develop an appropriate
graphic. After both of you have finished drawing a suitable graphic, exchange your graphics. Peer-edit the graphics. Discuss with your classmate the rationale for choosing that particular type. Was the text well written? Was the graphic labeled properly? Did it appear to
be an accurate representation of the text? Did the graphic include any extraneous information that embellished but did not clarify?

EXERCISE 11.4

Using the chart-making tool in your word processor, create and label graphics that illustrate the following data:

1. Percentage of households that own bicycles:
   England 32%
   France 26%
   Holland 56%
   Pakistan 42%
   United States 29%

2. An organizational flowchart depicting the hierarchical structure of a business, government agency, school, or religious institution. Either use one with which you are already familiar, or make one up. Make certain that your flowchart is logical and demonstrates levels of responsibilities.

3. A table of your own creation that represents precise numbers or statistics. Acquire your data from an outside source to create your table. Consider using information from a business or corporation with which you are already familiar.

SUMMARY

Everyone appreciates illustrations, drawings, charts, graphs, and other forms of graphics that clarify and add meaning to the text. Knowing how and when to use them is an important responsibility of technical editing. If the data are complex or the reader needs to visualize the data, graphics can add the necessary dimension to complete the reader's understanding. Also, well-chosen, accurate graphics can expedite production, reduce costs, and even save lives.

Editing for Effectiveness

4

This unit assumes you have a good working knowledge of the first two levels of edit: editing for correctness and editing for visual readability. Units One, Two, and Three cover the traditional conventions of the English language. We ask you to learn and practice using these conventions so that you will be able to apply them effectively as an editor. Essentially, then, you are acquiring skills that you will increasingly improve upon as you gain more editing experience.

This unit has two primary functions: first, to help you refine what you already know about editing, and second, to prepare you for a type of learning that is difficult to teach: critical thinking. Here, our focus concentrates on *editing for effectiveness*, where you must critique the value and quality of other individuals' writing and then demonstrate how to improve it. It's not hyperbole to suggest that editing on this level requires one to actually enter the thought processes of writers, for better or for worse. This editing will require you to make many judgment calls about style, organization, completeness, balance, and timeliness of technical documents. For instance, you will have to assess whether the author's approach to the topic is logical and coherent. In short, you will have to rely on your own ability to think critically and determine what is the most effective way to communicate someone else's ideas to a target audience.

Editing for effectiveness is the most comprehensive level of editing, requiring the editor to exhibit sound judgment, knowledge of the conventions of usage, attention to detail, a bit of imagination, and above all, insight into the dynamics of the English language. Furthermore, we maintain that editors need to be well-informed on the theories of learning and the psychology of human interaction, as well as possess some common sense.

The four chapters in this unit provide you with initial insights into all of these areas as they apply to technical editing. However, the theories of learning, style, organization, and logical presentation of ideas won't help unless you apply what you have learned. After you have finished with this unit, we invite you to turn a critical eye to magazine articles, textbooks, instruction manuals, web pages, and journal articles to judge the effectiveness of others' written words. A lot of bad and good writing is out there; we invite you to begin to distinguish the difference.

CHAPTER 12

Editing for Effectiveness: Where Art and Craft Come Together

At the conclusion of this chapter, you will be able to

- discuss why technical editors should know learning theory;
- apply to editing five cognitive strategies of learning.

Previously, we discussed two categories of editing: editing for correctness and editing for visual readability. This chapter concentrates on a third level, editing for effectiveness. Each level incorporates specific editorial tasks and responsibilities and, just as importantly, different types of editorial judgments. For example, Unit Two explains how editing for correctness (also called copyediting, light editing, language editing, minimum editing, screening editing, etc.) requires the editor to assess the correctness and accuracy of a document's grammar, spelling, punctuation, and other elements of mechanics based on conventional standards of usage. This level of editing is reasonably straightforward and judgment calls are minimal: either a subject correctly agrees with its verb or it does not. Because many of your clients in business and the government are themselves competent at detecting grammatical inconsistencies, it's natural for them to infer that missed or unedited surface errors in a document reflect poor writing, sloppy editing, and technical accuracy, creating a poor image of your agency.

> Nothing, not love, not greed, not passion or hatred, is stronger than a writer's need to change another writer's copy.
> —*Arthur Evans*

Similar to editing for correctness, editing for visual readability relies on conventional standards as well, but additional editorial judgment is often necessary. As you recall, editing for visual readability requires the editor to follow appropriate document design guidelines, which may be prescribed by the client, the government, your agency, or in some cases the nature of the document itself. Primarily, though, the editor's responsibilities for this level address issues of visual readability such as document design, document formatting, and the accuracy and completeness of illustrations and graphics.

This chapter examines the third and most comprehensive level: editing for effectiveness. Editing a technical document for effectiveness, as you will discover, often resembles an art form. Editorial decisions must incorporate instinct and judgment with experience and knowledge of conventions of usage. The editor must continually analyze, synthesize, and evaluate while staying focused on the big picture. Because each document is unique, the editor must balance the desires of the writer with the needs of the reader, which change with each new document. Among the editor's responsibilities at this level are planning the document with the writer; determining issues of suitability and readability for particular audiences; organizing information logically; reviewing for technical accuracy; shaping the tone, diction, and ease of reading into a pleasing style; and complying with legal and ethical issues.

Because editing for effectiveness incorporates so many variables and degrees of complexity, this chapter addresses only four aspects: theories of learning, application of style, principles of organization, and logic that creates effective content. Other issues, such as visual design, audience analysis, production, scheduling, writer-editor relationship, collaboration among team members, online editing, and ethical and legal issues, are treated in other chapters. Our focus in this chapter, therefore, is to consider principles of organization, style, and logic.

Although copyediting may concentrate on the sentence level, editing for effectiveness concentrates on the context and purpose of the document and whether it achieves the desired goals. For example, suppose you were hired to work cooperatively with an engineer to design an online help menu for a new software program. The engineer clearly has the technical knowledge to design and program the online help menu but lacks your expertise to translate that knowledge into a clearly written, logical progression of steps suitable for the skill level of the intended user. Although some naive individuals may believe this would be a relatively simple undertaking, just ask anyone old enough to remember using the first DOS operating system manuals and how difficult it was to follow the instructions created by engineers. Fortunately for many first-time DOS users, savvy entrepreneurs quickly realized the users' frustrations in attempting to understand and apply overly technical, ambiguous directions. Soon, a whole series of bestselling "DOS for Dummies" manuals appeared everywhere, even in grocery stores. What was the secret of their success? Nothing more than reconstituting the original instructions into an organized, audience-appropriate series of easy-to-follow instructions.

> An editor is someone who separates the wheat from the chaff and then prints the chaff.
> —*Adlai Stevenson*

Perhaps your achievements as a technical editor may not be as dramatic or as profitable as the previous example, but your knowledge of how language works and how to translate language into meaningful prose for specific audiences will always be in demand. However, before examining the specific issues that comprise editing for effectiveness, we provide an overview of theories of how people learn and why you as an editor or editor-writer can profit from a brief introduction into cognitive psychology.

THEORIES OF LEARNING

Anyone who has read a textbook, crammed for an exam, or sought to acquire a complex skill undoubtedly has wondered if there weren't more efficient or effective ways to learn. Recent research in cognitive psychology, which endeavors to understand the nature of the mind through empirical inquiry, strongly suggests there are cognitive strategies that make learning easier. *Cognition* means "coming to know" and focuses on the internal mental processes humans undertake while learning. If, as an editor or writer, you were told that by incorporating certain learning strategies or techniques into technical documents you could greatly enhance the probability of the reader's understanding the material, you would, of course, revise the material. Therefore, before proceeding to the specific issues of editing for effectiveness, we provide a brief overview of the theories of learning and suggest how you can apply those theories to create more effective technical documents that will greatly enhance your readers' ability to comprehend.

The Function of Schemata

Have you ever sat in a classroom or office and wondered why the person at the next desk seemed to easily grasp how to solve a calculus problem or write a well-organized, cogent document, but you struggled just to understand the fundamentals? Or perhaps you have been that rare individual who effortlessly solved complex math or physics problems and ruined the grading curve for the rest of the students. What constitutes the difference between those who easily understand certain concepts or problems and those who struggle to understand? Another way to consider this issue is to ask, "What do experts and novices do differently?" or "Is there anything a person can do to become an expert more quickly?" These are the essential questions raised by cognitive science. Although no magic pill yet exists to aid learning, through empirical research, cognitive science has uncovered theories of how experts process information and the strategies they employ that separate them from novices.

The structure of consciousness, the processes of thinking, and how we come to know are problems cognitive psychologists have worked on throughout the twentieth century. One of the earliest pioneers in this field, the Swiss psychologist Jean Piaget, theorized that mental maturity and the ability to problem-solve evolve from an individual's development of increasingly logical and complex schemata. Other cognitive researchers who followed Piaget, such as the German Gestalt psychologists, the Englishman Sir Frederic Bartlett, the Russian Lev Vygotsky, and the Americans Jerome Brunner and linguist Noam Chomsky, as well as many others, have made considerable advances in schemata theory.

> What I said never changed people. What they understood did.
> —*Author unknown*

This theory asserts that schemata are mental data structures or bundles that the mind stores as knowledge in the form of patterns or *data schemata*. When we recall something, we retrieve some of the data from that experience that have been appropriately stored in our brains as a type of data file. Also, similar to computer programs, we possess *process schemata* that allow us to amplify and organize information for later retrieval. When we want to know *about* something, we use our data schemata, and when we want to know *how*, we use our process schemata.

For example, when we drive to school or work each day, we activate both data and process schemata. The data include items such as our acquired knowledge of keys, doors, the radio, ignition switch, speedometer, and hundreds of other pieces of information we have previously committed to our long-term memory. Our process schemata works in conjunction with this data to: (1) turn the key in the lock, (2) open the door, (3) get behind the steering wheel, (4) move the seat-belt buckle over our body into the lock, (5) insert the key into the ignition, (6) turn the ignition switch forward, and so on. After we have been driving a short period of time, we do all of this without consciously thinking about it; this experience is called *automaticity*. However, recall your first attempts at driving. No doubt you felt nervous and awkward until you had sufficiently practiced the process.

So, in simplified terms, think of a technical document as supplying new information to a reader who must assimilate and store that information internally first as data schemata. However, if the document asks the reader to know anything beyond mere information, such as comprehending logical procedures, organizing concepts, or associating new information with information already acquired, the reader must activate previously learned process schemata to organize and assimilate this new learning. The cognitive strategies we provide in the next few pages represent examples of process schemata that help the reader process and organize information more effectively.

Cognitive research tells us that schemata actively direct and construct our perception of the world around us. At any moment, we are inundated with an incredible amount of information through our senses, all of it competing for attention; however, what limited amount of information our mind finally elects to perceive and what meaning we give to those perceptions are driven by our schemata. In part, this selectivity is a function of what we already know from previous data schemata we have processed and retained. New learning, then, combines or "wraps" already learned and stored schemata with our processing of new information that, if retained, becomes stored as new schemata.

Schemata have two principal functions: they direct our perception, and they make learning possible. Our perception of a new event is activated through our association to a previously learned schemata or sequence of events, which in turn assimilates the new event and constructs a new event context. Once that process occurs, the new event becomes meaningful as new schemata. However, if no previously learned schemata (or *concepts*) are available for developing a context, cognitive psychologists tell us that no learning will take place. For example, if we say "one, zero, one, one, zero, zero, one, one," you would have no way to determine whether this code is meaningful until we provide the previously learned schemata, *computer*. "Ah, now I know!" you might say. "It's binary code: computer language."

However, what happens to individuals who, through no fault of their own, have difficulty separating all the surrounding stimuli and limiting the perceptions they attend to? In essence, they are bombarded with many schemata that simultaneously compete for their attention. Many bright children who develop behavior problems have attention deficit disorder, or ADD. For example, in a room filled with people talking, an individual with ADD might attend to several nearby conversations while simultaneously carrying on a perfectly normal conversation with someone else. In their classrooms, children with ADD attempt to attend to many different stimuli occurring simultaneously and divide their attention between teacher-directed learning and what other children are doing around them. In short, they have too many competing schemata to assimilate and process.

> Education is an admirable thing, but it is well to remember from time to time that nothing that is worth knowing can be taught.
> —*Oscar Wilde*

TIP FOR TECHNICAL EDITORS

When editing any document that requires the audience to have prior knowledge of data or to be able to organize data for later retrieval, ask if sufficient information or directions have been provided to create appropriate schemata. Simply stated, have the writers considered the audience's level of sophistication?

COGNITIVE STRATEGIES AND EDITING

Schemata may be very interesting, but what do they have to do with editing? If we return to our original question about experts versus novices, and if strategies exist to enhance learning, we then can begin to understand in terms of perception and schemata. Although there are additional aspects that affect learning, the most important is the mind's ability to locate and represent previously learned schemata in order to provide a context for learning new material. For example, as you work as a copy editor, your ability to correct grammatical and mechanical errors depends on whether you have previously acquired schemata to perceive corrections that may be necessary. You possess the data schemata that help you identify concepts such as comma, noun, participle, etc., as well as process schemata that help you organize information and tell you to change a comma into a semicolon.

As an editor, or as is often the case, editor-writer, you will frequently be called upon to help plan technical documents. Your task is to present the information as efficiently and effectively as possible while anticipating the reader's needs. Psychologists have identified at least five cognitive strategies that facilitate these writing objectives and are simple to incorporate into technical documents. Using these strategies helps the reader organize information into process schemata, making learning easier by reducing the reader's "cognitive load." For example, if you return to the beginning of this chapter, you will notice that the first two paragraphs provide overviews of the two previous units, whereas the third paragraph explains what to expect in this chapter. This is a cognitive strategy called the *advance organizer.* It reminds you what you have previously learned and prepares you for new learning. In other words, it reinforces previously learned schemata while preparing a context for new information.

The remainder of this section presents cognitive strategies useful for writers and editors. We have selected the cognitive strategies we feel are most appropriate for inclusion in technical documents, but determining which to use for what content and type of reader is an important judgment call you will have to make. Remember, these strategies are intended to activate systematically readers' internal cognitive processes and help them organize the technical information presented in the document.

Cognitive Strategies

- *Frames, Type One*
- *Concept Maps*
- *Advance Organizer*
- *Analogy and Simile*
- *Imagery*

Frames, Type One

Frames, type one is a commonly used cognitive strategy that organizes large amounts of information spatially. It's useful to present both *declarative* knowledge (knowledge that is stated) and *procedural* knowledge (knowledge that initiates a process). More importantly, this strategy provides a succinct overview of the "big picture" by providing a contextual framework for the reader. Usually, it consists of a grid or matrix divided into rows and columns to represent knowledge and show relationships among the main ideas. It is an excellent graphic for displaying ideas, concepts, facts, procedures, and processes. Figure 12.1 is an example of a frame, type one.

TIP FOR TECHNICAL EDITORS

During the planning phase of a document, consider incorporating a frame, type one into the document as a graphic representation, especially if you are part of an editorial team and have the time and resources to design it. Otherwise, graphics of this nature are the responsibility of the technical writer, but many are not aware of how to create such a frame or why it is effective. First decide whether the document contains enough information to warrant a frame, and then determine if that information can be displayed to show relationships such as the diagram in Figure 12.1.

Concept Maps

Similar to framing, *concept maps* (also known as *semantic maps* or *graphic organizers*) are an excellent means to graphically display information. However, unlike frames, graphic concept maps create a visual representation of the relationships between concepts. Technical documents frequently use three types of concept maps: spider maps, hierarchy maps, and chain maps (Figure 12.2). *Spider maps* simply show relationships among concepts and are particularly good

Era	Geologic time period	Geologic events	Characteristics	Formations
Archeozoic	Unknown	Earliest forms of rocks metamorphosed by intense heat and pressure	Long periods of erosion; no life present	Crest of Rocky Mountains; Scandinavian Peninsula
Proterozoic	?–520 million years ago	Shallow seas form	Life begins in the seas; trilobites, mollusks	Glacier National Park; Baltic Shield; Ural Mountains
Paleozoic	520 million–210 million years ago	Accumulation of sediment/faulting begins	Elevation of land; extensive erosion; divided into seven periods	Appalachian Mountains; continents occupied by shallow seas; Great Lakes
Mesozoic	210 million–60 million years ago	Sandstones, limestones; Europe covered by epeiric seas	Land plants and invertebrate animals; later, dinosaurs; divided into three periods	Grand Canyon; Gondwanaland; Australian Shield
Cenozoic	60 million years ago–present	Glacial ice covers northern continents; seas begin to recede	Europe, Asia, North Africa covered by seas; mammals evolved; divided into two periods	Himalayas; Alps; Cascade Mountains of Oregon

FIGURE 12.1
Example of Frame, Type One: "Geologic Evolution."

for illustrating declarative knowledge. *Chain maps* indicate stages or series of chronological events and are frequently used to show procedures. *Hierarchical maps* generally denote relationships between levels of importance. Undoubtedly you have seen forms of these maps before, probably in the form of a flowchart. If possible, you should consider integrating them with technical documents during the planning stage, and don't be afraid to create hybrid versions.

Envisioning the "big picture" is often difficult for those who are not familiar with a particular concept that can be understood only in context of an organizational framework. Ask while editing if creating a chart in the form of a concept map would make the document more readable. Would a concept map also aid the intended audience?

EXERCISE 12.1

You are editing an executive summary for Hydroburn, Inc., a relatively new company that designs fuel cells for autos and now plans to go public with stock options. The writer has done a good job composing the summary and now plans to develop a PowerPoint presentation from it. However, written as text, you feel that the company's organizational structure seems too confusing, that the benefits of fuel cells would create a greater impact as a graphic illustration, and the process of how a fuel cell converts hydrogen into water and oxygen and finally electricity is too technical for the intended audience. Decide what type of concept map is appropriate for each problem, and then create a concept map that corresponds. Don't worry that you lack expertise in this area; just be creative with your mapmaking.

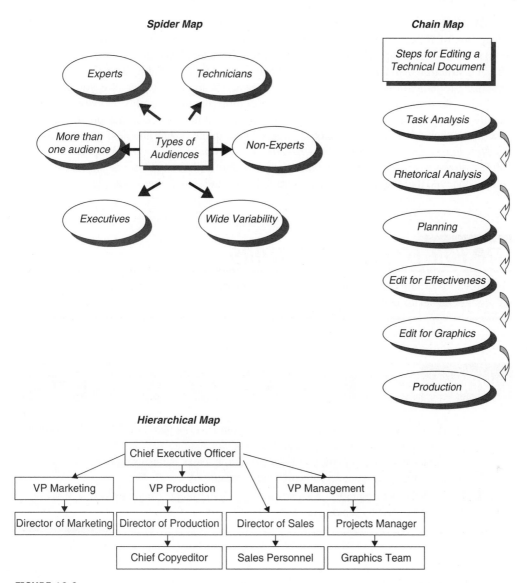

FIGURE 12.2
Types of Concept Maps.

Advance Organizer

The advance organizer is a cognitive strategy that functions as a "bridge" between what readers already know and knowledge they will transform into new schemata. That is, it prepares readers for the new information and learning they are about to read. This strategy is particularly useful for technical writers and editors when they want to express similarities between previously learned knowledge and new knowledge that is particularly complicated or abstract. Think of this strategy as an introduction, much like the opening three paragraphs of this chapter. Notice how our introduction to this chapter reminds you of the concepts we discuss in Units Two and Three to prepare you for the new concepts in this chapter. The advance organizer encourages the reader to make connections and transfer knowledge; it may take the form of several paragraphs, a few sentences, or, if well written, a single sentence.

Although the advance organizer usually works best as one or two complete paragraphs, constraints of time, length, or appropriateness may prevent you from fully developing it. In fact, most short technical documents do not require such elaboration. For example, if you edit an instruction manual where you are only concerned with providing directions for putting something together, a detailed advance organizer is not appropriate because you are not interested in the reader learning, just successfully completing the assembly. However, consider the effectiveness of using a sentence or two to reinforce what the reader has previously completed, and then what the reader should expect to complete during the next phase of assembly: *Now that you have completed the wheel assembly by bolting sprockets A and B onto wheel hub H, you are ready to begin assembling the gears.* Notice how just one transitional sentence unites concepts, making the document more readable and comprehensible.

TIP FOR TECHNICAL EDITORS

Use advance organizers as transitional devices to connect ideas and concepts or to make a process easier to comprehend.

EXERCISE 12.2

1. Locate a piece of your own writing or find your favorite textbook. Determine where you think it would be appropriate to add an advance organizer, and then write at least one advance organizer in the form of a paragraph. Next, using a single sentence as an advance organizer, connect three different paragraphs.
2. Find at least two examples of an advance organizer in a textbook or technical document. (Hint: start by looking at chapter introductions.)

Analogy and Simile

Similar to the advance organizer, analogies, similes, and imagery are cognitive strategies that connect previous knowledge with new knowledge. However, unlike advance organizers, these three strategies make cognitive connections through connotation rather than denotation. That is, analogies and similes connect knowledge through figurative language and what words *suggest,* not what words literally mean. Analogies express relationships in the form of an *as if* or *is to,* whereas similes create relationships using *like, as,* or *is similar to.*

For example, *The gyroscope was spinning so fast that it appeared as if it were motionless* and *A schematic diagram is to an engineer as a recipe is to a chef* are examples of analogies. To explain Einstein's theory of relativity, scientists employ a simile that says space and time are like a fabric that wrinkles when it approaches a large mass with gravity. (Don't worry; we don't fully understand Einstein's theory either, but using the fabric simile does help.)

Providing an occasional analogy can often make technical documents more understandable, especially for those who are less familiar with the specific information. As a technical editor, you may want to suggest to writers that they include an occasional analogy to make documents more readable.

Imagery

You may recall from high school or college reading short stories and novels where the author created memorable visual images of characters and scenes. No doubt you can still bring some of these images to mind even though you probably have long forgotten the plot and the principal characters. For example, if you have read Poe's "The Pit and the Pendulum," it's impossible to forget the image of a huge, razor-sharp pendulum swishing back and forth across the neck of the supine prisoner. Or perhaps you can conjure the image of a young, freckled boy sailing a raft down a muddy Mississippi River with a runaway slave in Twain's *The Adventures of Huckleberry Finn.*

Writers know that well-crafted images add interest and affect a reader's long-term memory. Cognitive psychologists have determined that using imagery enhances learning declarative or procedural knowledge and works best when the image is represented as a concrete idea or concept.

TIP FOR TECHNICAL EDITORS

Use analogies, similes, and imagery judiciously. Generally, technical documents call for specific, denotative language, and so none of these strategies may be appropriate. However, documents containing abstract concepts and intended for an audience of non-experts may benefit from incorporating an occasional analogy, simile, or image to clarify the writer's meaning. For example, suppose you were collaborating with a writer to develop a brochure explaining gastric banding surgery. You might write, "The stomach functions similar to a balloon that expands when full and contracts when empty. Gastric banding surgery is like tying a rubber band around the stomach to restrict the amount of food that can be absorbed."

EXERCISE 12.3

Technical editors work on documents from many different fields, including engineering, education, computer science, business, government, and medicine. Pretend you have been asked to collaborate on constructing a website for any three of these fields. Your client wants the website to communicate technical information to a global audience of non-experts. Develop a one-paragraph proposal for each field (e.g., explaining a surgical procedure, an engineering design for a new product, how a computer organizes information on its hard drive, etc.) and include an example of an analogy, a simile, and an image in the proposal you would use on the website.

SUMMARY

Theory underlies our assumptions about *why* something works the way it does. How often have you been asked to learn something without knowing why what you were learning was important? Did it seem irrelevant at the time? This chapter provides theories of learning that technical editors can employ to make documents more effective and relevant to the reader. Knowing how individuals learn best is certainly worthwhile information for editors and writers alike.

13

Style: Making Language Effective

At the conclusion of this chapter, you will be able to

- avoid the Official Style;
- relate why writing style is important to effective editing;
- assess sentences for clarity and variety;
- distinguish logical relationships within sentences;
- revise sentences in the passive voice;
- analyze sentences for unnecessary nominalizations, excessive verbiage, and improper diction.

At the beginning of Chapter 1, we suggest that editing for effectiveness is as much an art as a skill. We maintain that editing at this level requires not only astute discernment of what makes writing effective but also a high degree of sensitivity and intuition. Sure, anyone with a little understanding of the English language can create complete sentences and combine them into paragraphs devoid of grammatical or mechanical errors. However, what separates merely technically correct writing from writing that seemingly flows effortlessly and engages the reader? What makes writing appropriate for an intended audience? How does writing maintain the reader's attention while moving from one idea to the next?

Editing grant proposals, online help, websites, memos, brochures, instruction manuals, program analyses, and any number of other technical writing genres is not like editing a fast-moving detective novel that holds readers spellbound until the last page where they find out "whodunit." Instead, the technical editor concentrates on helping the writer develop a document that explains, instructs, or solves a problem in a clear, coherent style. How successful the writing style is depends on editorial choices: in word selection, sentence length, sentence type, tone, and voice.

> Editing should be, especially in the case of old writers, a counseling rather than a collaborating task. The tendency of the writer-editor to collaborate is natural, but he should say to himself, "How can I help this writer to say it better in his own style?" and avoid "How can I show him how I would write it, if it were my piece?"
> —*James Thurber*

Have you ever attended a party, reception, or meeting where you committed a fashion faux pas, where you were underdressed for the occasion and everyone seemed to be staring at you? How were you to know that 6:00 p.m. on the wedding invitation meant suiting up in a coat and tie or an evening dress instead of blue jeans? (Fortunately for you, the fashion police weren't working the wedding that night.) Who makes up these rules, anyway?

Language is like attire; you select it according to the occasion. Certainly you would not wear a tuxedo to go swimming, nor would you use slang expressions when interviewing for a new job. Custom dictates what is appropriate, and in the case of writing, choosing the proper style of writing for the occasion greatly influences how the reader will respond to the document. As an editor, you will have to make critical judgments regarding the style of the document: should it be formal, casual, warm, objective, informal, or sincere? In short, you will determine the appropriate "attire" for the document.

Figures 13.1 and 13.2 are two examples of letters sent to a potential employer. The first is deliberately exaggerated for illustrative purposes. First, notice how the style of writing reflects the author's attitude toward the subject matter. This attitude is known as *authorial tone.* Blayne's letter is full of careless surface errors, including misspelled and misused words. Furthermore, Blayne's tone is too casual and personal, which would offend many employers. Simply stated, Blayne has poorly chosen what to say and how to say it.

Meisha, on the other hand, writes in a more formal, businesslike style. Her tone is warm, friendly, but professional. Her writing reflects her confidence in herself and her appreciation for the opportunity to interview. Unlike Blayne, she understands that it would be inappropriate at this stage of their relationship to address Mr. Pisula by his first name. Meisha is much more likely to get the copyediting job, while Blayne will continue to draw unemployment (unless Mr. Pisula is Blayne's uncle).

Good writing begins by asking, *Who are our readers?* and *What do our readers need to know?* about the particular subject. After answering these essential questions, other considerations emerge, such as, *How do we organize what the readers need to know?* and *How much detail is necessary?* to make the ideas clear. Once these considerations and issues of grammar and mechanics have been met, one additional question remains: *How can the writing become clearer and more direct?*

Unfortunately, attention to clarity and issues of style often are forgotten, overlooked, or just pushed aside in the rush to create a final document. Generally, all writing, especially writing of a technical nature, needs to be unambiguous and easy to comprehend. However, you surely have read many documents that seemed overwritten, vague, and even impenetrable. Consider the property tax code from the state of Texas (Figure 13.3). This document sounds like language on steroids.

No doubt you have run across similar prose in rental or loan contracts, tax manuals, government documents, and even academic journals and textbooks. However, inflating language

---

Dear Richard,

Hey, its great to finally here from you. I can hardly wait for my interview next week. I really, really want the copy editing job and I know I'll be good at it. As you will discover when you meat me, I'm very good with grammer and know the ends and outs of working with people. Also, I'm a conscious employee. Unfortunately I can't supply the letter of recommendation from my former boss that you requested. He got mad because I was late for work a few times. I wouldn't have got fired if my boss would have cut me a little slack about being on time every day. I told him I was having issues with my girlfriend, but he didn't seem to care. Anyway, all the other guys at the pizza store like me and say I was a good worker, (except for Meisha who nobody can get along with). So, I'll see you next Thursday, perhaps we can do lunch—your treat, of course (just kidding) —Blayne

---

FIGURE 13.1
Sample of Poor Authorial Tone.

May 14th, 2005

Dear Mr. Pisula,

I appreciate the opportunity to meet with you next Thursday at 11:30 a.m. to discuss the position of technical copy editor. As I noted in my résumé, I have had two years' experience as copy editor for a small newspaper and have enjoyed the challenges this job presented. However, I am now interested in working for a larger firm such as yours and entering the technical editing field. I am familiar with your company's operations and believe I can make a significant contribution as a technical copy editor. I look forward to meeting with you next week.

Sincerely,

Meisha Roberts

FIGURE 13.2
Sample of Professional Authorial Tone.

APPLICABILITY OF TITLE.

This title applies to a taxing unit that is created by or pursuant to any general, special, or local law enacted before or after the enactment of this title unless a law enacted after enactment of this title by or pursuant to which the taxing unit is created expressly provides that this title does not apply. This title supersedes any provision of a municipal charter or ordinance relating to property taxation. Nothing in this title invalidates or restricts the right of voters to utilize municipal-level initiative and referendum to set a tax rate, level of spending, or limitation on tax increase for that municipality.

FIGURE 13.3
Applicability of Title Document.
Source: State of Texas.

with pompous, gassy prose impresses no one, especially the reader who must try to make sense of it. Embellished, overblown style is known by several different names: *legalese* for contracts and judicial documents, *bureaucratese* for government documents, and *academese* for academic articles and textbooks. Richard Lanham collectively calls all three the Official Style, a style that has been adopted by the U.S. government, the workplace, and certain disciplines. It is a style that is pretentious and frequently tries to be intimidating but unfortunately only leads to frustrating the reader. We agree with Michael Crichton, who points out that convoluted language is often used to "gussy up" simple ideas or disguise the fact that no ideas exist at all.

But before we go pointing fingers, ask yourself if you have tried to imitate what you thought was the Official Style of academe by looking up big words in the dictionary and inserting them into a paper. You probably didn't know what all the words meant, but you thought they surely would impress the teacher. Well, they didn't. Readers don't want to sit with a dictionary on their lap to look up every other word in a document—in fact, they won't.

> The Lord's Prayer is 66 words, the Gettysburg Address is 286 words, there are 1,322 words in the Declaration of Independence, but government regulations on the sale of cabbage total 26,911 words.
> —*David McIntosh*

AVOIDING THE OFFICIAL STYLE

The Official Style nominalizes, uses passive voice, and inflates language:

Clear: The technician corrected the computer error.
Official Style: A correction was made by the technician.

(The verb *corrected* was transformed into a noun. Also, the subject of the sentence, *technician,* is the object of a prepositional phrase.)

Clear: The management team designed the illustrations for the magazine.
Official Style: The illustrations for the magazine were designed by the management team. (passive voice)

Clear: Because we received the graphics late from the editor, the magazine was not ready for production.
Official Style: In so far as the illustrations of a graphic nature were detained as a result of a managerial impediment, the release, termination, and conclusion of said publication by the aforementioned individuals was slow to develop.

The tendency by some writers to inflate their style can be traced back more than four hundred years to the Renaissance in England. Eager to distinguish English as a language separate from Latin and French, many writers developed a flowery prose and engaged in writing contests to see who could be the wittiest. However, with the advent of the Royal Society in 1660 and the Scientific Revolution, precision and clarity replaced embellished, flowery language. The Industrial Revolution of the eighteenth and nineteenth centuries galvanized the growing need to develop writing that was crisp, direct, and technical in nature. Figure 13.4 provides an additional example of contemporary prose style that you don't want to imitate; it is followed by the same text edited for clarity.

From: Johnson Space Center Handbook (before editing)

207.1 Introduction

For JSC emergencies, call x33333. For Ellington Field emergencies, call x47231. For White Sands Test Facility emergencies, call x5911. Off site, call 911. The first priority of a person on site who observes an outbreak of fire or other situation that constitutes an emergency is to immediately contact the emergency phone numbers for assistance. In many cases, immediate evacuation of a facility or area may be necessary. Notification and evacuation are to have priority over attempts by untrained personnel to combat the emergency. It is JSC policy that employees not "fight" fires that cannot be safely extinguished with a hand held fire extinguisher unless such employees are certified members of a fire brigade or any formally organized fire department. Employees should not try to fight a fire unless they first notify the fire department, are trained in fire extinguisher use, and can safely extinguish the fire.

207.2 Purpose

The purpose of this chapter is to provide guidelines for reporting an emergency and details what shall be contained in emergency action plans that give information and guidance to building personnel after an incident has been discovered.

207.3 Scope

This chapter applies to the reporting of medical and fire emergencies and emergencies involving a hazardous substance anywhere on JSC property, including Ellington Field and the Sonny Carter

---

FIGURE 13.4
Example of the Official Style and Editing It for Clarity.

Training Facility. Also included are guidelines for general emergency action plans in each of the JSC and Ellington Field buildings.

207.4 Responsibility

207.4.1 Directors

Directors of organizations are responsible for the following actions:

a. Ensuring that emergency action plans are developed for their assigned facilities.

b. Appointing and providing for the training of facility managers for the building and/or facility.

c. Ensuring that hazards in or near the workplace are identified and properly mitigated.

d. Ensuring that the employees in the facility are aware of hazards of the workplace and are properly trained in the appropriate actions to take in emergency situations.

From: Johnson Space Center Handbook (after editing)

This could be you . . .

Several people could have been exposed to a toxic gas because they left their buildings after the gas release. They didn't know what to do.

A fire in an area with hazardous materials burned longer than it should have while firefighters tried to find out if they could use water on the fire. Little planning had been done for this emergency.

A computer area was flooded when a water line broke. The emergency plan didn't cover water leaks.

1. Who must follow this chapter?

You must follow this chapter if you work at JSC or a JSC field site. If you are a supervisor, facility manager, or director, Paragraph 16 of this chapter lists your responsibilities.

2. What does this chapter cover?

This chapter tells you what to do in an emergency of any kind and what emergency planning JSC must do.

Emergency action

3. What emergencies must I report?

You must report any emergency that you see. This includes any fire, no matter how small. Report fires that have been extinguished. They may still be smoldering and could reignite.

Remember, your emergency numbers are x33333 at JSC, x44444 at Ellington Field, 911 at any off-site location, and x5911 at White Sands Test Facility.

You must call your emergency number if you see an emergency.

FIGURE 13.4 (*continued*) Example of the Official Style and Editing It for Clarity.

> You must keep the emergency scene as undisturbed as possible. If you don't, valuable evidence for the investigators could be destroyed.
>
> 4. How will I be told that there is an emergency?
>
> You could be told of an emergency in two ways:
>
> > a. A building fire alarm.
>
> > b. The JSC employee alarm system. See the diagram at the end of Paragraph 5 below.
>
> 5. What must I do in an emergency?
>
> If you are involved in an emergency, you must take the actions in this table for the kind of emergency it is.

FIGURE 13.4 (*continued*)   Example of the Official Style and Editing It for Clarity.

EDITING FOR STYLE

Depending on your clientele, you as a technical editor may encounter writing that is overwritten, awkward, and in need of extensive surgery. (Sometimes the author's style is so dense and turgid, as in the previous two examples, that the documents must be returned to the author for a translated, simplified version.) Editing for style requires the editor to make many judgment calls at the sentence and word level. Unlike grammar and mechanics, there are no prescribed rules, just guidelines. However, keeping these few stylistic guidelines in mind as you edit can significantly improve the readability of the document. In fact, some recent empirical research indicates that the improvement may be as much as 30%.

"Have something to say and say it clearly," states Matthew Arnold. This is good advice. But as simple as this counsel seems, you know that saying something clearly involves many factors, the most important of which is good, qualified judgment. Before surgeons operate on their patients and recommend a specific procedure, they qualify their recommendations by running tests. Yet despite all the tests they run, their final diagnosis is an educated opinion as to the best course of action.

Although technical editing may not be brain surgery, effective editing does require running tests, making a diagnosis, and determining how to make a document say something clearly. Therefore, we can say that editors of style test sentences and words for effectiveness, and editorial choices evolve from those results. Generally, stylistic analysis consists of asking the following ten questions:

1. Does the style serve the writer's purpose?
2. What assumptions can be made regarding the reader's initial level of understanding?
3. Will the style be suitable for an international audience?
4. Are the sentences varied in length and type?
5. Do the sentences convey a sense of logic?
6. How well do the ideas flow from sentence to sentence?
7. Does the structure of each sentence convey meaning clearly and accurately?
8. Is the choice of words appropriate for this particular audience?
9. Are the verbs active?
10. Does the overall style create a clear sense of readability?

Although this list may appear a bit daunting at first, while working through this section you will discover you already know most of the fundamentals. Now, however, we want to take you beyond mere sentence exercises and encourage you to apply your knowledge to working models. We devote the remainder of this section to identifying the problems associated with poor writing style and to providing specific strategies you can employ to resolve them. Of course, nothing can substitute for extensive on-the-job experience, but by following these strategies, your editing skills should dramatically improve. Issues we address in this section include sentence structure, length, and type; diction; active and passive voice; and coherence.

> Have something to say, and say it as clearly as you can. That is the only secret of style.
> —*Matthew Arnold*

SENTENCE STRUCTURE: WHERE STYLE BEGINS

Style, as it pertains to sentences, goes further than merely following a formula for sentence construction, such as subject + verb + complement, or subject + verb + object; it concerns the *syntax*, or the arrangement and relationship of the individual words, phrases, and clauses to communicate meaning effectively.

However, before discussing syntax and the more sophisticated aspects of sentences, we need to briefly review basic sentence structure. First, remember from Unit Two that a sentence is an independent clause, which means it contains a subject and a verb (S + V) and completes a thought. Direct objects with accompanying modifiers may be added to the basic sentence core to form a second sentence pattern (S + V + O). A third sentence pattern (S + LV + C) contains a subject, linking verb, and subject complement such as a noun, noun substitute, or an adjective.

Sentence Review from Unit Two

Basic sentence pattern   (Subject + Verb):

Melissa works. (S + V)

Melissa works quickly. (S + V with the addition of the adverb quickly)

Melissa and the new copy editor work quickly. (compound S + V + adverb quickly)

Basic sentence pattern plus object   (Subject + Verb + Object):

She copyedits technical manuals. (S + V + O)

She and two other editors copyedit technical manuals for the Lexington Corporation. (compound S + V + O with a prepositional phrase added)

She copyedits and illustrates technical manuals. (S + compound V + O)

Basic sentence pattern plus complement   (Subject + Linking Verb + Complement):

Ms. Rice is the new office manager. (S + LV + C)

Ms. Rice and Mr. James are our technical consultants. (compound S + LV + C)

The technical writers are busy with the management team now. (S + LV + C + prep. phrase + adverb)

Much of the writing you will encounter as an editor may not be as crisp and straightforward as the review examples. Generally, clear writing begins with a concrete subject and an active verb. However, sometimes the subject and the verb get lost among strings of prepositional and participial phrases, dependent clauses, vague pronoun references, and connotative words. For example: *Due to their inability to deliver a determination for the lack of fiduciary shortfall, our engineers are constrained to apply their understanding as how to proceed with the project.* Huh? Revised for readability, the meaning behind the sentence becomes clear: *Until we determine how we ran out of money, the engineers cannot continue with the project.* As demonstrated by this example, the subject and verb may be difficult to locate because the first version of the sentence is overwritten; the author is trying to appear erudite.

Many other problems exist with this sentence and it requires too much interpretation to be clear. For example, did you notice the two prepositional phrases *for the lack* and *of fiduciary shortfall*? An astute reader might believe that the engineers had plenty of money because they *lacked* a fiduciary shortfall. The point is this: whenever the reader must struggle to interpret the meaning behind a document, the responsibility falls back upon the writer. However, editors should make suggestions for revision. Also, if the writer's intended meaning is unclear, and misinterpretation of the document could result in harm, then the editor has an ethical obligation to contact the writer for clarification.

TIP FOR TECHNICAL EDITORS

When you encounter dense, ambiguous sentences that need radical surgery, always begin by finding the verb first and then the subject. Be careful, though, not to confuse a verb with a participle or a gerund (see Unit Two). After you have located the verb, say "who" or "what" and then state the verb. Return to the previous example. Did you identify the verb *are constrained*? (Also, did you notice that the verb is passive?) Next, ask "who" or "what" *are constrained,* and the answer is *engineers*. You now have the sentence core; the remaining task is to filter out the jargon and the unnecessary phrases. Finally, ask "why" are the engineers constrained—because of *their inability to deliver a determination for the lack of fiduciary shortfall*; that is, because they cannot understand why they are out of money.

EXERCISE 13.1

Determine the author's meaning in the following sentences by first locating the verb(s) and then the subject(s) that are the source of that action. After locating the subject and verb, revise the sentences for clarity. Compare your sentences with other members of your class.

1. Due to the fact that the interpretation of the electrical system inherently relies upon knowledge of a schematic diagram, the electricians must perform a demonstration of their electrical acumen if they are to be hired by our company.
2. Attesting to the small but contented number of males unencumbered by bans of prodigious self-actualization within an institute of employment is a similar number of females who would like to reverse the males' direction.

3. It is without doubt that a certifiable condition pertaining to the circumlocution of blood with the circulatory system commonly known as "tired blood" is attributable to the failure of the individual who possess that circulatory system to voluntarily ingest plant material prosperous with iron nutrients.

SENTENCE TYPE: VARIETY IS KEY

In the previous section, we reviewed what constitutes basic sentence structure. We suggested that effective editing requires revising overwritten, convoluted sentences. By first determining the core sentence's subject and verb, you can in most cases successfully translate poorly written sentences into meaningful prose. This section expands that proposition by considering the stylistic advantages of using a variety of sentence types.

The English language consists of four sentence types: simple, compound, complex, and compound-complex. The amount of information a sentence conveys depends on the writer's or editor's selection of sentence type. From Unit Two, you recall:

- *Simple sentences* convey one unit of meaning. They contain a subject and predicate, which may have modifiers to enhance or provide additional detail about the subject. Simple sentences are frequently used to focus the reader's attention on a particular idea before proceeding. (Note: A simple sentence can contain a compound subject, a compound predicate, or both.)
- *Compound sentences* commonly convey two elements of information that parallel each other in relationship. They consist of two independent clauses, and they are joined by a coordinate conjunction or a conjunctive adverb. Compound sentences provide a break from overusing simple and complex sentences.
- *Complex sentences* demonstrate logical relationships between ideas. These sentences contain an independent clause and a subordinate clause that create logical relationships of time, place, choice, cause, contrast, condition, or purpose. In addition, when used properly, complex sentences create coherence within a paragraph.
- *Compound-complex sentences* generally are formed by two independent clauses; one of these independent clauses is a complex sentence. They provide more information than the other three types of sentences. Compound-complex sentences should be used sparingly, usually to move the reader rapidly through uncomplicated but necessary information.

TIP FOR TECHNICAL EDITORS

Sometimes, documents do not require much sentence variety. Simple sentences are often preferred when:

- A step-by-step procedure needs to be followed, such as a technical manual that describes how to put a product together (e.g., an infant's portable crib).
- The document is going to be translated into several different languages.

SENTENCE LOGIC: KEEPING INFORMATION COHERENT

It should now be clear that each sentence type has a function. The simple sentence creates meaning directly through a subject and verb with possible modifiers. The compound sentence combines two simple sentences that are related in meaning to either coordinate or contrast ideas. Another type, the compound-complex sentence, is nothing more than a complex sentence added to a simple sentence with a conjunction, making it compound as well.

However, the complex sentence is distinctive for another reason: it expressly demonstrates the logical relationship between ideas. The inherent structure of a complex sentence subordinates one idea to another. That is, complex sentences consist of a subordinate clause (a dependent clause introduced by a subordinate conjunction) logically conjoined to an independent clause (usually a simple sentence). Following are some examples of complex sentences that demonstrate logical relationships of time, place, contrast, reason, choice, condition, purpose, or result. Study the examples and review the subordinate conjunctions.

Examples of complex sentences

(subordinate clause)   *(independent clause)*

Before we take a break, let's finish the cover design for the brochure.

or

(independent clause)   *(subordinate clause)*

Let's finish the cover design for the brochure before we take a break.

First, notice that these two complex sentences convey the same meaning; however, the subordinate clause in the first sentence has been moved behind the independent clause in the second sentence. Also, notice that an introductory subordinate clause is separated from the independent clause with a comma. Finally, notice the logical sequence of events—X will happen before Y—which creates a chronological sequence. Examine the following examples for additional logical relationships. Words in boldface are the subordinate conjunctions that introduce the full subordinate clause.

Logical Relationships Created by Subordinate Conjunctions

Time:

Since the new editor joined our team, she has completely revised our style guide.
When you finish with the Edison account, fax them a copy of the plans.

Place:

Wherever you list your technical editor qualifications on your résumé, be certain that you list the dates employed as well.
She left the layouts **where** Mr. Evans told her.

Contrast:

Although technical editors do not write the documents, they can help the author by giving advice on style and organization.

Even though technical editors work primarily with written documents, they may also assist in preparing speeches.

Reason:

Because lawyers, government employees, and engineers frequently have trouble writing clear, coherent reports, technical editors are often hired to assist them.

Choice:

We must turn in the report tomorrow *whether* you have edited it or not.

Condition:

If you do not finish copyediting the engineers' report, you will have to stay tonight and complete it.

Technical editors know that they will need to justify their corrections *even if* they must take more time.

Result:

We need to determine a method of fixing the software *so* we can complete the graphics on time.

In order that we complete the Radisson account on time, you will need to send the feasibility report to copyediting today.

Subordinate Conjunctions: Indications of Logical Relationships

Time	Place	Contrast	Reason	Choice	Condition	Result
after	where	although	because	whether	as if	in order that
as	wherever	even though	as		as though	so
before		though			if	so that
since		whereas			even if	
until					unless	
when						
whenever						

TIP FOR TECHNICAL EDITORS

Although complex sentences may initially appear confusing, you already use them fluently in your speech and probably in your writing. When editing for effectiveness, consider using a complex sentence to subordinate one simple sentence to another to create a logical union of ideas. Your readers expect documents to make logical connections. Furthermore, because of their unique functions, complex sentences add coherence and readability.

Editors need to recognize all four types of sentences and know how each type functions to create meaning. In addition, they should be able to take existing sentences within a document and refashion them to create a more fluent style. See how much you have learned so far by completing the following exercises.

EXERCISE 13.2

The research department at your organization has supplied you with a list of facts on global warming. As an editor-writer, you have been asked to compose a short news item for the science/technical section of your daily paper. Write a short article on global warming using all four types of sentences.

- Scientists predict 2.7 to 11 degree Fahrenheit rise in temperature during next fifty years.
- Could cause increased flooding, storms, agricultural losses.
- Carbon dioxide levels have increased 30% during previous 100 years.
- Some regions have already warmed by 5 degrees Fahrenheit.
- World is experiencing rapid buildup of greenhouse gases.
- Cars, sport utility vehicles, and light trucks create 20% of carbon dioxide pollution.
- Power plants release 36% of all CO_2 emissions.
- Global warming can cause:

  Major shifts in precipitation;
  Shifting ranges in infectious disease and increasing infection around the world;
  Sea level rise four to ten inches; destruction of beaches;
  Drastic habitat shifts for plants and animals;
  Altered migratory animal behavior;
  More common and severe winter floods and summer droughts.

EXERCISE 13.3

You've just been handed the following article to edit (Figure 13.5). It's going to appear tomorrow in the food section of a major regional newspaper. After skimming it, you determine it does not need major grammatical revisions, but it does need more sentence variety. Unfortunately, you have been given only twenty minutes to edit this article before it goes to the production office. Using copyediting marks, improve the style as much as possible in the time allowed. Concentrate primarily on the four sentence types but also try to revise any sentences with passive verbs.

The worst breakfast you can eat are two egg McMuffins and two hash browns. This is according to a study from the State University of New York at Buffalo. The hash browns and bread from the McMuffins inflame the arteries until lunchtime. Total calorie intake is 930. The fats from these foods release free radical molecules in the blood cells. The free radical molecules trigger the inflammation.

One of the best breakfasts a person can eat is cold cereal. This is according to a research study from the University of Toronto. Children who consumed more than eight cold servings of cereal over a two-week period had less body fat. Children who eat cold cereal weigh on-average 10 pounds less than other children their age. Cereal provides nutrients as well as increases metabolism. Increasing metabolism helps burn fat consumed later in the day.

Eating almonds is a good way to lose weight. They also help reduce cholesterol. This is the result of a study published in *The International Journal of Obesity*. Almond eaters in the study reduced their systolic blood pressure by 11 percent. Their Body Mass Index readings dropped 18 percent. Eating almonds produces a 35 percent decrease in LDL. LDL is the bad cholesterol. Almonds are rich in vitamin E.

Broccoli is an excellent health food. A research study conducted by Northeastern Ohio University's College of Medicine shows that broccoli, cabbage, and brussel sprouts inhibit the herpes simplex virus. These plants contain indole-3-carbinol. This is a key protein. Studies conducted on monkeys showed that indole-3-carbinol blocked the virus 99.9 percent of the time.

FIGURE 13.5
What's for Breakfast?

NOMINALIZATIONS

Earlier, we explained that the Official Style of bureaucrats, academicians, and many corporations and businesses is frequently overblown, pompous, and attempts to add credibility to what is being said. An analysis of these types of documents indicates a common denominator: all of them use *nominalizations*. The term *nominalization* means that the verb and possibly adjectives within a sentence have been made into nouns. Three things happen when this occurs. First, the action conveyed by the verb gets lost. Second, the reader becomes confused as to the writer's intended subject. And third, nominalizations create abstract nouns, which in themselves can create confusion. For example, examine the following sentences for nominalizations.

Clear: The documents will be forwarded to the copy editor after the graphics team approves them.

Nominalization: The **forwarding** of the documents to the copy editor will occur after **approval** by the graphics team has been obtained.

Clear: The manager suspended production of the copy until she had time to reassess it.

Nominalization: **Suspension** of production of the copy will be maintained until **reassessment** by the manager.

Clear: The engineers have resourcefully reduced the project's costs.

Nominalization: The resourcefulness of the engineers has resulted in a **reduction** of the project's costs.

Notice in each example how the initial verb has been nominalized and abstracted. Also, the length of each nominalized sentence has been increased by including unnecessary words. Do you agree that readers prefer straightforward, clearly written sentences over sentences containing nominalizations? Study the examples below to gain a further understanding of how simple verbs and adjectives become nominalizations. Notice that many of the verbs and adjectives end with similar suffixes that make them abstract nouns (*-tion, -ment, -tance, -dance, -ness, -ence, -ability, -sity*).

Verbs	Nominalized form	Adjectives	Nominalized form
oppose	opposition	casual	casualness
discern	discernment	reckless	recklessness
obstruct	obstruction	capable	capability
investigate	investigation	skilled	skillfulness
confront	confrontation	responsible	responsibility
confine	confinement	careful	carefulness
develop	development	creative	creativeness
instigate	instigation	special	specialization
evaluate	evaluation	extraordinary	extraordinariness
document	documentation	confidential	confidentiality

Several additional points need to be made about nominalizations. First, not all nominalizations are bad. For example, a nominalization might be used to replace what would otherwise be an awkward sentence.

*Due to the fact that he **recommended** the **removal** of all electric typewriters from the office, this **resulted in** a loss of productivity* becomes more straightforward when written as: *His **recommendation** to **remove** all electric typewriters resulted in a loss of productivity.* Notice that the verb *recommended* has been nominalized as *recommendation* to make the sentence less wordy and clearer. Also, notice that the nominalized verb *removal* has been made active by using *remove*.

Generally, though, restating active verbs and adjectives as nominalizations that create abstract concepts confuses readers, especially if they are not familiar with the concepts. Common nominalizations such as *liberation* (instead of *liberate*), *taxation* (*tax*), *assessment* (*assess*), *documentation* (*document*), and so on are acceptable, but editors must continually make judgment calls about whether a nominalized form is appropriate for the intended audience. For an in-depth treatment of this topic, we recommend Joseph Williams's *Style: Ten Lessons in Clarity and Grace.*

EXERCISE 13.4

Locate the nominalizations in the following passages. Compare your revision with a classmate's; your version may differ. Be able to defend your revision. We begin with an example:

The evolution of a method for the embedment of module function objects into code designed to run on handheld computers has the effect that we will see many more "smart" distance area cell phones, remote controls, and personal assistants.

Revision: As we work out ways to embed functions as objects in code that runs on handheld computers, we will see many more remote controls, personal assistants, and "smart" cell phones aware of their location.

1) It is our request that upon your return a review be conducted on all vendor submissions and a report prepared with recommendations for a two-vendor runoff competition.

2) Sustainability of these fishing areas is ensured through regional fishery management planning consultation, with the result that a Fisheries Management Plan is produced. The Coast Guard is responsible for FMP enforcement at sea as well as enforcement of laws for the protection of marine mammals and endangered species.

3) "The high-yield portfolio write-off and mark-down losses announcement today reflects the continued deterioration of the high-yield portfolio and losses associated with selling certain bonds," the CFO said.

4) The innovation history of our world-class fashion emporium includes the creation of the preppy look, all-wool worsted double-stretch fabric, and button-down collar shirt apparel.

(Courtesy of Lisa and Jonathan Price, www.WebWritingThatWorks.com.)

VOICE: ACTIVE OR PASSIVE?

Voice for most of us means the ability to speak or to declare something to be true. In terms of language usage, though, "voice" refers to the special relationship between the subject and its verb: whether the subject of the sentence is performing the action (active voice), or the verb in the sentence is acting upon the subject (passive voice). Unfortunately, the issue of voice is an often overlooked factor of writing, especially in technical documents.

Most of us grew up hearing our English teachers say, "Avoid the passive voice; write in the active!" This was good advice, but sometimes the passive voice may work better than the active. Nevertheless, readers of technical documents usually expect straightforward sentences with the subject first followed by an active verb and a direct object or subject complement.

As with nominalizations, use of the active or passive voice comes down to a judgment call based on the sentence's intended meaning. Normally, we speak in the active voice. Also, most writers, editors, and readers prefer the active voice because it enlivens the prose by clearly establishing who the subject is and what the subject is doing. For example, the sentence *Bill rebooted the computer* is active voice and unambiguous. What action Bill performs, *rebooted*, cannot be mistaken; furthermore, it clearly demonstrates who is responsible for the action: Bill. If we express the same idea in the passive voice, *The computer was rebooted by Bill*, the emphasis in the sentence shifts from Bill performing the action to the computer receiving the action. At first you may want to dismiss this seemingly subtle difference, but the accumulative effect of stringing passive sentences together within a document frequently creates confusion as to who is doing what. When writers mix nominalizations with numerous passive sentences, misunderstandings are almost certain to occur.

Many editor-writers fail to realize that voice doesn't just refer to the grammatical construction of a sentence; it also affects the tone of the document. For example, the active voice, where the subject clearly performs the action, generally enlivens a document's prose, creates a feeling that the author is confident about what is being said, and has command of the language. The passive voice, though, where the verb is acting upon the subject, can create prose that sounds flat and unsure. For example, compare the following sentences. Notice how the active-voice sentence seems more direct.

Self-esteem **is raised** when work skills **are learned**. (passive voice)
Self-esteem **rises** when people **learn** work skills. (active voice)

The passive voice *is created* when a form of the verb "to be" *is used* in conjunction with a regular verb (such as in this sentence). Also, the subject of a passive sentence receives the action from the verb. Here is another example that illustrates these points.

Mohammed investigated the harassment allegation is active voice because the subject, Mohammed, is performing an action. This sentence is straightforward; Mohammed is responsible for investigating the allegations. However, if we wrote this same sentence in the passive voice—*The harassment allegation was investigated by Mohammed*—notice the subtle changes. First, in the passive example, a form of the verb "to be" functions as an auxiliary verb. Second, the subject in the passive sentence changes to *allegation,* and now more emphasis is on the allegation itself than on Mohammed's investigating the allegation.

The passive voice, however, does have its place in writing, and a good editor will be able to determine if its use is justified. For example, some scientists use the passive voice because they feel it lends more objectivity to their writing by eliminating the first person. For example, instead of saying, *I analyzed the data*, a scientist may write, *The data were analyzed*. Notice that the pronoun *I* has been omitted in the passive sentence. Of course, everyone knows that some individual or group of individuals (or a computer!) actually analyzed the data, but the passive voice creates a tone or feeling that appears more objective by not calling attention to the researchers themselves. Also, you will commonly find that newspaper writers use the passive voice for the same reason.

Suit the action to the word, the word to the action.
—*William Shakespeare*

EXERCISE 13.5

Notice how the following sentences are awkward because they employ passive verbs. Use your editorial judgment to improve them. The passive verbs are boldfaced, but you may want to choose a different verb to revise the sentence. Again, we begin with an example:

Promotions are **earned** by editors with initiative and who have skills that are **learned** through experience. (passive-voice version)

Editors with initiative and skills earn promotions. (active-voice version)

1. A variety of physical formats are **found** to be available for technical manuals.
2. Identifying a list of materials and parts was **begun** as the first step in inventory.
3. An editor's reputation can be **affected** by the quality of the editor's correspondence with clients.
4. Editors are **required** to work closely with their supervisors so the costs of the publication can be **determined** by all publication members.
5. Your paychecks will be **withheld** in cases of noncompliance with all directives issued from this office.

TIP FOR TECHNICAL EDITORS

Although you may hear that good technical or scientific writing uses the passive voice to maintain objectivity, in practice you will find that many researchers, scientists, and technical writers actually use the first person. Editors should be aware of their clients' preferences and understand the pros and cons of using either voice. Also, we recommend that the decision regarding which voice to use should be negotiated during the early planning if the opportunity arises. All else being the same, stick to the active voice.

Here's a quick review of passive sentences:

- They usually contain a form of the verb *to be* plus a past participle.
- They do not identify who is doing the action.
- They can create a tone that lacks confidence and assertiveness.

NO GOBBLEDYGOOK!

During the Clinton administration, Vice President Al Gore resuscitated the National Performance Review agency and renamed it the National Partnership for Reinventing Government (NPR). Using plain, clear language when speaking and writing became one of the NPR's major objectives. According to the NPR, for too long government documents have contained *gobbledygook,* a term coined to designate jargon-filled documents containing drivel, meaningless expressions, or awkward sentences. In fact, the NPR annually bestows the "No Gobbledygook Award" to the government agency that has eliminated the most gobbledygook from its documents. We're not sure how NPR goes about deciding who receives this award, but sorting through all those government documents has to be an incredibly difficult editing job.

According to the NPR, most documents can be reduced in length by at least 25%. Also, a study done by naval officers determined that they took 17 to 23% less time to read plainly written documents than the original versions. Here's where the Navy took notice. If this savings in reading time is translated into money, the study concluded that the Department of the Navy could save over $250 million per year!

> Whatever isn't *plainly* stated the reader will invariably misconstrue.
> —*John Trimble*

EXERCISE 13.6

Figures 13.6 and 13.7 are two versions of the same document; the first is written in gobbledygook with a heavy emphasis on the passive voice. The second is a plain-language version. Avoid the temptation to look at the revised version now. Instead, locate the passive verbs in the original version and eliminate any unnecessary expressions. After you have completed this, write your shortened, plain-language version and compare it to the revision.

(Source: http://www.plainlanguage.gov/.)

THE LARD FACTOR

Unlike a mathematical equation that results in either a wrong or right answer, communicating through language with all its nuances and possible connotations simply cannot be as precise. However, language's incredible flexibility allows for countless approaches to communicating an idea and gives unprecedented power to writers and editors. As you can see from the previous examples, effective editing calls for numerous editorial decisions and judgments.

A major part of an editor-writer's power is to know when to quit—to quit adding words, that is. Some writers mistakenly believe that writing that sounds profound *is* profound, that

---

VETERANS BENEFITS ADMINISTRATION

OLD VERSION—Pension Medical Evidence

addressee
street:
city
state/zip

Dear addressee:

Please furnish medical evidence in support of your pension claim. The best evidence to submit would be a report of a recent examination by your personal physician, or a report from a hospital or clinic that has treated you recently. The report should include complete findings and diagnoses of the condition which render you permanently and totally disabled. It is not necessary for you to receive an examination at this time. We only need a report from a doctor, hospital, or clinic that has treated you recently.

This evidence should be submitted as soon as possible, preferably within 60 days. If we do not receive this information within 60 days from the date of this letter, your claim will be denied. Evidence must be received in the Department of Veterans Affairs within one year from the date of this letter; otherwise, benefits, if entitlement is established, may not be paid prior to the date of its receipt.

SHOW VETERAN'S FULL NAME AND VA FILE NUMBER ON ALL EVIDENCE SUBMITTED.

Privacy Act Information: The information requested by this letter is authorized by existing law (38 U.S.C. 210 (c) (1)) and is considered necessary and relevant to determine entitlement to maximum benefits applied for under the law. The information submitted may be disclosed outside the Department of Veterans Affairs only as permitted by law.

Sincerely,

Bill U. Friendly

FIGURE 13.6
Letter Written in Gobbledygook.

VETERANS BENEFITS ADMINISTRATION

NEW VERSION—Pension Medical Evidence

addressee
street:
city
state/zip

Dear addressee:

We have your claim for a pension. Our laws require us to ask you for more information. The information you give us will help us decide whether we can pay you a pension.

FIGURE 13.7
Plain-Language Version of Letter.

What We Need

Send us a medical report from a doctor or clinic that you visited in the past six months. The report should show why you can't work.

Please take this letter and the enclosed Doctor's Guide to your doctor.

When We Need It

We need the doctor's report by [date]. We'll have to turn down your claim if we don't get the report by that date.

Your Right to Privacy

The information you give us is private. We might have to give out this information in a few special cases. But we will not give it out to the general public without your permission. We've attached a form which explains your privacy rights.

If you have any questions, call us toll-free by dialing 1-800-827-1000. Our TDD number for the hearing impaired is 1-800-829-4833. If you call, please have this letter with you.

Sincerely,

Enclosures:
Your Privacy Rights
Doctor's Guide

FIGURE 13.7 (*continued*)   Plain-Language Version of Letter.

including big words and stringing long clauses together adds credibility to ideas. Nothing is further from the truth.

Technical writing is difficult enough to understand for most people without a document containing excessive verbiage and a pretense toward profundity. Simply put, bad writing is *lard*. Everyone knows what lard is to the body: unnecessary fat that weighs down an individual. It's the same for writing. Unnecessary words and phrases increase the amount of time it takes to read a document while reducing the reader's ability to comprehend. Written lard, which is unnecessary verbiage, must be eliminated, especially from technical documents. Remember, most readers have time to read a document or publication only once. They may not give it a second chance if the wording isn't clear and concise.

Wordy documents share three distinguishing characteristics. First, as we have previously discussed, nominalized verbs and adjectives often create lard by obscuring the sentence's subject and predicate, resulting in confusion and unnecessary prepositional phrases. Second, overuse of passive verbs and *to be* verbs generally yields weak, ineffective statements. Finally, prepositional phrases strung together in a series create confusion and can often be eliminated.

Although editors must follow budget and time constraints, effective editing requires taking time to eliminate a document's lard. Editing out the lard should not be an editor's responsibility when editing for correctness or appropriateness; it must be negotiated with the client as a separate function of editing for effectiveness or level-three editing. Why? Because depending on the length of the document to be edited, its lack of clarity, and the technical

The work was like peeling an onion. The outer skin came off with difficulty . . . but in no time you'd be down to its innards, tears streaming from your eyes as more and more beautiful reductions became possible.
—*Edward Blishen*

When in doubt, delete it.
—*Philip Cosby*

writer's abilities, you may be involved in an extensive, sentence-by-sentence rewrite. If you discover that a document does require extensive revision, and your company has not negotiated to edit on this level, you may return the document to the writer and ask for a revised draft. Here are four suggestions for eliminating wordiness or lard:

- *Select words that are concise, not abstract.* This particularly applies to nominalizations. Also, avoid redundancy, such as *another additional* (just use *another*), *eliminate altogether* (*eliminate*), *join together* (*join*), etc.
- *Avoid lengthy sentences.* Use a variety of sentence types, but in general, keep the length of the sentences relatively short. We recommend no more than twenty words per sentence if possible. Avoid stringing prepositional phrases together. Use active verbs when suitable.
- *Keep paragraphs relatively short.* Think of paragraphs as compositions in miniature. The topic sentence is similar to a thesis statement for an entire essay, while all subsequent sentences in the paragraph pertain to the topic sentence.
- *Consider using bullets or numbers for lists.* Bulleted or numbered lists make a series of items or sequential events easier for the reader to comprehend. Use them when appropriate.
- *Use transitions to connect sentences.* Transitions are words and phrases such as *first, second, however, in addition, consequently*, etc. that often eliminate entire sentences by linking ideas logically and coherently.

TIP FOR TECHNICAL EDITORS

Creating lists often reduces verbiage and provides information in a quick, aesthetically pleasing format. When making a list, use numbers (*1, 2, 3*) or letters (*A, B, C*) when you want to prioritize, show chronology, or importance. Use bullets when you want to create a list where all the elements are of equal importance or the sequence is not significant.

With minimal experience and by remembering these few rules, you can eliminate excessive verbiage. It's not difficult, but it's time-consuming. However, with experience, editors learn to combine ideas and purge documents of extraneous words. Begin by examining the examples below; then, try your editing skills on the sentences in Exercise 13.7. Afterward, revise the documents in Exercises 13.8 and 13.9 by eliminating unnecessary words and phrases and by combining ideas. You will be amazed how much lard you can remove. Think of yourself as a language surgeon performing liposuction on an obese patient.

The first two examples are taken from the U.S. government's website *Plain Language* (http://www.plainlanguage.gov/). We recommend that you familiarize yourself with this site for additional good tips on writing and editing, especially before you edit any government documents.

Example 1

Original version	**Plain-language version**
If you take less than your entitled share of production for any month, but you pay royalties on the full volume of your entitled share in accordance with the provisions of this section, you will owe	Suppose that one month you pay royalties on your full share of production but take less than your entitled share. In this case, you may balance your account in one of the following ways

no additional royalty for that lease for prior periods when you later take more than your entitled share to balance your account. This also applies when the other participants pay you money to balance your account.

without having to pay more royalty. You may either:

a. Take more than your entitled share in the future; or

b. Accept money from other participants.

Example 2

Original version	Plain-language version
For good reasons, the Secretary may grant extensions of time in 30-day increments for filing of the lease and all required bonds, provided that additional extension requests are submitted and approved before the expiration of the original 30 days or the previously granted extension.	We may extend the time you have to file the lease and required bonds. Each extension will be for a 30-day period. To get an extension, you must write to us giving the reasons that you need more time. We must receive your extension request in time to approve it before your current deadline or extension expires.

Notice that in the plain-language versions above, the essential information has been transformed into relatively short sentences, active verbs are used, and strung-together prepositional phrases are avoided.

Continue by examining the following examples of wordy sentences and the suggested revisions we provide. See how we calculate what Richard Lanham calls the "lard factor" in each sentence. Then, complete the document exercise to assess your ability to eliminate lard from writing.

When we came into the office and went over to Joseph's desk, we saw that he had not completed the markups he had been given to complete, and in my opinion, he is too inexperienced to continue with the project we all are working on. (original version, 46 words)

While at Joseph's desk, we determined he did not complete the markups and concluded he is too inexperienced to continue with the project. (revised version, 23 words)

$$\text{Lard Factor:} \frac{23 \text{ (words in revised version)}}{46 \text{ (words in original)}} = (.50 \text{ or } 50\%)$$

This means that the revised version is only half the length of the wordy version. We have reduced the lard by 50%. (Of course, it's important not to misrepresent the meaning of the original sentences for the sake of brevity.)

The memo on the office bulletin board stressed the need for strong compliance to the application process for submitting graphics to the editors, who need all the graphics turned in at the proper time designated on the submission form. (original version, 39 words)

The memo stressed that applications for submitting graphics be turned in to the editors at the designated time. (revised version, 18 words)

$$\text{Lard Factor: } \frac{18 \text{ (words in revised version)}}{39 \text{ (words in original)}} = (.46 \text{ or } 46\%)$$

This means 54% of the sentence (the lard) was eliminated.

EXERCISE 13.7

Revise the following wordy sentences, being careful not to change or misrepresent the author's intended meaning. Feel free to select new verbs that are more precise. After you are satisfied with your revision, calculate the lard factor. Challenge your classmates to determine who can eliminate the highest percentage of lard while remaining true to the original version.

1. I think my outdated computer, which is five years old, needs to be replaced by a new one that is much faster.
2. The new intern has a very nice personality and is attractive and everyone seems to be drawn to him.
3. Due to the fact that that many writers use a lot of jargon when they write, a workshop will be held Tuesday for the benefit of our writers and editors so they can address this issue in a timely manner.
4. My managing editor says she wants to go back to writing for the magazine she used to write for in Paris before she had this job because she misses the challenges that job called on her to perform.
5. As many people who read carefully are aware, statistics can often be manipulated to conform to a writer's agenda, and many times can be just plain nonsense. For example, to say that 50% of the legal marriages that occurred during the past 30 years included women is laughable and even flawed. Or, to say that a person who graduated number 50 in a class of 100 people graduated in the bottom 50% of the class instead of the top 50% of the class is to put a spin on the that person's class ranking.

EXERCISE 13.8

Revise this portion of a feasibility study for wordiness and calculate the lard factor. Compare your revision with others in the class. Who removed the most lard but did not "cut into the bone"?

Section 2.1 Data and Methods

Most students who attend universities cannot even afford to bring their cars to school because the parking fee is too much. Jennifer Johnson, a senior at State University College in Oneonta, who decided this year not to bring a car to campus, said, "The cost of registering a vehicle alone is too expensive." (*The Daily Star*) Most campuses raise the fees to try keep to the students from bringing their cars on campus. However, the problem only gets worse at the beginning of each new semester. From our research, we have found that some universities and colleges are trying to correct or improve the parking problem.

Many universities are now installing a parking system where students must park in certain areas or lots. Some universities have installed a bus system to transport students around campus. The University of Michigan has established a university parking service and transportation service that takes faculty and students all over campus. However, this is a bit expensive for University of South Alabama students. When we asked Chief Clay about what had been done in the past, he replied, "A long time ago, like early 80's, we had some restricted parking. Students living in Alpha, Beta, and Gamma dorms were restricted to their respective parking lots and were not allowed to park their cars anywhere else on campus. That changed when Delta was activated and Alpha discontinued being used as a dorm. Certain administrators felt the distance to Delta was to great for students to walk so they lifted the restrictions making all parking spaces for students white and blue for all faculty and staff."

THE TECH EDITOR'S CUBICLE

The following humorous example (Figure 13.8) of how the story of Little Red Riding Hood might be written by a technical writer appears on the U.S. government's *Plain Language* website. As an employer of thousands of technical writers, the U.S. federal government has recognized the need to have the documents it produces written in straightforward, plain language. As a technical editor, your task will often be to "detechnicalize" documents, especially after considering the document's intended audience. As an exercise, revise the following tale of Little Red Riding Hood but remember that your audience in this scenario is children.

At a previous but undetermined timeframe, a single-family domestic domicile was inhabited by a young girl, known as Little Red Riding Hood (LRRH), and her Maternal Parent (MP). The Maternal Parent (MP) had once provided for the fabrication of an article of clothing, a cloak in nature (including a "hood" or protective covering for the head of the wearer), that was RGB code [255,0,0] in hue (aka, "red"). As a result of this action, and the resultant repeated usage of the "hood," the young girl was always known as LRRH in substitution for the name identified on her birth certificate and other identifying documentation.

During one 24-hour interval, a request was issued by the MP for LRRH to deliver a package to the MP's Maternal Parent (MPMP) (genealogically identified as the Grandmaternal Unit (GU) with respects to LRRH). This package was to include:

- cheesecakes
- fresh butter
- one dozen (12) strawberries

Little Red Riding Hood (LRRH) optioned to accept the Task Order (TO). LRRH further sourced a package delivery vehicle with the proper functionality for the Task Order, selecting a wicker basket. After a

FIGURE 13.8
Little Red Riding Hood.

thorough and complete market survey, leveraging LRRH's experience with similar Task Orders in the past, cheesecake and fresh butter were acquired from the kitchen, whereas strawberries were acquired from the garden. While the latter item was not, strictly speaking, within the bounds of the Task Order, the marginal cost savings as compared to waiting for strawberries to grow in the kitchen appeared to be of great benefit to the MP in the completion of the Task.

With initial outsourcing complete, the journey was commenced by LRRH (see Appendix A: Proposed Map of Route Between the Domiciles of MP and GU). During a brief eleventh-hour meeting, MP issued a contract rider requiring the complete confidentiality of all personnel working the Task Order. LRRH assured MP that there would be no violation of this rider.

In the course of executing the Task Order, LRRH was approached by market competitor Old Grey Wolf (OGW). There were inquiries from OGW to LRRH regarding the nature of the Task Order, and in violation of the contract rider, LRRH disclosed sensitive and mission-critical data relating to the Task. Table 1-1 illustrates the nature of the information believed to have been compromised:

Table 1-1: Information Compromised by LRRH During Interactions With GW

Nature of Data	Disclosed To	Severity of Disclosure
Contents of Basket	Old Grey Wolf	Medium
Nature of Task	Old Grey Wolf	High
Destination of Journey	Old Grey Wolf	High

The identity of LRRH had been predetermined by OGW using standard practices of observation; therefore, that information was not compromised by the actions of LRRH.

It was the intent of OGW to compromise the functionality of LRRH, but the potential negative impact on its operations by the nearby presence of an organized unit of fully-functional Wood Cutters (WC) provided for the redirection of its action item to the domicile of GU.

Though LRRH had blatantly violated the terms of the contract rider, this violation went unreported to supervisory entities (i.e., MP) by the violator. LRRH continued to action the Task Order despite clear and compelling evidence that the integrity of the process had been disenfranchised by the OGW.

While LRRH continued to analyze its processes through the implementation of the Task Order, OGW leveraged its greater cumulative experience and used Best Practices to arrive at the GU client site in a more efficient and expedient manner than LRRH. Therein, the functionality of GU was impacted by the biorhythmic needs of OGW in a negative manner.

Upon the dissemination of information related to the pending closure of the Task Order assigned to LRRH, OGW engaged in an enterprise-wide analysis of situational readiness. Determining that there were vulnerabilities in OGW's methodology, OGW elected to redesign the external identifiers of OGW to better emulate those of GU, by means of garbing the nightgown generally associated with GU and altering the vocal patterns of OGW to align with precedents set by GU.

After completing the Task Order by delivering the deliverables:

- cheesecakes;
- fresh butter; and
- one dozen (12) strawberries

FIGURE 13.8 (*continued*) Little Red Riding Hood.

LRRH recorded observations of the host system. These observations included, but were not limited to:

- My what big ears you have!
- My what big eyes you have!
- My what a big nose you have!
- My what big teeth you have!

Upon receipt of the host system status analysis, OGW prepared and delivered a response regarding the functionality of the concerned functionalities, to include:

- This functionality leverages the soundwaves generated from other sources, such as LRRH, to amplify the positive audio signal from such sources for the end user.
- This functionality absorbs underutilized light emissions and their reflection from objects thereon, such as LRRH, to better provide for the identification of nearby entities by the end user.
- This functionality analyzes the available transient atmospheric particles against a matrix of known particle cultures, such as LRRH, to provide near-instantaneous and transparent supplemental feedback to the end user.
- This functionality greatly impacts the capacity of the OGW to reprocess physical assets related to LRRH in such a manner as to benefit the continued functional life-cycle of the OGW operations!!

Immediately thereafter, Old Grey Wolf (OGW) executed its asset plan action item and severely compromised the functionality of Little Red Riding Hood (LRRH).

FIGURE 13.8 (*continued*)   Little Red Riding Hood.
Source: U.S. Federal Aviation Administration.

EXERCISE 13.9

Figure 13.9 is another example of overcomplicated, technical jargon from a National Park Service manual. Revise this passage into plain English.

CHAPTER 5: OUTGOING LOANS

A. INTRODUCTION

The National Park Service enters into two types of loans, incoming and outgoing. Incoming loan transactions are considered accessions. Refer to Chapter 2, Accessions, Section C for procedures on incoming loans. The National Park Service makes outgoing loans to further its mission of preservation, education and research. Museum property sent out by a park to another park, repository, non-NPS institution or organization, or service-providing organization for exhibition, exhibit preparation, study, conservation, photography, collections management or storage, is considered an outgoing loan by the lending park. Outgoing loans are temporary assignments of custody (but not title) by the lender (NPS park or center) to the borrower.

In order to avoid possible liability, all conditions of the outgoing loan between lender and borrower must be agreed to by a signed outgoing loan agreement prior to the initiation of the outgoing loan. Outgoing loan terms and conditions provide legal protection to both the lender and the borrower and are included in this chapter.

FIGURE 13.9
Park Service Manual Extract.

The National Park Service enters into two types of outgoing loans: standard outgoing loans and repository outgoing loans.

1. Standard Outgoing Loans

Objects from the parks museum collection are loaned only for the purposes of exhibition, research, scientific preparation, analysis, photography, conservation, or other requested services. Loans are made to educational institutions (e.g., NPS park museums, non-NPS museums, historical societies, universities and other organizations); service-providing organizations (e.g., non-NPS and NPS conservation and analytical laboratories or exhibit preparation firms or contractors providing these services); and other National Park Service divisions, offices, or units. Only cataloged objects can be loaned to institutions for exhibit purposes. If objects loaned for research purposes are not cataloged, they must be adequately documented through another means (i.e., field specimen log for archeological collections). Objects loaned for conservation purposes must be cataloged, unless the conservation treatment is necessary to assist with the preparation of the material for identification purposes.

2. Repository Outgoing Loans

Repository outgoing loans are made for purposes of collections management (including cataloging and storage) or solely storage to non-NPS repositories, such as universities or research institutions, to NPS centers, and occasionally to other NPS parks. Arrangements with a non-NPS repository may be covered by a cooperative agreement (see Figure 5.10), although each loan transaction should also be documented with an outgoing loan agreement. Outgoing loan agreements must be completed and signed for each transaction reviewed regularly. Outgoing loans to non-NPS repositories should be renewed every ten (10)years.

3. Procedures for Documenting Objects Sent to NPS Conservation Treatment Facilities

Objects sent to a NPS conservation treatment facility (e.g., Division of Conservation, Harpers Ferry Center) for treatment or exhibit preparation are processed as a standard outgoing loan.

Objects to be loaned to the conservation treatment facility are identified by either an Object Treatment Request (OTR), initiated by the lending park, or selected by the curator and/or a conservator from objects identified by an approved exhibit plan and a separate list prepared.

An OTR must be submitted to the NPS conservation treatment facility, through the regional curator, for objects that are not included in an exhibit plan. The lending park prepares the outgoing loan agreement unless prepared by the NPS conservation treatment facility. Objects may be shipped only when the loan agreement is signed by both responsible officials. Shipment of objects will be documented with a receipt for property initiated by the lending park and signed by the park curator and an authorized official at the NPS conservation treatment facility. As objects are returned to the park, the same procedure applies, except the receipt for property is initiated by the authorized official at the NPS conservation treatment facility.

Exhibit projects, such as those planned at Harpers Ferry Center, often include a large number of objects requiring treatment, mounting, or other exhibit preparation, and may be sent in batches to the treatment facility over a period of time. A single outgoing loan agreement may cover all the shipments needed for an entire exhibit. After the loan agreement is signed by both parties, objects may be shipped to the facility at mutually agreed upon times. Each shipment to and from the lending park and the treatment facility will be documented with receipts for property as indicated above for the OTR process.

4.Documentation of Outgoing Loans

Information needed to make an outgoing loan is contained on the Outgoing Loan Agreement (Form 10-127 Rev.; Figure 5.2). The conditions governing outgoing loans are described in the Conditions for Standard

FIGURE 13.9 (*continued*)   Park Service Manual Extract.

Outgoing Loans (Form 10-127a Rev.; Figure 5.4), the Conditions for Standard Outgoing Loans (NPS Conservation Treatment Facilities) (Form 10-127b; Figure 5.5), the Conditions for Repository Outgoing Loans (non-NPS) (Form 10-127c Rev.; Figure 5.6) and Conditions for Repository Outgoing Loans (NPS) (Form 10-127d Rev.; Figure 5.7). Specific conditions, such as special handling or additional insurance conditions, to be met by the borrower should be noted in the special conditions section of the outgoing loan agreement.

The superintendent must sign the outgoing loan agreement, but other staff may process the loan. The standard outgoing loan agreement is generated by the lending park. Between NPS units, the lending park will follow outgoing loan procedures. The borrowing park may use the lenders outgoing loan agreement to document the incoming loan, but must follow all other incoming documentation procedures as outlined in Chapter 2 of this handbook. In this instance, the accession number assigned to the incoming loan must be placed on the lenders outgoing loan form by the borrower when filed. The NPS repository agreement may be generated by the NPS repository but must be signed by the superintendent. All non-NPS repository loan agreements must be generated by the lending park.

The lending park or center must update the status field in ANCS. An object temporary removal slip should be completed for all standard loans. For further details on documenting and tracking loans, see Chapter 5, Section E.

Flow charts for processing loans are found in Figure 5.11, Flow Chart for Standard Outgoing Loans; and Figure 5.12, Flow Chart for Repository Outgoing Loans.

FIGURE 13.9 (*continued*)   Park Service Manual Extract.
Source: U.S. National Park Service

DICTION: CHOOSING WORDS CAREFULLY

By now, you realize that editing for style consists of much more than simply recalling grammatical conventions and fashioning sentences together. A technical document's readability depends on how effortlessly and clearly the information "flows" to the reader. Surely you have encountered textbooks, academic articles, reports, and other forms of writing that are simply too dense, tangled, or complex to plow through. Editors encounter writing like this every day; the writers have a lot to say but don't always know how to say it. Effective editing frequently determines the success or failure of a piece of writing, especially writing of a technical nature.

Choosing the right word, or *diction,* is another major factor in creating an effective, clear style. But it's more than selecting a suitable synonym from a thesaurus. Other considerations must be taken into account to avoid misunderstandings. For example, when editing a document, an editor needs to be aware of whether any

- concrete terms can be substituted for unclear, abstract terms;
- words that have connotative meanings might be misinterpreted;
- euphemisms have been inserted to soften unpleasant information (e.g., *the departed* for two or more dead people);
- technical jargon or acronyms are inappropriate for the intended audience;
- slang terms, sexist language, or idiomatic expressions hinder readability.

Concrete versus Abstract

In theory, technical writing should be as specific as possible and use concrete words at the lowest level of abstraction. For example, *Southern oak tree* is much more specific than *plant* or *vegetation*. Or to take an example from computing, *OS/2* is more concrete than making general references to an operating environment.

> *Abstract:* The editors met for a long time, but they didn't get a whole lot done.
> *Concrete:* The senior editors met for two hours but failed to decide on Roger's proposal.
> *Abstract:* The duplicating costs for our department have seen a significant increase recently.
> *Concrete:* The maintenance costs for our department's copying machine have increased 20% since last February.
> *Abstract:* The commonality of the people found within the diversity of the multitude of cultures, indicates a need for a greater perspective of multiculturalism.
> *Concrete:* Although people may have diverse cultural backgrounds, they share common interests.

Some abstract words to avoid include *activities, aspects, concepts, devices, factors, functions, outputs, systems, things*, and *variables*.

TIP FOR TECHNICAL EDITORS

Generally, technical writing focuses on what is concrete and specific, whereas philosophical writing focuses on the abstract. Writing that intends to inform or persuade often uses both concrete and abstract words, depending on the subject matter and audience. We suggest that when editing technical documents and reports, you choose precise words and avoid abstractions. However, when you decide that abstractions are necessary, consider using examples to clarify them.

Connotative versus Denotative

One reason English has become an international language is its ability to convey ideas precisely and accurately. As a hybrid language that has become transformed through centuries of absorbing and adapting to other languages, English today consists of many words (commonly known as synonyms) that have similar yet slightly different meanings. For example, the generic term for one who fights, *fighter*, can be labeled as a *boxer, pugilist, combatant, soldier, warrior, rebel, mercenary, terrorist*, or *freedom fighter*, each term suggesting a different association or emotional response to a type of fighter. Depending on a writer's intentions, the English language affords a vast repertoire of synonyms that allow writers to focus precisely on their intended meaning. Ironically, though, English's capacity for conveying ideas precisely and accurately can also lead to significant misunderstandings.

The strict, literal definition of a word found in the dictionary is its *denotation*. For example, the word *house* denotes a dwelling, a place where someone resides. When you read the word *house*, nothing particularly comes to mind except its literal definition. However, suppose we changed the word *house* to *home.* Now do images of a cat curled up on an oval rug by a fireplace come to mind? Or are you reminded of the smell of your grandmother's hot apple pie drifting out her screen door to where you are swinging on her front porch? These images are called *connotations*. Many words in the English language, such as *home*, have denotative as well as connotative meanings that suggest associations to a reader beyond their literal definition. This is where a good technical editor needs to be careful and can help the writer. For example, *kick the bucket* is figurative language meaning *to die*. If you were editing medical software, the section on women's miscarriages should not use this figurative term with regard to fetuses.

Although all editors need to be conscious of diction and what words may imply to the reader, technical editors need to be particularly aware when a word connotes more than one meaning and might lead to confusion. By definition, technical writing provides technological and scientific information to an audience that expects the language to be precise and accurate. English affords writers the opportunity to be exact. Effective editing, however, often uncovers words and terms that writers overlook as having multiple meanings.

Euphemisms

When writers want to soften the impact of a situation or event, they often resort to euphemistic expressions. Politically correct language is a primary example. Handicapped people are *physically challenged*. Civilian deaths resulting from bombs are *collateral damage*. A person's death may be referred to as *passing on to her reward, his time has come, gone to the great beyond, passed away,* or *deceased.* Canadians refer to being fired as being *made surplus.*

When Janet Jackson's breast was exposed at the 2004 Super Bowl halftime performance, she explained that she had a *wardrobe malfunction.* When Caspar Weinberger was the secretary of defense and American troops were withdrawn to ships offshore of Lebanon, he explained, "We are not leaving Lebanon. The Marines are merely being deployed two or three miles to the west." Jimmy Carter affirmed that the failed rescue attempt of American hostages in Iran was *an incomplete success.* According to Lewis Thurston, who was New Jersey's chief of staff, staff members do not have chauffeurs but *aides who drive.* When a major automobile manufacturer had to lay off 50,000 employees, it justified its move by saying it had "initiated a career alternative enhancement program."

Technical editors need to be aware of euphemisms. Depending on the intended audience, they should call the writer's attention to indirect language and suggest more precise terminology or eliminating the euphemism altogether.

> Academic Bad Writing is indeed old news, and no secret. But it is also on-going: a thriving, flourishing, burgeoning industry with all too much product. The market is saturated, indeed the water is up over the second floor windows, but the rain keeps falling. The vampire keeps waking up every night to find fresh blood, so all we can do is keep pounding away on the stake through the heart.
> —*George Orwell*

EXERCISE 13.10

Try to match the euphemisms and double-talk expressions with the plain English version (list continues on next page).

1. _____ Aerodynamic personnel decelerator	A. Toilet plunger
2. _____ Pre-dawn vertical insertion	B. Road signs
3. _____ Hydro blast force cup	C. Violent peace
4. _____ Permanent pre-hostility	D. Flea
5. _____ Limited armed conflict	E. Fan
6. _____ Manually powered fastener-driving impact device	F. Nut
7. _____ Grain-consuming animal units	G. Parachute
8. _____ Mounted confirmatory route markers	H. Refugees
9. _____ Forcible ejection of the internal bomb components	I. Junk

10. ____ Hemotophagous arthropod vector	J. Smoke
11. ____ Hexiform rotatable surface compressor unit	K. Hammer
12. ____ Universal obscurant	L. Pigs and cows
13. ____ Previously owned parts	M. Invasion
14. ____ Environmentally operable panel	N. The bomb blew up
15. ____ Ambient noncombatant personnel	O. Peace

Answers:
1. G   6. K   11. F
2. M   7. L   12. J
3. A   8. B   13. I
4. O   9. N   14. E
5. C   10. D   15. H

Jargon, Acronyms, Slang Terms

Manuals just slow you down and make you feel stupid. The directions are too slow, too detailed, and use to much abstract, arcane or academic language—like "boot up" instead of "turn on the red switch in the back."
—*Neil Fiore*

It's a safe assumption that most readers hate jargon, and some technical writers have a bad reputation for incorporating jargon into their writing. *Jargon* refers to words, terms, acronyms, and abbreviations used by a specific profession or group of individuals within their area of specialty. Government officials, the military, social workers, educators, engineers, doctors, and scientists are just a few of the professions who use jargon to make themselves sound knowledgeable. Expressions such as *point in time, context-specific, economic-wise,* and *meaningful experience* represent just a few fancy terms used as *catchwords* or *buzzwords*.

The use of jargon is audience-specific (another example of jargon). A good technical editor will always be conscious of the intended audience and question if the writer's terminology fits the knowledge level of the reader. For example, suppose the directions accompanying a new pair of sunglasses said, "The refractory coating cannot be subjugated to prolonged exposure to the impact of loose particulate matter initiated by air-borne currents." Huh? Stated without the jargon, the directions might read, "During windy conditions, dust can damage the coating on your sunglasses."

However, jargon in the proper context can be useful. For example, individuals already familiar with computers understand terms such as *boot up, file allocation, RAM, CD drive,* and so on. Certainly, a technical manual for novices would have to explain each term in plain language and present technical elements in a simplified, straightforward manner. Businesses, corporations, agencies, and institutions develop their own *corporate culture,* and individuals within a specific corporate culture will understand jargon that outsiders would not. Again, the point is to assess the needs and expectations of the intended audience.

The same is true for *acronyms,* abbreviations formed from initial letters. Common acronyms, such as UNICEF or NASA, generally do not require an explanation. However, editors should not assume readers know certain acronyms, especially in technical and scientific fields. Most acronyms need to be spelled out first and followed by the acronym in parenthesis. For example, a report from Mothers Against Drunk Driving (MADD) or a report on a superconducting quantum interference device (SQUID) needs the acronym spelled out initially. After the acronym has been introduced, though, it may be substituted throughout the remainder of the document. Do not fail to notice that the acronym is stated in all capital letters.

Finally, avoid using slang and clichés in technical documents. Slang expressions create extremely informal language, and clichés suggest a lack of original thinking. When writers use slang or clichés, they run the risk of not being understood, especially by international audiences.

Both forms of speech are short-lived. Expressions that were popular twenty years ago are rarely used now: *sleazebag, at this point in time, yuppie, bug off, far out, bottom line, number-cruncher.* Select language that is appropriate for your audience. Speech filled with slang and clichés may be acceptable among close friends but wouldn't be appropriate in more formal situations. You wouldn't want us to refer to you as *dudes* and *dudettes,* now would you? A *cliché* is an overused expression such as *butterflies in the stomach* for nervousness or *strong as an ox* for strength.

Here are just a few examples of poor diction, clichés, and jargon:

Diction:

> A twenty-two point two cubic foot frost-free refrigerator-freezer
>
> X-dot-Desktop (*the pronunciation of X.Desktop, a real-life Unix-based software product from SCO*)

Clichés:

> The Irish should just play Notre Dame football.
>
> "There's no excuse . . . they whipped us on both sides of the ball . . . give credit to the New York Giants. They were ready to play and they got after us. They took it right to us . . . we got into a hole we couldn't dig ourselves out of." (*Dave Campo of the Dallas Cowboys*)
>
> We'll reach across party lines.
>
> A race car driver's no good without a pit crew.
>
> We're here to serve the American people.
>
> We'll fight for working families.

New jargon:

> Red States/Blue States.
>
> Enemy combatant (*as opposed to a* friendly *combatant?*)
>
> Wardrobe malfunction (*sure, blame the wardrobe*)
>
> Blog (*a monster from the 1950s?*)
>
> Body wash (*a.k.a. soap*)

EXERCISE 13.11

Take ten minutes of class time to brainstorm as many clichés, slang terms, euphemisms, and jargon words as you can. Identify each type. Ask other class members which ones they haven't heard before. Ask international students to provide similar terms from their native language.

SUMMARY

This chapter demonstrates how the elements of style can either create effective writing or, in the case of the Official Style, make it unintelligible. Although choosing the right word, eliminating unnecessary words from sentences, avoiding the passive voice, using transitions, creating coherence, and other elements that constitute style may seem secondary to the other editing considerations, proper writing style has been shown to be one of the most important aspects of effective communication.

CHAPTER 14

Organization: Where Form and Function Come Together

At the conclusion of this chapter, you will be able to

- define the essential principles of effective organization;
- distinguish eight organizational patterns of writing;
- edit to create well-organized technical documents;
- recommend important organizational changes for readability.

Form cannot be separated from function. From our experience, this principle of good writing is greatly underemphasized. The form of a document—the manner in which the author arranges verbal and visual information on a page—contributes to the reader's comprehension and understanding of the author's intent. Too often, inexperienced technical writers and editors focus on grammatical and mechanical issues while neglecting organizational considerations. Of course, following the conventions of usage is essential to good writing; however, effective editing also encompasses analyzing a document's content for a logically coherent, well-organized presentation. This chapter emphasizes the different organizational principles associated with textual information.

Determining the arrangement of a document's content coincides with collecting the information about its subject. Both aspects should be tentatively resolved during the planning stage. Unfortunately, editors cannot always collaborate with writers to plan documents, and upon receiving the final draft they may have to recommend a comprehensive revision of the structure. Although the content and organization framework for the document are the writer's responsibility, technical editors must evaluate the document's function and the tasks the readers are being asked to undertake. Editors must consider if the organizational framework of the material is suited to the needs of the audience and if it effectively communicates the writer's intentions.

> Then anyone who leaves behind him a written manual, and likewise anyone who receives it, in the belief that such writing will be clear and certain, must be exceedingly simple-minded.
>
> —*Plato*

For example, a manual for a word processing program might be organized differently for beginners than for those already familiar with basic word processing functions. Considering the needs and expectations of each type of audience, the technical writer would probably arrange the beginner's manual in a chronological fashion by first presenting basic skills such as how to open a new document, edit typing errors, change font size and style, format margins, and so on. More complicated procedures such as setting macros, merging documents, and formatting graphics would follow later. However, a manual for a more experienced audience might simply categorize each word processing function according to its use. By organizing the functions into categories, then, the technical writer creates a reference manual more suited to the experienced user.

You may recall our previous discussion of learning theory and the two principal functions of schemata: to direct the reader's perception and to make learning possible. We also discussed two types of schemata: *data schemata,* which provide new information, and *process schemata,* which allow us to organize that information into meaningful scripts or processes. Can you now begin to see why the organizational structure of information is so essential? Our new word processing user must confront new information (data schemata) while learning how to apply it (process schemata). An effective manual for the beginner must consider both aspects. However, the experienced user is already familiar with many of the technical characteristics of word processing and may merely seek supplemental knowledge of how to apply a specific function. Both levels of users will be introduced to new data and process schemata; the difference is in their arrangement and the level of detail.

THE PRINCIPLES OF EFFECTIVE ORGANIZATION

As an editor, you can subject any document to one simple test to determine if it is well organized: construct a *working outline* from the document's content. That is, determine if the document has a clear thesis and if the document can be organized into logical parts that support that thesis. This is not as daunting as it may sound. First, most documents already follow easily recognizable, standard patterns of organization known as *templates.* Second, many documents introduce their main points through headings and subheadings. Third, some documents provide a table of contents that reveals the intended structure. And fourth, the topic sentences of individual paragraphs should reflect a specific point that supports the thesis sentence. A logically organized document generally will resemble Figure 14.1.

Notice that all the main points in this outline delineate separate aspects of the thesis. Also notice that all of the supporting details develop a main point. An outline reveals whether a document follows a logical progression, and whether the writer provides sufficient support for each main idea. Analyzing a document for organization is an important editorial task. For example, examine the sample outlines in Figures 14.2 and 14.3. These outlines introduce two types of organizational patterns. Observe their similarities and differences.

EXERCISE 14.1

Locate two journal or magazine articles that are technical in nature. Create an outline from each article similar to the examples in Figures 14.1, 14.2, and 14.3. Be certain to locate the thesis and the main points that support the thesis. Determine from the thesis and the text if the article is attempting to inform or persuade the reader. How well does the article fulfill its purpose?

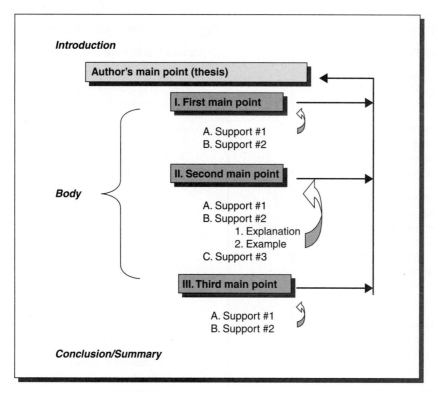

FIGURE 14.1
Generic Outline.

Just as there are no prescribed rules for writing style, the same holds true for organizing a document. Writers and editors must continually make judgment calls in both cases. However, style and organization follow basic principles and patterns that readers recognize, especially patterns of organization. For example, scientific journals usually organize empirical studies in a prescribed format:

1. Abstract
2. Introduction of the problem
3. Review of the literature
4. Methodology
5. Results of the experiment
6. Conclusions

Popular "help" magazine articles frequently begin with a brief personal anecdote in which someone confronts adversity. Usually a discussion follows the personal anecdote as the writer elaborates on the problem and how it affects many other people. Finally, the writer provides an annotated list of what the reader can do when faced with similar circumstances. This is a typical *problem-to-solution* organizational pattern that readers immediately recognize.

Most of us are familiar with reports or articles that compare specific types of products or services. Here technical writers frequently structure their articles according to a different

---

Beginner's Word Processing Manual

Introduction to Word Processing

Thesis: This manual provides a step-by-step approach for beginners who desire to learn basic word processing commands.

I. Step One: Getting Started
  A. Loading the software
  B. Learning basic terms
  C. Accessing the help menu

II. Step Two: Page Setup
  A. Setting the margins
    1. Top and bottom margins
    2. Side margins
  B. Determining the paper size
  C. Creating paragraphs
    1. Setting indentations
    2. Line spacing

III. Step Three: Creating Fonts
  A. Determining the font type
  B. Determining the font style
  C. Determining the font size

IV. Step Four: Beginning to Type
  A. Opening a new document
  B. Beginning to type
  C. Saving the document
  D. Printing the document

FIGURE 14.2
Sample Outline: Chronological Structure.

Reference Manual for Word Processing

Introduction to Word Processing

Thesis: This manual is intended as a reference for experienced users who desire to advance their word processing knowledge.

I. Autocorrect Function
  A. Defined
  B. What it does
  C. How to use it
    1. Selecting from the menu
    2. Formatting as you type
    3. Creating exceptions

II. Spelling and Grammar Function
  A. Defined
  B. What it does
  C. How to use it
    1. Applying to a document
    2. Setting grammar options
    3. Using the Fleisch-Kincaid scale
    4. Looking up synonyms

III. Track Changes Function
  A. Defined
  B. What it does
  C. How to use it
    1. Highlighting changes
    2. Accepting or rejecting changes
    3. Comparing documents

FIGURE 14.3
Sample Outline: Classification Structure.

---

organizational pattern: *comparison and contrast*. Generally, these writers compare features such as cost, durability, ease of use, and feedback from previous users. Then they compare the pros and cons of each aspect. Some magazines, such as *Consumer Reports,* derive their entire purpose from comparing and contrasting products, mutual funds, and vacation resorts.

THE MOST COMMON ORGANIZATIONAL PATTERNS

Familiarize yourself with the eight most common patterns of organization shown in Figure 14.4. These eight patterns can be classified into two major types: *sequential* and *categorical*. Note that many technical documents combine patterns.

Because sequentially organized information progresses from one aspect or event to the next, readers easily anticipate the next logical step. It's important that you as an editor check sequentially organized documents for missing steps or logical gaps in the process.

Categorical patterns of organization may not be as readily discernible as sequential patterns because the categories chosen by the writer are not necessarily arranged logically but by function. Also, categorical patterns frequently incorporate sequential patterns. For example,

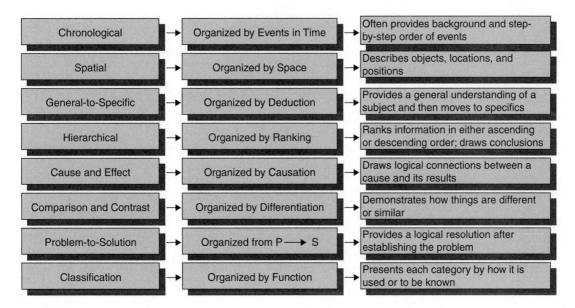

FIGURE 14.4
The Most Common Organizational Patterns.

a reference manual for experienced word processing users may be arranged by categorizing similar word processing functions. However, each function may include a sequential pattern to demonstrate a step-by-step approach to utilizing that function.

Sequential Patterns of Organization

- Chronological
- Spatial
- General-to-specific
- Hierarchical

Categorical Patterns of Organization

- Cause and effect
- Comparison and contrast
- Problem-to-solution
- Classification

Chronological

The chronological organizational pattern sequences events. It is frequently used to narrate a procedure, a process, or step-by-step instructions and is most commonly employed in assembly manuals, reports, empirical studies (methodology section), ethnographic studies, introductions, and résumés.

Figure 14.5 uses the chronological pattern. The audience has some prior experience with telescopes. In the margin, the technical editor has made organizational suggestions to the writer.

Figure 14.6 also uses the chronological pattern. The audience may have had little experience with 35mm cameras. The presentation is clear, organized, and easy to follow.

Chronological organization is often the easiest to develop because the narrative follows a natural time sequence. However, editors need to make certain that the writer has not made huge jumps that may confuse the reader. Most technical documents ask for a sequence of events, not an explanation. Discussion or judgments of the events are either totally excluded or left to a separate section. For example, the *methodology* section of a scientific article narrates the process a researcher uses to evaluate the hypothesis. The *conclusion* section generally is reserved for making judgments regarding the study.

Telescope Assembly

Use the following steps to assemble your telescope. Note: Section headings list which LX200 model (7", 8", 10" or 12") is covered under that heading.

1. The Field Tripod (7", 8", 10", and 12" LX200 Models)

The Field Tripods (Figs. 1 and 2) for Meade 8", 10", and 12" LX200 telescopes are supplied as completely assembled units, except for the spreader bar (4, Fig. 1) and the 6 lock knobs (2 knobs for each of the 3 tripod legs) used to adjust the height of the tripod. These knobs are packed separately for safety in shipment.

For visual (*i.e.* non-photographic) observations, the drive base (17, Fig. 3) of the telescope's fork mount is attached directly to the field tripod.

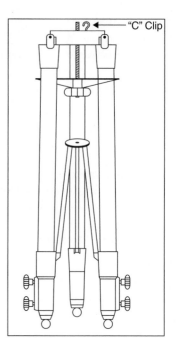

(1) Tripod Head (5) Lock Knobs
(2) Threaded Rod (6) Extension Strut
(3) Tension Knob (7) Tension Hub
(4) Spreader Bar

The telescope in this way is mounted in an "Altazimuth" ("Altitude-Azimuth," or "vertical-horizontal") format. The telescope in this configuration moves along vertical and horizontal axes, corresponding respectively to the Declination and Right Ascension axes (explained later in this manual) in an astronomical observing mode.

Alternately, the field tripod can be used in conjunction with the appropriate optional equatorial wedge (see Appendix A for instructions of the use of the equatorial wedge) for long exposure astrophotography. The equatorial wedge permits alignment of the telescope's Polar Axis with the Celestial Pole (or North Star). After removing the field tripod from its shipping carton, stand the tripod vertically, with the tripod feet down and with the tripod still fully collapsed (see Fig. 2). Grasp two of the tripod legs and, with the full weight of the tripod on the third leg, *gently* pull the legs apart to a fully open position.

Thread in the 6 lock-knobs (2 on each tripod leg) near the foot of each tripod leg. Refer to Fig. 1. These lock-knobs are used to fix the height of the inner, extendible tripod leg sections. ***Note:* "Firm feel" tightening**

FIGURE 14.5
Chronological Organization.

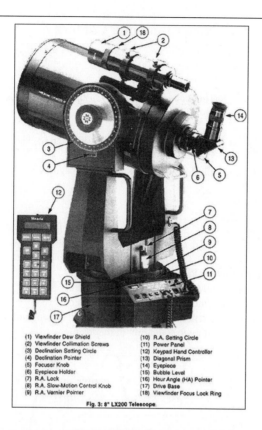

Enlarge graphic of scope—some parts are difficult to see.

(1) Viewfinder Dew Shield
(2) Viewfinder Collimation Screws
(3) Declination Setting Circle
(4) Declination Pointer
(5) Focuser Knob
(6) Eyepiece Holder
(7) R.A. Lock
(8) R.A. Slow-Motion Control Knob
(9) R.A. Vernier Pointer
(10) R.A. Setting Circle
(11) Power Panel
(12) Keypad Hand Controller
(13) Diagonal Prism
(14) Eyepiece
(15) Bubble Level
(16) Hour Angle (HA) Pointer
(17) Drive Base
(18) Viewfinder Focus Lock Ring

Fig. 3: 8" LX200 Telescope

is sufficient; over-tightening may result in stripping of the knob threads or damage to the tripod legs and results in no additional strength.

The spreader bar (4, Fig. 1) has been removed for shipment. To replace, first remove the threaded rod (2, Fig. 1) from the tripod head (1, Fig. 1); a small piece of plastic holds the threaded rod in place. Remove the small plastic bag that is stapled to the threaded rod. This bag contains the "C" clip retainer (used below) and an extra clip.

Slide the spreader bar onto the threaded rod (note the correct orientation as shown in Fig. 1) and position the threaded rod back through the tripod head. Place the clip retainer (a "C" clip) into the slot in the threaded rod. This clip holds the threaded rod in place. See Fig. 2.

FIGURE 14.5 (*continued*)   Chronological Organization.

Used with the permission of Meade Instruments Corporation.

Spatial

A spatial organizational pattern describes "what is where." Most commonly, spatially organized documents describe objects or physical sites. The writer moves around the subject matter in a logical direction, perhaps from top to bottom, or from one quadrant to another. Although description is primary, often technical writers include spatial analysis as well. For example, an urban planner not only would describe urban sprawl around a city but also would include the reasons for the sprawl and the processes that led to it. Typical types of documents that incorporate spatial patterns include feasibility studies, proposals, journal articles, and any document where a close description is needed, such as photographs, organizational charts, drawings, blueprints, brochures, and instructions. Figure 14.7 uses the spatial pattern

---

Operation of a Camera

Automatic Flash

To use the automatic flash, switch the camera into Auto Flash mode. The flash will pop and fire automatically for dark or backlit subjects.

WARNING: Keep fingers away from flash area to avoid obstruction of flash.

NOTE: If you do not want to use flash, use Normal mode. See page 44.

Red-Eye Reduction Feature

The red-eye reduction feature uses the camera's red-eye reduction lamp. The lamp gently shines into the subject's eyes to decrease the size of the pupils, reducing chance of red-eye occurrence.

To turn on the Red-Eye Reduction Feature:
1. Press the mode button until it shows the letter R on the LCD panel.
2. Wait for red light, which signals the red-eye feature is activated.
   *To turn off the red-eye reduction, simply press the mode button until the LCD panel clears. The red light will turn off, signaling deactivation of the feature.

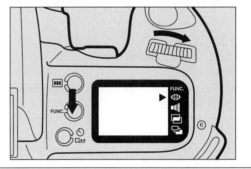

FIGURE 14.6
Chronological Organization.

of organization to describe a well-arranged campsite. Comments on the organization are included in the margin.

Organizing information spatially is an effective means to describe the relationship or juxtaposition among elements that comprise a whole. The brochure selection in Figure 14.7 incorporates a spatially designed diagram that helps the reader to envision how a primitive campsite can be set up properly. When appropriate, writers should use signposts (*in the upper left, toward the center, above,* etc.) to move the reader through the description. Signposts are particularly important when a writer must describe a complicated or multifaceted component or physical site. Finally, good editing requires the editor to assume the position of the reader and determine if the spatial organization successfully achieves the writer's objectives.

General-to-Specific

General-to-specific is a sequential organizational pattern that presents the reader with a broad overview of the subject before addressing specific details. This pattern functions as an

Description of a Primitive Campsite

Seasoned veterans know there is much more to selecting and laying out a primitive campsite than just finding a beautiful location. Other factors, including safety, comfort, wildlife, and the environment, affect a good site selection.

> The description begins with a general thesis of points to be covered.

#1 Select a level site: If you are camping in the mountains, finding a level site may be difficult. However, if you want to sleep well and wake up refreshed, sleeping on level, dry ground is critical. If you cannot find level ground, don't sleep laterally but position your head above your feet. This is important for a good night's rest. You don't want to roll toward the fire during the night.

Begin early in the afternoon scouting for a site before everyone is too tired. Don't forget to look up. Setting up a tent or building a fire under trees is asking for trouble. Falling branches can be dangerous.

Photo courtesy of Ben Williford

> The organizational pattern includes four categories spatially arranged.

#2 Camp near water: Locating a site near a source of water is also important. The water may not be potable, but you must have a means of extinguishing your campfire. Too many forest fires have been caused by careless campers who thought their fires were out. It's best to forgo making campfires altogether and take portable stoves to cook on. Also, locate your tent at least 300 feet from the source of water. Rivers rise rapidly during a thunderstorm.

> The graphic illustrates the arrangement of the campsite—provides a visual aid.

Wind direction

2

3

1

4

#1 Level campsite for tent and campfire
#2 Source of water
#3 Location of campfire
#4 Pantry

Photo by Barry Nowlin

FIGURE 14.7
Spatial Organization.

#3 Build your fire safely: Nearly 25% of wildfires are the result of human carelessness. If you must have a fire, construct a fire-ring around the fire at least 20 feet in diameter. Burn only fallen branches and avoid making sparks that could set the woods on fire. Locate your fire downwind from your tent. Extinguish it completely when disembarking. Make certain you leave only your footprints.

#4 Secure your food: Nothing is more frightening and even dangerous than wild animals showing up in the middle of the night at your campsite. The last thing you want is a bear, skunk, raccoon, bobcat, or coyote foraging around your tent at two in the morning. Take precautions and keep all your food together in a sealed knapsack. Suspend it between two trees with a rope at least ten feet high and away from branches. This becomes your pantry. Remember, most animals can climb trees without any trouble. Take care to position your pantry at least 50 yards downwind from your campsite. Bears have been known to be attracted to toothpaste and photographic film.

Conclusion: Most people find primitive camping an exceptionally satisfying way to reenergize. With a little common sense and preparation, camping in the wilderness will create memories for a lifetime. Preparing a good campsite will go a long way in making those memories happy.

FIGURE 14.7 (*continued*) Spatial Organization.

advance organizer to provide background information and orient the reader. This pattern's most common usage includes introductions and body paragraphs in technical documents, such as memos, feasibility studies, journal articles, and reports. Typically, paragraphs begin with a general statement (the topic sentence), and then proceed to discuss that topic. Figure 14.8 illustrates the general-to-specific pattern.

The general-to-specific pattern of organization is indispensable for technical writing. It acquaints the readers with background information and supplies a contextual basis for the entire document. Virtually all documents, whether memos, instructional manuals, feasibility studies, or scientific journal articles, use some form of the general-to-specific pattern. A good editor understands that readers often need the big picture, and when editing a document should call the writer's attention to documents that do not already contain a general context.

Introduction to the World Wide Web: Setting Up Your Own Website

Introductions usually begin with general information and work toward the specific purpose of the paper, called the thesis.

If you randomly asked people to name the most revolutionary invention of the twentieth century, many would say "the computer." Although the first computer can actually trace its origins back to the abacus in 500 BC, the computer's ability to communicate worldwide has evolved only during the past forty years. Partnership and cooperation among the government, academia, and industry have all led to the creation of this unprecedented new technology—the Internet.

Use of this new technology captured everyone's imagination in 1969 when Tim Berners-Lee conceived of integrating a web browser into the NeXTStep operating system. This was the birth of the World Wide Web (WWW). Soon more user-friendly, secure browsers emerged, including Mosaic, Netscape, and of course Microsoft's Internet Explorer. Now web browsers are capable of transmitting streaming video and radio, as well as allowing individuals to download their bank statements and MP3 music files. In essence, the WWW consists of billions of files and programs that allow anyone with a computer and internet access to unlimited amounts of information.

Today anyone with some basic knowledge of computers can create a personal website within a weekend. Small businesses, too, can create a simple homepage that provides information about themselves and their services. Many software programs are available that make web page creation easy even for the beginner.

FIGURE 14.8
General-to-Specific Organization.

<u>This pamphlet provides the basic guidelines for creating an uncomplicated personal homepage. After introducing you to some basic terminology of web design, it tells you how to get started: design headings, text and graphics, and how to connect to the Internet.</u>

> Thesis statement (our underline).

I. Getting Started

<u>Before you actually begin to design your web page, you should know how the Internet works and some basic terminology.</u> First, the *World Wide Web* is a system of computers throughout the world that are linked together. They are able to communicate to each other through a transfer protocol called HTTP, which sends out a request to all computers containing the desired information. A typical HTTP request contains a web page address, such as http://www.technicalediting.com/htm. The computer (*server*) holding that specific address then opens to the requested web page.

> Notice that each paragraph begins with a topic sentence.

II. How the WWW Works

<u>Accessing a particular website is a wonder of recent technology.</u> To begin, all the information for a website is stored as files on computers commonly called servers. After your computer locates a particular website using the HTTP protocol, it reverts to a browser. Being able to read a web page requires the use of a web browser such as Netscape Navigator or Internet Explorer. Browsers make sense of the special markup codes used to design web pages. These codes are written in a language called HTML, which means hypertext markup language. This pamphlet introduces you to HTML coding in the next section.

> Topic sentences introduce the general idea of the paragraph.

III. How to Write HTML Codes

Writing HTML codes for a web page is not difficult if . . .

IV. Designing Your First Web Page

Now that you know basic HTML codes, it's time to practice . . .

V. Putting Your Web Page on the Internet

It's time to look at your web page on the internet by . . .

> The body of the document incorporates sequential steps.

Conclusion

FIGURE 14.8 (*continued*)   General-to-Specific Organization.

Hierarchical

Hierarchical is a sequential organizational pattern that presents material in either ascending or descending order of importance. Often, writers use this pattern to persuade readers to accept the writer's viewpoint. For example, a memo sent to the engineering department recommending that it design an anti-rollover device for SUVs might begin with the most persuasive argument: safety. Next, the memo might address another but less important issue: the cost of installation. Finally, the least important issue—a small but growing customer demand for this device—would complete the memo. Hierarchical patterns of organization meet the readers' need and expectation for a logical progression of ideas. Randomly organized ideas give the impression that the writer isn't credible or in control.

The hierarchical pattern of organization is an effective means of developing many types of reports, memos, feasibility studies, journal articles, and proposals. However, it is not enough to rank information from most important to least important or vice versa. The writer also needs to explain why one aspect is more important than another. Keep this in mind when you edit documents with a hierarchical pattern of organization. (See Figure 14.9.)

> *Edwards, Edwards, and Edwards*
> *Technical Editing Associates, Inc.*
> *1800 Madison Avenue*
> *New York, NY 12345*
>
> July 23, 2005
>
> To: Ms. Amanda Avery
>
> From: Frank Grammar,
>   Chief Technical Editor
>
> Re: Edit of *How to Design a Web Page*
>
> Dear Ms. Avery,
>
> Please find attached a markup copy of your instructional manual *How to Design a Web Page* sent to us for comprehensive editing as per our contract agreement. Although the manual as a whole provides a comprehensive and thoughtful approach to web design, I suggest three critical revisions.
>
> First and most importantly, I feel that the manual's organizational pattern needs revision. *How to Design a Web Page* follows a sequential, step-by-step approach to web design; however, I recommend swapping step 4, "Creating Titles," with step 5, "Creating Headings." Reordering these two steps will appear more logical to your readers and decrease the opportunity for confusion. Also, please reconsider your introduction. First, it seems a bit brief. Your readers, as I understand it, are new to web design and may need more general information on how the web operates as a function of the Internet. If the readers are able to conceptualize the general workings of the web, they'll grasp web design quicker.
>
> In addition to these organizational matters, I'd like to suggest several stylistic changes as well. First, consider changing the tone of the manual. As written, it's a bit too formal and academic. Consider writing from a more personal point of view (first person); your readers will find the material more engaging. Also, you use a considerable amount of technical jargon that I'm afraid beginning web designers may not understand. Perhaps a separate glossary or text box with explanations would be helpful. Second, consider using more transitions. I have marked them for you and believe they will provide your manual with more coherence.
>
> Finally, although relatively minor, I suggest you align three of your graphics, Fig. 10, Fig. 11, and Fig. 14, closer to the text that discusses them. These occur on pages three, five, and six and are marked on your attached copy. Also, I have made a few other suggestions on the copy for you to consider.
>
> We appreciate your business and hope to hear from you within the week. If you have any questions, please call. Or, if we need to meet, I'm available as well.
>
> Sincerely,
>
> Frank Lee Grammar

Annotations (left margin):
- The memo begins with brief introduction, including a thesis.
- The paragraphs are arranged in descending order of importance.
- The body paragraphs begin with topic sentences.

FIGURE 14.9
Hierarchical Organization.

Cause and Effect

Cause and effect is a categorical organizational pattern that demonstrates causal relationships. Technical writers frequently use cause and effect to argue that a particular result will occur if a certain action is instituted. For example, engineers might wonder: if they replaced metal ball bearings with nylon ball bearings in tire hubs (cause), what would be the impact on performance (effect)? If taxes are decreased (cause), what would be the overall impact on the economy (effect)? Not all reasoning moves from cause to effect, though; sometimes individuals must reason inferentially by first examining the effect and then trying to determine the cause. You find your goldfish floating belly up (effect), and you wonder what was the cause: old age, disease, dirty water, hungry cat? A manager may notice that employee morale is low (effect) and wants to determine the cause.

Cause-and-effect arguments are often integrated with other types of organizational patterns. For example, an environmental report might begin by discussing the detrimental effects of allowing a road to be built adjacent to a city's reservoir. The report may organize the effects in a hierarchical order from most to least important. Other patterns within an overall cause-and-effect organizational framework are also widely used.

The federal government employs many technical writers and editors and occasionally subcontracts work out to civilian agencies. At some point in your career, you are likely to be involved with documents that originate from a governmental agency or bureau. Figure 14.10 represents the type of governmental document you are likely to encounter. Notice how the document's organizational structure combines two different patterns: cause and effect and general-to-specific. Also notice that this document is one governmental agency's (NOAA's) reaction to the conclusions drawn by two other governmental agencies, the Army Corps of Engineers and the Government Accountability Office (GAO).

The issue here is whether the Army Corps of Engineers should build a system of jetties at Oregon Inlet on North Carolina's Outer Banks. From this NOAA document, it's clear that the project is controversial, and this agency questions some of the positive benefits that would derive from building these jetties. Read the document carefully for its organizational pattern and style. Do you believe you could improve on either?

Cause and effect is a common organizational pattern used by technical writers. Effective editing of such a document requires the editor first to ask, "Is this truly a cause-and-effect situation?" That is, are the effects the writer claims will result from a particular cause actually causally linked? Or do other considerations and variables exist that also contribute to the effects but were not considered by the writer? For example, a high school teacher attempts to start her car after school and determines that the battery cable has come loose. She saw some students around her car earlier that day, so she hastily concludes that they are the cause of the loose cable. Are there other possibilities the teacher has not considered? Is this faulty reasoning about cause and effect?

The lack of sufficient supporting evidence to establish a causal relationship is another weakness of many cause-and-effect documents. As editor, ask a second question regarding the *amount* and *type* of evidence the writer offers as justification for claiming the causal link. Does the writer provide facts, examples, authoritative statements, statistics, or other types of evidence for establishing the connection between cause and effect? Once you have determined

**RESOLUTION OF THE OCTOBER 16, 2001
NATIONAL OCEANIC AND ATMOSPHERIC ADMINISTRATION REFERRAL TO THE
COUNCIL ON ENVIRONMENTAL QUALITY
OF THE FINAL ENVIRONMENTAL IMPACT STATEMENT
FOR THE ARMY CORPS OF ENGINEERS' MANTEO (SHALLOWBAG) BAY PROJECT.**

The purpose of the project is to improve navigation at Oregon Inlet, located in the northern region of the barrier islands known as the Outer Banks. The FEIS evaluates the Corps' proposal to construct two jetties, deepen the navigation channel from the present authorized depth of 14 feet to a design depth of 20 feet, and conduct ongoing dredging to maintain the channel.

> The document provides a brief introduction.
>
> The second paragraph provides the thesis (our underline).

NOAA strongly supports the goal of assuring safe navigation for the commercial and recreational fishing vessels using Oregon Inlet. NOAA is concerned, however, that the proposed jetties would cause unacceptable environmental harm to larval fish and their habitat with corresponding negative impacts on commercial and recreational fishery resources. NOAA also questioned whether the economic analysis supported a determination that project benefits would outweigh project costs. NOAA believes that other dredging and navigation aid alternatives could achieve the project's goal in a more cost-effective and environmentally acceptable manner.

I. Background

The Outer Banks are dynamic barrier islands of shifting sands with a history of opening and closing inlets between the Atlantic Ocean and the estuary known as Pamlico Sound. The Oregon Inlet is subjected to the most severe wave climate on the East Coast of the United States. Since its opening during an 1846 hurricane, Oregon Inlet has migrated over 2 miles to the south in response to the wave climate's transportation of considerable quantities of sand along the beaches of Bodie Island to the north and Pea Island to the south.

> The background provides the reader with a chronology of Corps involvement.

The proposed project is based on the fact that Oregon Inlet has substantial vessel traffic because it is currently the only navigable inlet between Rudee Inlet, near Cape Henry, Virginia, 85 miles north, and Hatteras Inlet, North Carolina, 45 miles south. In 1950, Congress authorized the Corps to dredge and maintain a 14-foot deep by 400-foot wide channel through Oregon Inlet and 12-foot deep by 100-foot wide channels through Pamlico Sound to the towns of Manteo and Wanchese. The Corps has worked to maintain the channel at Oregon Inlet since 1962. Between 1962 and 1965, the Corps maintained the channel using hopper dredges, which remove dredged sand from the area. Between 1965 and 1982 the Corps maintained the channel primarily by using sidecast dredges, which redeposit sand adjacent to the dredged area. During the early 1990's, the Corps maintained the ocean bar channel using ocean certified pipeline dredges that deposited the dredged material on Pea Island between 1 and 2 miles south of Oregon Inlet.

II. Core Issues Evaluated During the Referral

A. **Economic Analysis**

> The evaluation of the effects of building jetties begins here with the first effect: economic cost-benefit analysis.

The GAO review of the cost-benefit analysis for this project found that it did not provide a reliable basis for deciding whether to proceed with the project. GAO at 38. The GAO highlighted three main concerns: 1) certain data was outdated and incomplete, 2) certain assumptions did not have adequate support, and 3) the analysis did not adequately examine the inherent risk and uncertainty in key variables that could significantly affect the project's benefits and costs. In its report, the GAO stated that they "did not assess the net effects of all the limitations we found with the economic analysis because obtaining the necessary data would take an inordinate amount of time and expense." *Id.* at 3. The Department of the Army responded

FIGURE 14.10
Cause-and-Effect Organization.
Source: NOAA News Releases 2003.

that the GAO conclusions appeared reasonable and acknowledged that economic reanalysis would be required before starting construction. *Id*. at 76. Issues identified include the following:

First, the analysis did not adequately address the implications of changes in commercial and recreational fishing use of Oregon Inlet. The authorized 20-foot channel was designed in 1970 to accommodate increased fishing by deeper draft commercial trawlers and continuous navigation access across Oregon Inlet's ocean bar. This projected fishery never materialized and the fishery currently is overcapitalized. NOAA and the Corps agree that under existing conditions available fish can be taken by vessels that transit the ocean bar. GDM 8-5, EIS 4-16. Thus, the fishery resources offshore of Oregon Inlet are fully exploited, in some cases over-exploited, by vessels navigating a channel that the Corps maintains at 14 feet between 15 and 23 percent of the time (depending on weather conditions and Corps funding). The existing value of the fishery is valued at $17,986,000 in both the no action alternative and the preferred alternative (GDM 7-17, 7-22), while the preferred alternative is estimated to provide $7,237,000 in additional benefits based largely on additional recreational boat trips and increased "fishing efficiency savings" of increased certainty in ocean access via the inlet. GDM 8-2 – 8-9.

> Notice the use of introductory topic sentences for each paragraph.

While the Corps no longer justified the project on the basis of additional fish landings, the EIS statement of purpose and need maintains that the project is necessary to allow large trawlers to navigate the inlet safely. The Corps projected approximately $2 million in annualized benefits based on cost savings from reducing the number of trips that would be forced to detour by conditions in the inlet. However, data on trawler use of Oregon Inlet dates from the mid-1980s, when 234 trawlers worked the inlet, instead of North Carolina Division of Fisheries data from 1999–2001 showing that this number decreased to only 97 trawlers. During that same period, smaller recreational fishing vessels increased nearly ten-fold while the total fish landings (trawlers and smaller vessels combined) remained relatively constant.

Thus, the use of mid-1980s data misapplied the cost savings associated with deep draft vessels to a fleet increasingly composed of smaller vessels that do not appear to be as affected by inlet conditions. Adjusting the analysis based on the more recent and reliable North Carolina data, and excluding fixed costs that should not have been included, the GAO found that commercial fishing benefits would be reduced by about 90 percent, from about $2 million to $194,000. This cost recalculation is important to the project's cost-benefit ratio. It also affects the balance between the need for the project and the potential environmental costs and risks entailed. The cost-benefit analysis also may overstate the costs (and effort) of the "no action" alternative and understate the costs (and risks) of the proposed action. Updating Corps expenditures for fiscal years 1997 through 2001, the GAO found that the net costs of the jetty project increased by over $1 million. The Corps also corroborated NOAA's concern that the Corps record of jetty projects indicates that the sand effects of the proposed project could vary by as much as 40 percent. This uncertainty is of greater importance than indicated in the EIS in light of the lack of any solid estimate of the project's costs and benefits, the history of failing to fully fund the current dredging program, and the importance of the sand by-pass plan to the project's mitigation of fisheries impacts.

B. **Potential Effect on Fishery Resources**

The potential effect of the jetties and sand management scheme upon the commercial and recreational fishery is a subject that is particularly important for weighing the costs and benefits of the project. The larval fish habitat at issue has been designated as Habitat Areas of Particular Concern by the regional Fishery Management Council because of their high value to the fisheries and vulnerability to degradation. The EIS concluded that "the overall impact of the recommended plan on larval [fish] will be minimal," EIS 6-17, but further noted that "potentially significant impacts by no means can be ruled out." EIS 6-16. These conclusions were not based on any field data on the effects of the Corps jetty design, data admittedly difficult to produce. NOAA maintains that even a 10 percent reduction in larval fish accessing the estuary would be significant, and believes that a reduction of up to 60 percent is "within reason" based on a Corps study using a scale model of Oregon Inlet. NOAA Ref. at 6. Based on NOAA data valuing

> The second effect of jetties: impact on larval fish.

FIGURE 14.10 (*continued*) Cause-and-Effect Organization.

Source: NOAA News Releases 2003.

commercial fish landings at $13.6 million per year, a 10 percent to 60 percent reduction in the fishery could cost the project's intended beneficiaries between $1.36 million and $8.6 million per year.

C. **Safety**

> The third effect: safety. Do you believe the writer should have arranged the importance of the effects in descending order?

Using data that is no longer available, the Corps projected that 14 lives would be lost under the no action alternative and that all would be saved under any of the action alternatives. Yet, the safety record of vessels using the inlet has improved significantly in the last 20 years. The GAO analysis of available accident information suggests that loss of life at the inlet in the past often was due to factors such as navigational errors, alcohol consumption, or the lack of life vests, which jetties would not influence. There is evidence in the record before CEQ of some incidents in the inlet, but questions remain as to whether jetties might create their own hazards, or might increase risky activity by giving less experienced boaters a false sense of safety, or whether the construction of hardened shorelines might make running aground more dangerous, particularly given the extreme weather and wave climate of the Outer Banks. The eastern seaboard has jetty projects in a number of locations and there are no data that demonstrate, one way or the other, whether the existence of jetties induces riskier behavior.

D. **Potential Effect on FWS and NPS Lands**

> The fourth effect: use of the National Park System and National Wildlife Refuge System.

The Secretary of the Interior also asked CEQ to consider the Department's concerns regarding the proposed project's effects on the Cape Hatteras National Seashore and the Pea Island National Wildlife Refuge. Interior has been concerned for some time about its ability to authorize the proposed use of National Park System and Refuge lands for this project, because to do so may conflict with its management authorities. This is a threshold issue for the feasibility of the Corps proposal that has never been resolved.

The National Wildlife Refuge System is a national network of lands and waters that is administered for the conservation, management, and where appropriate, restoration of the fish, wildlife, and plant resources and their habitats within the United States for the benefit of present and future generations of Americans. 16 U.S.C. § 668dd(a)(1). Under this statute, each refuge must be managed to fulfill the mission of the System, as well as the specific purposes for which each refuge was established. FWS is authorized to permit the use of any area within the Refuge System for any purpose whenever the agency determines that such uses are "compatible with the major purposes" for which the Refuge was established. In 1982, FWS determined that the construction of jetties is not compatible with the purposes for which the Pea Island National Wildlife Refuge was established. EIS App. A, DOI comments at 2. FWS concluded that the jetties would create conditions that will significantly alter the refuge environment, resulting in deterioration or elimination of wildlife habitat, and certain wildlife-related public use. *Id.* FWS and the Corps have coordinated on this issue for many years and the Corps has proposed several jetty design changes to respond to the compatibility concerns. However, FWS has not been able to conclude that the project is compatible with the major purposes of the Pea Island National Wildlife Refuge.

Successful resolution of these compatibility and impairment concerns would require significant additional efforts. Further expenditure of public resources to do so is not advisable in light of the project's potential impacts on fishery resources and questions about the outdated economic analysis.

> The conclusion summarizes the need for further analysis if this project is going to continue.

III. **Conclusion**

For the reasons stated above, unresolved issues remain concerning both the environmental impacts and economic benefits of the proposed project. Further analysis is unlikely to resolve these issues or yield a sufficient margin of benefit to justify the potential impacts of the project. Accordingly, after further consultation with the agencies, and based on currently available information the Corps has agreed to cease work on and further funding of the proposal for the construction of jetties at Oregon Inlet. The Corps will

FIGURE 14.10 (*continued*)   Cause-and-Effect Organization.
Source: NOAA News Releases 2003.

use the information developed through the NEPA process and work closely with NOAA to develop alternative approaches for improving navigation around the Outer Banks, including its Channel Widener Project. The Corps intends to conduct depth surveys of the inlet's navigational channel at more frequent intervals, especially during the peak commercial fishing season. The surveys will be conducted at least monthly and after major storm events. The Corps intends to make the data available to the public in the internationally accepted S57 format. To assist in this effort, NOAA is committed to enhancing safety and vessel operational efficiency by identifying and posting critical chart corrections in advance of Coast Guard weekly notices to mariners and to making these data available in association with NOAA's Electronic Navigational Charts.

FIGURE 14.10 (*continued*)   Cause-and-Effect Organization.
Source: NOAA News Releases 2003.

that the writer intends to demonstrate a cause-and-effect correlation, judge the worthiness of the writer's reasoning and supporting evidence.

Comparison and Contrast

Comparison and contrast is an organizational pattern that illustrates similarities and differences. It is especially useful in technical writing to compare products and processes in feasibility studies and proposals. The writer's principal objective is either to *present* or to *argue for* a particular option. For example, when determining which new computer to purchase, you might want to compare and contrast the performance of an AMD microchip to an Intel microchip. Some technical documents would simply present the data about each chip without drawing conclusions; others would argue that one is better than the other. Regardless, both documents must first establish the criteria used to make the comparison, such as speed, reliability, cost, overall performance, resistance to heat, etc.

Comparison-and-contrast documents can be organized by one of two methods: whole-to-whole or part-to-part (see Figures 14.11 and 14.12). For example, a technical document using the whole-to-whole method would completely discuss the AMD chip first and then proceed to compare and contrast the Intel chip afterwards. However, editors should note that when the writer begins discussing the second option (the Intel microchip in this case), it's important to refer back frequently to the first option (the AMD microchip) and remind the reader what is being compared. This consideration is not necessary using the second method, part-to-part, because each aspect is compared on a point-by-point basis.

Comparison and contrast is one of the most useful organizational patterns for technical documents. By distinguishing similarities and differences, readers can more readily evaluate the advantages or disadvantages of what is being compared. Rather than asking readers to form their own conclusions, writers may argue for the adoption of a particular plan of action or for instituting a particular process by comparing and contrasting them. For example, a feasibility study may compare two types of computers by describing the advantages and disadvantages of each type without expressing an opinion as to the better choice. However, many feasibility studies express a preference for the adoption of one computer over the other. The resulting comparison, then, becomes the "evidence" that supports that preference.

Undoubtedly you have dealt with competing ideas based on an educated comparison and contrast of their merits. For example, you may have decided to buy a Honda instead of a Kia because you compared their safety and reliability records. Or you may have chosen a particular college curriculum, professor, or class through the same process of weighing the pros and cons.

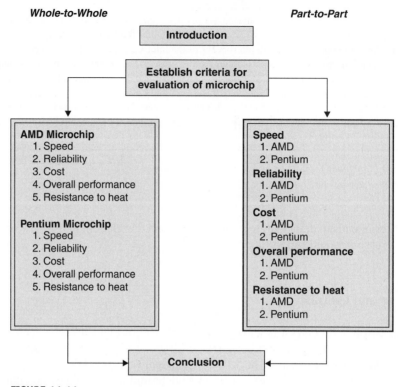

FIGURE 14.11
Comparison-and-Contrast Organization.

Parents: Are You Treating Your Children Right?

The introduction provides some background and identifies the intended audience: parents.	There is a good reason it's called the *common* cold. Research indicates that on average, a child can expect to have eight to ten colds before reaching the age of two. Although the frequency of colds declines as children grow older, it remains the number one reason why children miss school and parents miss work. Colds are the most infectious disease known to medical science.
	Unfortunately, children often are improperly treated for what parents assume is "just another cold." Their child might complain that their nose is stuffy, their ears hurt, and they may have nasal secretions and a little fever. This sounds like another cold to be treated with one of hundreds of over-the counter medications. However, what many parents don't realize is they may be treating their child for a cold when in fact the child may be suffering from allergies or sinusitis, which must be treated differently. Parents need to acquaint themselves with the underlying causes of each type of nasal infection, and learn how to treat and prevent them. Too often parents and doctors treat common colds with antibiotics, which do not help, but cause the child's body to build up a resistance to antibiotics. Also, many parents and doctors treat children for low-grade fever, which actually is the body's way of fighting viruses by keeping them from reproducing. Being able to distinguish between a cold, sinusitis, and allergies and learning to properly treat each type of infection will reduce many trips to the doctor's office and unnecessary time a child is ill.
Notice that the thesis states the purpose and establishes the point-by-point comparison.	
The main topics are delineated with easy-to-follow headings in parallel form.	**What is it?**
	Colds: Colds are created by a wide constellation of viruses. Sneezing, coughing, vomiting, and diarrhea are all means our bodies use to attempt to eject these viruses. Fevers that accompany colds indicate our bodies are attempting to fight off an infection.

FIGURE 14.12
Comparison and Contrast (Part-to-Part).

Sinusitis: The sinus cavities are located just below our eyes and on either side of our nose. When the mucous membranes lining the cavity become swollen and fill with fluid, the result is sinusitis. Unlike the common cold, though, sinusitis is the result of an acute, subacute, or chronic bacterial infection, not a viral infection. Each type of bacterial infection requires different treatment.

Allergies: When a child's cold symptoms just won't go away, does the child suffer from allergies or just successive colds? This is a common problem faced by many parents. The lining of our noses are filled with millions of mast cells whose job is to prevent harmful particles from entering our bodies. Many of us have hypersensitive mast cells that react to pollen, dust, and pet dander. When the mast cells detect a foreign particle such as pollen, they release histamine and many other chemicals that cause our noses to sneeze, itch, and swell.

> Point-by-point, systematic comparisons and contrasts run throughout the document, meeting the reader's expectations for a logically sequenced presentation.

Who gets it?

Colds: Colds are the most common childhood infection, and children usually acquire them from other children. Cold viruses are easily transmitted in daycare facilities when a child carrying a new virus strain spreads the virus by sneezing. Colds can occur year-round, especially during the winter or the rainy season in tropic climates. Moms are a bit more likely to contract a cold than dads, usually one more per year on average.

Sinusitis: Anyone can contract a sinus infection, particularly those exposed to secondhand smoke from cigarettes. Usually a sinus infection begins as a cold or allergy. Children, especially boys, who have ear infections, cystic fibrosis, problems with their immune systems, and deviated septa are more prone to sinusitis. Swimming and breathing cold air have also been known to cause sinus problems.

Allergies: Two components are responsible for allergies—genetics and the environment. Allergies are family related. For unknown reasons, firstborn children whose parents have nasal allergies are more susceptible than younger siblings. Allergic rhinitis is the immune system's overresponse to environmental conditions. Children raised on farms have significantly reduced incidences of allergies; apparently their immune systems learn to respond to the widespread stimuli that trigger allergic reactions. Although allergies are common among children aged two to seven, younger children have been known to contract them as well.

Is it contagious?

Colds: Most people contract colds from others who sneeze or blow their nose, but not likely from coughing as many people believe. Live viruses within the nasal passage are expelled during a sneeze and easily infect anyone within close proximity. Also, wiping around your nose or eyes after touching an object someone has contaminated with nasal secretions often leads to contracting a cold.

Sinusitis: Generally, sinus infections are not contagious. However, the colds that lead to sinusitis are easily contracted by other individuals.

Allergies: Most research indicates that allergies are not contagious, with one exception—hives. Some people who have received blood transfusions have contracted hives as a result.

How is it diagnosed?

Colds: The common cold is identified by its symptoms and time of duration. Generally, colds are a short-term illness with little or no fever, sneezing, and some nasal emission.

Sinusitis: The only sure method of diagnosis for sinusitis is a medical exam, which is often accompanied by an X-ray. Also, a medical history of sinus infections is a good indicator of recurring sinusitis.

Allergies: Like sinusitis, allergies are best diagnosed through a physical examination and medical history of the individual and the biological parents. Nasal secretions must be sent to a laboratory to confirm allergies.

How is it treated?

Colds: Colds usually last approximately one week. Antibiotics should not be used unless the cold persists for more than two weeks, which suggests sinusitis. Many over-the-counter cold "remedies" are effective in treating the symptoms; however, some new antiviral medications have been shown to be

FIGURE 14.12 (*continued*) Comparison and Contrast (Part-to-Part).

| | effective. Grandmother's chicken soup has helped many for over 2,000 years because the combination of heat, hydration, and salt actually seems to fight infection.
  Sinusitis: Two forms of sinus infections, acute and subacute, are bacterial in nature and take two to three weeks to run their course. Antibiotics may be prescribed but must be continued even after the symptoms abate. Saline nose drops have been found to help heal the mucous lining of the nose; however, decongestants and antihistamines often retard the healing process. Sometimes sinusitis is misdiagnosed. Various types of fungus as well as allergies also display the same symptoms. Chronic sinusitis may call for X-rays and lab tests. |
|---|---|
| Final comparison and conclusion. | |
| | **Conclusion:** Parents should be aware of the similarities and differences between these three illnesses and treat their children accordingly. The most important lesson, though, is not to use antibiotics for common colds caused by viral infections. Increasingly adults are showing resistance to antibiotics because they may have been overmedicated as children. So learn the symptoms and the treatments. It can save parents and children unnecessary trips to the doctor and time away from work and play. |

FIGURE 14.12 (*continued*)   Comparison and Contrast (Part-to-Part).

Feasibility studies, proposals, memos, and brochures are the most common types of writing that employ comparison and contrast. Think about how many times you have seen or read advertisements that compare one product to another. Often, technical writers incorporate both whole-to-whole and part-to-part patterns in a document. First, as an introduction, they provide a general overview in the form of a whole-to-whole pattern. Then in the body they may provide specifics through a comparative analysis of individual aspects.

As an editor, you will encounter the comparison-and-contrast organizational pattern frequently. You will discover that many writers incorporate other organizational patterns we have discussed into the comparison-and-contrast structure. For example, the specific items compared may be arranged from most important to least important, or they may be arranged according to classification. You will quickly get used to variations. However, as an editor, you will need to ask if the organizational framework is suited to the needs of the particular audience and if the writer makes the comparisons clear.

Problem-to-Solution

Problem-to-solution is a type of categorical organizational pattern that defines a current problem and then offers a solution to that problem. Technical writers use this three-part pattern to illustrate the resolution for a perceived difficulty. Most often, the document, which may take the form of a proposal, journal article, memo, report, or even a feasibility study, begins with an overview of "what is not working" (see Figure 14.13). After describing the problem, the document expands upon the method(s) that were taken to analyze, rectify, or improve on it. Finally, the third section, the solution, calls for a course of action based on the findings.

Editing a problem-to-solution document will require you to consider carefully several aspects. First, you need to make certain that the problem section is clear to all readers. For example, if engineers who are intimately involved in the project wrote the report, they may make invalid assumptions about how much their audience already knows. So first ask if the writing is *audience-appropriate*. Second, determine if the methods section clearly explains the analysis or the process that was undertaken to resolve the problem. Are all the steps explained, or do gaps in logic exist? Finally, ask if the solution is reasonable or promises too much. You may need a subject matter expert (SME) to help you determine this. If the engineers have claimed their solution will make a product 7% more efficient within three months, you may need to question that claim.

Project Overview

ODOT and the Port of Portland have begun a project to improve the flow of Airport Way traffic accessing I-205 north at the Airport Way interchange.

The Airport Way interchange provides a critical connection for travelers and the movement of regional goods. Future congestion at this interchange was flagged as an issue in the late 1990s. That future became reality in 2004, when both east- and westbound traffic on Airport Way started experiencing delays that today can stretch to 45-minutes. During the evening rush hour, there is more northbound traffic getting on I-205 at Airport Way than at any other I-205 interchange in Oregon. Costly slowdowns are projected to increase if nothing is done, with negative consequences for the regional economy.

> The document begins with an overview of the problem.

ODOT and the Port made a commitment to the Federal Aviation Administration to have a solution to the interchange congestion problem in place by the end of 2014. They are beginning the project with a study that will identify appropriate solutions to the congestion on Airport Way and the associated problems with the northbound turning movement, while preserving options for future improvements on I-205.

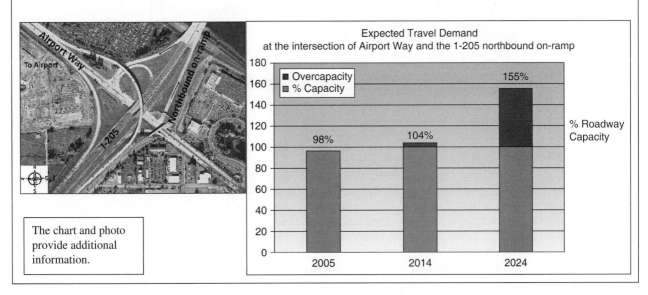

> The chart and photo provide additional information.

FIGURE 14.13

Problem-to-Solution Organization.

Source: State of Oregon.

Solving the Problem

ODOT and the Port are committed to developing a solution to the Airport Way interchange congestion problem. The problem solving begins by collaborating with community stakeholders and affected agencies to examine the extent of the interchange problem, identify key constraints and alternative solutions, and select which alternatives merit further consideration. Project partners will study the selected alternatives in a process required by the National Environmental Policy Act (generally an environmental impact statement—EIS, or environmental assessment—EA) and then select a locally preferred alternative.

The following committees are providing direction for the project:

- **Public Involvement Team**
  The Public Involvement Team is comprised of staff from ODOT and the Port, and members of the consultant team. This team is responsible for planning and implementing public involvement and outreach with people who may be impacted by the problem and proposed solutions. This team coordinates online surveys, open houses, project newsletters and website, and stakeholder interviews, briefings and advisory committee meetings.
- **Project Development Team**
  The Project Development Team is responsible for coordinating the project and finding consensus amongst partners on critical issues. The members of the Project Development Team represent the Port, ODOT, the City of Portland, Metro, Washington Department of Transportation (WSDOT), Federal Highway Administration (FHWA), and project staff.
- **Stakeholder Advisory Committee**
  The role of the Stakeholder Advisory Committee (SAC) is to advise the Project Development Team and Policy Group to ensure that the interests, issues, knowledge, and recommendations of the local community are considered in project decisions. Committee participants are also tasked with sharing project information with their various constituencies. Look here for a list of the committee members and their affiliations, proposed meeting schedule and meeting notes.

> An overview of the solution is followed by naming those who will solve it.

Airport Way Interchange Project Schedule

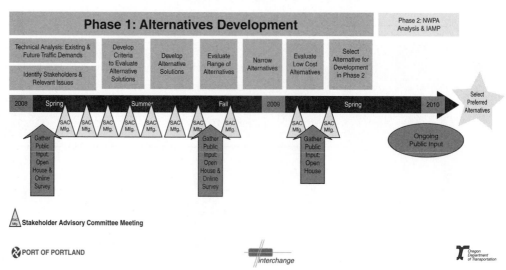

Steps in Developing Alternatives:

Step 1: Technical Analysis: Existing and Future Traffic Demands
The project team studied the current extent of the interchange congestion problem and will forecast what traffic will be like in 2028 in order to develop solutions that account for future regional growth.

FIGURE 14.13 (*continued*)   Problem-to-Solution Organization.

Step 2: Identify Stakeholders and Relevant Issues
This step, also known as "scoping," involved the project team conducting an Issues Workshop with the Stakeholder Advisory Committee, interviewing more than 30 potentially impacted individuals, hosting a public open house, and conducting an online survey. The project team used this information to develop a baseline understanding of the issues to be considered in the project analysis and to identify potentially affected individuals and groups whose opinions should be taken into account.

A step-by-step approach to the problem and how it was handled.

Step 3: Develop Criteria to Evaluate Alternative Solutions
The project team developed evaluation criteria to evaluate and compare alternative solutions. The criteria are based on the project Goals and Objectives—desirable project outcomes that are derived from issues identified by the Stakeholder Advisory Committee and members of the public via interviews, an open house, and an online survey.

Step 4: Develop Alternative Solutions
For this step, the project team held a three-day Value Planning Workshop in July 2008 with representatives from ODOT, the Port of Portland, WSDOT, Metro, FHWA, City of Portland, transportation engineers and planners, and the Stakeholder Advisory Committee. During the workshop, participants brainstormed 65 solutions to improve the flow of Airport Way traffic accessing I-205 north at the Airport Way interchange. After evaluating the advantages and disadvantages of each solution, workshop participants identified 22 solutions, including three non-engineered actions (Traffic Demand Management, improvement to public transportation and land use management) to advance to the next evaluation process (Step 5).

Step 5: Evaluate Range of Solutions
August 2008, the Stakeholder Advisory Committee and Project Development Team began a qualitative evaluation of the 19 engineered alternatives based on the criteria that reflect the project Goals and Objectives (Step 3). Most of the 19 alternatives are not stand-alone solutions, and in many cases, compliment one another in addressing the project's three identified key constraints. As a result, proposed solutions have been combined into ten (10) "alternative packages," each of which may be combined with any of the three non-engineering solutions. Of the ten alternative packages, five (1, 2, 3, 4, and 6) appear to perform the best based on evaluation criteria. Members of the public were invited to review and comment on the 10 alternative packages and evaluation process at open houses and online surveys.

Step 6: Narrow Alternatives
After considering public comment from the online survey, open house and stakeholder briefings, and evaluation results, the Stakeholder Advisory Committee and Project Development Team will select up to five alternative packages that most adequately address the project's core purpose and need. The project team will conduct traffic modeling, visual simulations, and quantitative evaluations for these selected alternatives.

Step 7: Select Alternatives for Development in Phase 2
The Project Management Team and Project Development Team will consider comments from the Stakeholder Advisory Committee and members of the public before selecting an alternative to advance to Phase 2. In this next phase of the project, the Project Development Team will thoroughly study the likely effects that each alternative will have on the natural and built environment. ODOT will publish the results of this analysis and encourage public input to thoroughly evaluate this alternative.

Links to additional information regarding this project.

Background and Context

Look here for project history, Frequently Asked Questions, and links to additional resources.

FIGURE 14.13 (*continued*) Problem-to-Solution Organization.

Classification

Classification is an organizational pattern that partitions information into categories. For example, types of editing might be partitioned into three major categories: correctness, readability, and effectiveness. Moreover, each of these major categories can be partitioned into additional categories that define their function in more detail. Categorizing items helps us make sense of our world. When you walk into a large store, items are categorized by function: housewares, electronics, sports, garden supplies, clothes, and so on; clothes are further categorized as men's, women's, children's, petite, plus, and so on.

Remember, readers have expectations that information will be presented logically, especially if it is arranged in categories. You should not randomly mix organizational types; it confuses the reader. For example, returning to the store analogy, you know from experience that your idea of where something belongs may not be the same as the store manager's. Why, you may ask, is the pet food located between the pharmacy and the garden supplies? Shouldn't it be categorized as a type of food and be located near the food section? Why are the light bulbs located in housewares instead of home repairs? When you replace a light bulb in your home, aren't you repairing? Effective editing requires the same considerations to avoid mixing categories and to follow the reader's expectations.

NOAA's Geostationary and Polar-Orbiting Weather Satellites

After a brief introduction, the article classifies the two types of weather satellites.

Operating the country's system of environmental (weather) satellites is one of the major responsibilities of the National Oceanic and Atmospheric Administration's (NOAA's) National Environmental Satellite, Data, and Information Service (NESDIS). NESDIS operates the satellites and manages the processing and distribution of the millions of bits of data and images theses satellites produce daily. The primary customer is NOAA's National Weather Service, which uses satellite data to create forecasts for the public, television, radio, and weather advisory services. Satellite information is also shared with various Federal agencies, such as the Departments of Agriculture, Interior, Defense, and Transportation; with other countries, such as Japan, India, and Russia, and members of the European Space Agency (ESA) and the United Kingdom Meteorological Office; and with the private sector.

NOAA's . . . operational weather satellite system is composed of two types of satellites: geostationary operational environmental satellites (GOES) for short-range warning and "now-casting" and polar-orbiting satellites for longer-term forecasting. Both types of satellite are necessary for providing a complete global weather monitoring system.

A new series of GOES and polar-orbiting satellites has been developed for NOAA by the National Aeronautics and Space Administration (NASA). The new GOES-I through M series provide higher spatial and temporal resolution images and full-time operational soundings (vertical temperature and moisture profiles of the atmosphere). The newest polar-orbiting meteorological satellites (that began with NOAA-K in 1998) provide improved atmospheric temperature and moisture data in all weather situations. This new technology will help provide the National Weather Service the most advanced weather forecast system in the world.

Geostationary Operational Environmental Satellites (GOES)

The first type of satellite is classified as to its purpose and types of operations.

GOES satellites provide the kind of continuous monitoring necessary for intensive data analysis. They circle the Earth in a geosynchronous orbit, which means they orbit the equatorial plane of the Earth at a speed matching the Earth's rotation. This allows them to hover continuously over one position on the surface. The geosynchronous plane is about 35,800 km (22,300 miles) above the Earth, high enough to allow the satellites a full-disc view of the Earth. Because they stay above a fixed spot on the surface, they

FIGURE 14.14
Classification Organization.

provide a constant vigil for the atmospheric "triggers" for severe weather conditions such as tornadoes, flash floods, hail storms, and hurricanes. When these conditions develop the GOES satellites are able to monitor storm development and track their movements.

GOES satellite imagery is also used to estimate rainfall during the thunderstorms and hurricanes for flash flood warnings, as well as estimates snowfall accumulations and overall extent of snow cover. Such data help meteorologists issue winter storm warnings and spring snow melt advisories. Satellite sensors also detect ice fields and map the movements of sea and lake ice.

NASA launched the first GOES for NOAA in 1975 and followed it with another in 1977. Currently, the United States is operating GOES-10 and GOES-12. (GOES-9, which is partially operational, is being provided to the Japanese Meteorological Agency to replace their failing geostationary satellite.) GOES-11 is being stored in orbit as a replacement for GOES-12 or GOES-10 in the event of failure.

GOES-10 and GOES-12

The United States normally operates two meteorological satellites in geostationary orbit over the equator. Each satellite views almost a third of the Earth's surface: one monitors North and South America and most of the Atlantic Ocean, the other North America and the Pacific Ocean basin. GOES-12 (or GOES-East) is positioned at 75 W longitude and the equator, while GOES-10 (or GOES-West) is positioned at 135 W longitude and the equator. The two operate together to produce a full-face picture of the Earth, day and night. Coverage extends approximately from 20 W longitude to 165 E longitude. This figure shows the coverage provided by each satellite.

The main mission is carried out by the primary instruments, the <u>Imager</u> and the <u>Sounder</u>. The imager is a multichannel instrument that senses radiant energy and reflected solar energy from the Earth's surface and atmosphere. The Sounder provides data to determine the vertical temperature and moisture profile of the atmosphere, surface and cloud top temperatures, and ozone distribution.

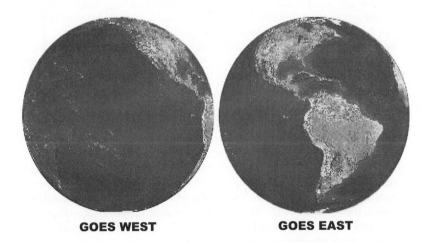

GOES WEST   **GOES EAST**

Other instruments on board the spacecraft are a Search and Rescue transponder, a data collection and relay system for ground-based data platforms, and a space environment monitor. The latter consists of a magnetometer, an X-ray sensor, a high energy proton and alpha detector, and an energetic particles sensor. All are used for monitoring the near-Earth space environment or solar "weather." GOES-12, the newest satellite also carries a Solar X-Ray Imager (SXI).

GOES-10 Characteristics

Main body:	2.0m (6.6 ft) by 2.1m (6.9 ft) by 2.3m (7.5 ft)
Solar array:	4.8m (15.8 ft) by 2.7m (8.9 feet)

FIGURE 14.14 (*continued*)   Classification Organization.

Weight at liftoff:	2105 kg (4641 pounds)
Launch vehicle:	Atlas I
Launch date:	April 25, 1997 Cape Canaveral Air Station, FL
Orbital information:	Type: Geosynchronous Altitude: 35,786 km (22,236 statute miles) Period: 1,436 minutes Inclination: 0.41 degrees
Sensors:	Imager Sounder Space Environment Monitor (SEM) Data Collection System (DCS) Search and Rescue (SAR) Transponder

> A table provides data in a straightforward manner.

The United States reaps many benefits from the new series of GOES satellites as they aid forecasters in providing better advanced warnings of thunderstorms, flash floods, hurricanes, and other severe weather. The GOES-I series provide meteorologists and hydrologists with detailed weather measurements, more frequent imagery, and new types of atmospheric soundings. The data gathered by the GOES satellites, combined with that from new Doppler radars and sophisticated communications systems make for improved forecasts and weather warnings that save lives, protect property, and benefit agricultural and a variety of commercial interests.

For users who establish their own direct readout receiving station, the GOES satellites transmit low resolution imagery in the WEFAX service. WEFAX can be received with an inexpensive receiver. Highest resolution Imager and Sounder data is found in the GVAR primary data user service which requires more complex receiving equipment. More information about establishing receiving stations can be obtained from the Email contact at the bottom of the page.

For more detailed information about the GOES satellites, see the GOES I-M DataBook, Revision 1, published 4 January 1997 by Space Systems-Loral. The most recent pictures received directly from the NOAA GOES satellites can be found at the NOAA GOES Server.

Polar-Orbiting Satellites (POS)

Complementing the geostationary satellites are two polar-orbiting satellites known as Advanced Television Infrared Observation Satellite (TIROS-N or ATN), constantly circling the Earth in an almost north-south orbit, passing close to both poles. The orbits are circular, with an altitude between 830 (morning orbit) and 870 (afternoon orbit) km, and are sun synchronous. One satellite crosses the equator at 7:30 a.m. local time, the other at 1:40 p.m. local time. The circular orbit permits uniform data acquisition by the satellite and efficient control of the satellite by the NOAA Command and Data Acquisition (CDA) stations located near Fairbanks, Alaska and Wallops Island, Virginia. Operating as pair, these satellites ensure that data for any region of the Earth are no more than six hours old.

> The second classification mirrors the first.

A suite of instruments is able to measure many parameters of the Earth's atmosphere, its surface, cloud cover, incoming solar protons, positive ions, electron-flux density, and the energy spectrum at the satellite altitude. As a part of the mission, the satellites can receive, process, and retransmit data from Search and Rescue beacon transmitters, and automatic data collection platforms on land, ocean buoys, or aboard free-floating balloons. The primary instrument aboard the satellite is the Advanced Very High Resolution Radiometer or AVHRR.

Data from all the satellite sensors is transmitted to the ground via a broadcast called the High Resolution Picture Transmission (HRPT). A second data transmission consists of only image data from two of the AVHRR channels, called Automatic Picture Transmission (APT). For users who want to establish their own direct readout receiving station, low resolution imagery data in the APT service can be received with inexpensive equipment, while the highest resolution data transmitted in the HRPT service utilizes

FIGURE 14.14 (*continued*)   Classification Organization.

a more complex receiver. Additional information about establishing receiving station can be obtained from the Email contact below.

NOAA-15 Characteristics

Main body:	4.2m (13.75 ft) long, 1.88m (6.2 ft) diameter
Solar array:	2.73m (8.96 ft) by 6.14m (20.16 ft)
Weight at liftoff:	2231.7 kg (4920 pounds) including 756.7 kg of expendable fuel
Launch vehicle:	Lockheed Martin Titan II
Launch date:	May 13, 1998 Vandenberg Air Force Base, CA
Orbital information:	Type: Sun synchronous Altitude: 833 km Period: 101.2 minutes Inclination: 98.70 degrees
Sensors:	Advanced Very High Resolution Radiometer (AVHRR/3) Advanced Microwave Sounding Unit-A (AMSU-A) Advanced Microwave Sounding Unit-B (AMSU-B) High Resolution Infrared Radiation Sounder (HIRS/3) Space Environment Monitor (SEM/2) Search and Rescue (SAR) Repeater and Processor Data Collection System (DCS/2)

The polar orbiters are able to monitor the entire Earth, tracking atmospheric variables and providing atmospheric data and cloud images. They track weather conditions that eventually affect the weather and climate of the United States. The satellites provide visible and infrared radiometer data that are used for imaging purposes, radiation measurements, and temperature profiles. The polar orbiters' ultraviolet sensors also provide ozone levels in the atmosphere and are able to detect the "ozone hole" over Antarctica during mid-September to mid-November. These satellites send more than 16,000 global measurements daily via NOAA's CDA station to NOAA computers, adding valuable information for forecasting models, especially for remote ocean areas, where conventional data are lacking.

Currently, NOAA is operating five polar orbiters. A new series of polar orbiters, with improved sensors, began with the launch of NOAA-15 in May 1998 and NOAA-16 on September 21, 2000. The newest, NOAA-17, was launched on June 24, 2002. NOAA-12, NOAA-14, and NOAA-15 all continue transmitting data as stand-by satellites. NOAA-16 and NOAA-17 are classified as the "operational" satellites.

FIGURE 14.14 (*continued*)   Classification Organization.

Organizing information by classification of categories is particularly important to scientists, engineers, and technicians. For example, it would be impossible to understand the biological sciences without classifying plants and animals into kingdom, genus, species, etc. Categories can also be subdivided with partitions or subcategories. A brochure for a particular national forest might first categorize the types of plants found within the forest. Under the broad category of *trees*, the brochure would then subdivide the trees into the various types. Finally, each type would be described in some detail.

Classification is useful for almost any type of technical document, from police reports (e.g., classification of perpetrators by gender, ethnic group, age, etc.) to feasibility studies (e.g., types of suitable building sites). When analyzing a document that has been organized by classification, you should make certain that the classifications are consistent, inclusive, and follow logically. See Figure 14.14 for an example.

ORGANIZATIONAL MODELS

You can expect some of your editorial duties to include editing information that, from an organizational standpoint, is prepackaged and formulaic. That is, the technical information itself is arranged according to the audience's preconceived organizational expectations. These organizational models usually take the form of an *informal* report, or larger, more detailed *formal* reports, or *proposals* and *feasibility studies*. Technical writers as well as editors should be familiar with the accepted standards for these organizational models, which can be found within the various style guides as well as many technical writing texts. Although it is not our intention here to provide a detailed description of all the organizational models, we do present a brief overview of the three primary ones.

Informal Reports

Informal reports usually follow one of two organizational formats: *memos* or *letters*. Generally, they are distinguished from longer, formal reports by their depth of coverage. Memos are most often circulated within an organization, whereas business letters are usually sent outside. Although the intended audiences may be different, both formats typically address the same issues involving training recommendations, problem analyses, progress reports, equipment testing, and company policies.

Informal reports, whether designed to inform, persuade, recommend, or analyze, have a general organizational structure: (1) an *introductory* element that states the purpose and problem to be discussed; (2) a *body* that provides necessary background, description, data, and possible consequences of the problem; and (3) a *concluding* element that restates the purpose of the report and suggests a plan of action. Audiences find headings that separate these elements useful. See Figure 14.15.

Letters are also used as a type of informal report. Similar to the memo, letter reports should include headings pertinent to the issue being discussed while maintaining the same overall organizational structure: introductory summary, body, and conclusion. A letter report includes more specifics than a simple business letter, but it will not address the degree of complexity or detail of a formal report.

Whether you are editing a memo or a letter report, keep the organizational aspects in mind. Judge whether the writer has provided enough introductory background, details, and reasons for change. Determine if the headings are parallel in form and if they are relevant to the audience. Consider if the headings should ask a question rather than make a statement. Finally, decide whether the writer has sufficiently developed the headings with appropriate and sufficient details, examples, reasons, or data.

Formal Reports

Formal reports, as the name suggests, concern more complex projects than informal reports, and as you would expect, are much longer. Commonly, the audience for a formal report is more diverse than its informal counterpart. Formal reports may be intended for people

MEMORANDUM

TO: Mr. Colin Smith
FROM: Frank Glover
DATE: June 24th, 2006
SUBJECT: Computer Equipment

Introductory Summary

Last week we upgraded all the computers in the Research Division with the latest version of SPSS. The upgrade was long overdue. However, Research's current computer hardware is not adequate to run the SPSS upgrade seamlessly and we need to consider several alternatives.

Problems with the Current Computer Equipment

Twelve individuals in our Research Division currently share six computers that were purchased over five years ago. Only two of these computers have more than one gigabyte of hard drive storage and none has more than 64 megs of memory. Most importantly, four of these computers continue to operate using a beta version of Windows 98.

As you know, our company has recently signed a contract with the Navy to test their G-4 guidance systems for the F-68 drones. Mr. Smith, our present computers simply cannot handle the type and amount of statistical data this contract will require. Several individuals in Research worry that their present computers may crash while trying to perform the statistical analysis on the G-4.

Recommendations

I believe we have two alternatives. First, we can upgrade our current computers by installing more memory, adding a greater hard drive capacity, and purchasing the latest Windows operating system. I have asked William Fields, our tech specialist, to conduct an equipment analysis, and he has determined that upgrades to our current computers would run close to $2,250. Mr. Fields believes the upgrades will initially do the job but will render the computers obsolete in two to three years.

The second alternative is to replace the existing computers with new ones. After checking prices and comparable systems, Mr. Fields recommends we replace all six computers with six Bell 2400s which, with our discount, will cost $8,426.

Conclusions

Given the two alternatives, I recommend we replace the older computers with Bell 2400s because they are:

- considerably faster than our current computers
- less likely to need service and cause delays
- prepackaged with the necessary Windows operating system
- capable of being networked throughout the company
- designed to be easily upgradable

In addition, Bell Komputers offers a payment plan that would allow us to pay for the computers over the next three years without interest.

Summary

After weighing the two options, I recommend we replace the older computers with the Bell 2400s. Although the initial cost will be considerably higher, the savings in time and efficiency will certainly pay for itself. If you have additional questions, please feel free to contact me.

FIGURE 14.15
Sample Recommendation Memo.

inside or outside the organization, technical or nontechnical individuals, stakeholders, or just interested parties. Formal reports usually provide a detailed analysis of a problem or provide a recommendation by frequently incorporating supporting statistics, charts, graphics, and data to supplement the discussion section. Following is a description of the various organizational sections for a formal report. We have provided a few guidelines; however, before you begin editing any formal report, be certain you have read your company's style manual. Each of these nine elements requires its own format.

The Nine Elements of a Formal Report

1. *Cover page:* includes the title of the project, the client's name, the name of your organization, and the date.
2. *Transmittal letter:* a brief, nontechnical letter or memo that introduces the report and invites the audience to read further.
3. *Table of contents:* an outline that provides the audience with the structure of the report, how it is sectioned, what is emphasized, and where the information can be located.
4. *List of illustrations:* a sequential listing of the illustrations in the report.
5. *Executive summary:* a short, usually single-page overview of the entire report for the decision-makers to consult before reading the specifics.
6. *Introduction:* an abstract of the report providing its objectives, its purpose, and a description of the project or problem.
7. *Discussion:* the longest portion of a formal report that, on a technical basis, is partitioned into sections and subsections that illustrate the collected data, demonstrate how the data were verified, draw conclusions from the data, and recommend a plan of action based on the conclusions.
8. *Conclusion:* a comprehensive and exhaustive assessment of all conclusions and recommendations.
9. *Appendix:* a section after the conclusion where the reader can locate supporting information, charts, and other forms of data from the discussion sections that would otherwise interrupt the flow of the report.

Proposals and Feasibility Studies

Similar to reports, proposals can be either be formal or informal, depending on their scope. As a general rule, informal proposals are five or fewer pages and cover less depth than a formal proposal. Regardless of its length, a proposal attempts to persuade an audience, either in-house colleagues or external clients, that a particular course of action should be taken. It may take the form of adopting a particular concept, initiating a new method for achieving a specific outcome, devising a new product, creating a new service, receiving a grant, and so on.

A feasibility study, on the other hand, illustrates the practical application of a proposal. It addresses specifics such as the costs, methods for implementation, effectiveness, expected results, and possible alternatives for the recommended changes initiated by a proposal. Like a proposal, a feasibility study can be directed toward an in-house audience or to clients outside the organization.

Although the content will differ, proposals and feasibility studies follow the same general organizational format of formal and informal reports: summary introduction, body, and conclusion. Also, they include most of the same nine elements listed for the formal report. Remember, though, whereas reports are primarily informational (although they may contain recommendations), proposals and feasibility studies argue for some type of major change.

As a technical editor, you can expect to make significant contributions to these types of standard documents.

COMPUTER DOCUMENTATION

The computer industry has developed its own models for typical documents written for internal and external readers. Technical editors need to be familiar with these document models as well as some of the terminology specific to the industry. For example, the industry typically refers to product-related documents as *documentation*.

Documentation Sets

Typical *internal* documentation includes product requirements, specifications, test cases, and other documents used by the engineers. They are usually written by business analysts and engineers or programmers for use within the company, follow a format and outline prescribed by the company, and contain much jargon. These documents may be reviewed by a technical editor but often are not. Some of them are used by technical writers as source material when writing external documentation and marketing materials.

Typical *external* documentation for software includes installation guides, administrators' guides, user guides, reference manuals, and online help. These materials are aimed at users of the software. They are typically written and edited by technical writers and editors and follow a task-oriented model, which combines the chronological (how-to instructions) and general-to-specific (introductions and overviews) organizational patterns described earlier.

Task-Oriented User Documentation

Hardware and software *user guides* for both beginners and experienced users should take a task-oriented approach. The tasks covered and the level of detail provided vary with the audience's level of knowledge, but the principle is the same: identify what the user needs to accomplish (the task), and then explain how to do it. Often this involves describing more than one function in the hardware or software and explaining the sequence of steps and the consequences of choices the user needs to make along the way. Chapter 22 discusses task analysis for user guides.

Reference manuals typically take a more function-oriented approach in the descriptive material but are often organized by task as well.

Figure 14.16 shows the task-oriented contents list of a typical user guide for a word processor.

EXERCISE 14.2

Select any two of the documents from this chapter and create an outline for each. Identify the thesis statement first. Then write major headings for each of the main points. After you

List of Chapters

1. Introducing Writer
2. Setting Up Writer
3. Working with Text
4. Formatting Pages
5. Printing, Faxing, Exporting, and Emailing
6. Introduction to Styles
7. Working with Styles
8. Working with Graphics
9. Working with Tables
10. Working with Templates
11. Using Mail Merge
12. Creating Tables of Contents, Indexes, and Bibliographies
13. Working with Master Documents
14. Working with Fields
15. Using Forms in Writer
16. Customizing Writer

Partial Contents List for Chapter 3, Working with Text

Selecting Text
Cutting, Copying, and Pasting Text
Finding and Replacing Text and Formatting
Formatting Paragraphs
Formatting Characters
Creating Numbered or Bulleted Lists
Using Footnotes and Endnotes
Checking Spelling and Grammar
Hyphenating Words
Undoing and Redoing Changes

FIGURE 14.16
Contents of User Guide for a Word Processor.

are satisfied that you have listed the main points, determine if each is fully developed by listing the minor points under each one. Return to the generic outline (Figure 14.1) if you have questions about how to organize an outline.

EXERCISE 14.3

Locate a journal article, feasibility study, how-to manual, or any other technical document and create an outline similar to the previous exercise. Begin by identifying the type of organizational pattern. Then write a one-sentence statement describing who you believe to be the intended audience. Next, write out the thesis statement, and then construct an outline from the document. (You do not need to outline introductions and conclusions.) Make copies of your outline and the original document and distribute them to your classmates for their feedback.

EXERCISE 14.4

Select any of the eight organizational patterns and write your own 500- to 750-word article. Begin by developing an outline, and then write the article. Make copies of your final draft and distribute them to classmates. Ask them to reconstruct an outline from the article you composed. After they have completed their outline, collect a copy of their outline and compare it to your original. If their outlines greatly differ from yours, discuss any discrepancies.

SUMMARY

Regardless of the function of any technical document, it cannot be successful without the best form or organizational pattern that complements it. Readers expect information to be organized in certain ways. Technical editors must know the eight essential patterns for organizing information and be able to apply them to the proper situation.

CHAPTER 15

Coherence: Making Sentences and Paragraphs Flow

At the conclusion of this chapter, you will be able to

- differentiate effective from ineffective paragraphs;
- recognize the importance of paragraphs that specialize;
- edit technical documents for cohesion;
- analyze documents for logical reasoning;
- decide the overall effectiveness of a well-written technical document.

The ability to reason logically is derived in part from life experiences, but it can also be consciously acquired through study and the application of specific skills. Technical editors acknowledge that their own ability to reason logically (at least on paper) first began by learning the standard conventions of writing and editing and then by applying that knowledge. While editing, you must engage the thought processes of the writers (their *process schemata*) and try to comprehend what the writer is trying to say. If the style, organization, or logic of a document is poorly crafted, your endeavor to comprehend may turn into a struggle. You will find that such mental gymnastics may be physically exhausting as well.

> Words and sentences are subject to revision; paragraphs and whole compositions are subjects of prevision.
> —*Barrett Wendell*

For example, Chapter 14 discusses organizational patterns for presenting information. While reading that section, did you consider that these patterns are logically structured as well? From experience, readers already know these patterns and maintain expectations that the writer will follow them. Surely a writer would not want to begin with the body of a document before introducing it. Readers naturally expect background information first. Or if a document is organized chronologically, readers do not expect the writer to sequence events or steps randomly. What happens to readers (and editors) as they try to make sense of a poorly planned, ill-constructed document? In most cases, they simply won't read it.

If you completed the exercises in Chapter 14 that asked you to outline the examples, you can analyze your reconstructed outlines to determine if the document progresses logically from one main idea to the next. Remember, readers of technical documents do not like the unexpected, especially digressions or skips in reasoning. If you have not had the opportunity to work as a technical editor, you might be surprised to learn that educated professional writers lapse into irrational thought more often than we like to admit.

BODY PARAGRAPHS: COMPOSITIONS IN MINIATURE

Writing can be classified into four principal modes: *narrative, descriptive, expository, and persuasive*. Each mode has a specific function, and each initiates specific reader expectations. When writers choose to develop their ideas through narration, the reader expects to be told about a sequence of events, often in the form of a story. Description, on the other hand, directs the reader's attention to the physical surroundings of a setting, object, or a person's state of mind. Expository writing focuses on informing the reader, whereas persuasive writing seeks to convince the reader to accept a particular point of view. Most often, writers combine two or more of these modes of writing within their documents.

Although they use the narrative mode less often, technical writers frequently combine descriptive, expository, and persuasive writing to advance their ideas. For example, a chronologically organized document may begin with a general description of product or process and then develop the body of the document through exposition. Or a problem-to-solution feasibility study may first present a problem using description, then use exposition to inform the methods used to resolve the problem, and finally turn to persuasion to argue for a course of action. Determining which mode or combination to use depends on the intended audience and the purpose of the document. Again, this is where the technical editor's experience and expertise become important. Editors make organizational suggestions that significantly affect the effectiveness and readability of a document.

Modes of discourse are developed as paragraphs, which resemble miniature compositions. Both contain a stated purpose and information that develops that purpose. Whereas a composition has a thesis statement to introduce the central idea for the entire composition, the paragraph has a topic sentence that introduces the central idea for the paragraph. And whereas the composition relies on paragraphs to develop the thesis, the paragraph uses sentences for its development. In short, each is a logical unit of thought; the composition is just on a larger scale.

Before you read further, take a guess as to how long the modern paragraph has been around. It may surprise you to learn that the paragraph as we know it is not much over 100 years old. In 1866, Alexander Bain, a high school English teacher in the United Kingdom, became increasingly disappointed in the quality of his students' writing. As he describes it, their efforts rambled, lacked cohesion, and generally failed to sustain a rational approach to a topic. Out of his frustration was born the modern paragraph, which contains a topic sentence and subsequent sentences that connect logically to each other. Combined, the sentences develop into a single unified thought. Soon, specialized paragraphs with specific functions followed: paragraphs that define, compare, contrast, describe, explain, classify, inform, or show causal relationships.

Examine the following paragraph from the Centers for Disease Control (CDC). Notice that the first sentence establishes reader expectations for all subsequent sentences. This is the topic sentence: what the paragraph will discuss. Do you agree that the paragraph is unified throughout and develops the topic sentence?

Editing is the same as quarreling with writers—same thing exactly.
—*Harold Wallace Ross*

When the 2009 H1N1 flu outbreak was first detected in mid-April 2009, CDC began working with states to collect, compile, and analyze information regarding the outbreak, including the numbers of confirmed and probable cases and the ages of these people. The information analyzed by CDC supports the conclusion that 2009 H1N1 flu has caused greater disease burden in people younger than 25 years of age than older people. At this time, there are relatively fewer cases and deaths reported in people 65 years and older, which is unusual when compared with seasonal flu. However, pregnancy and other previously recognized high risk medical conditions from seasonal influenza appear to be associated with increased risk of complications from this 2009 H1N1. These underlying conditions include asthma, diabetes, suppressed immune systems, heart disease, kidney disease, neurocognitive and neuromuscular disorders, and pregnancy.

TIP FOR TECHNICAL EDITORS

Good technical writing will follow at least one mode of discourse. Which mode(s) the writer uses should be readily identifiable; otherwise the document will appear illogical and confusing. Except for specialty paragraphs such as introductory, concluding, and transitional paragraphs, all body paragraphs will generally have a topic sentence and several sentences that support it. Usually, paragraphs of a technical nature are comparatively short: four to six sentences. But this is not a rule, just a general guideline. Also, all supporting sentences in a body paragraph must specifically address the topic sentence. Otherwise, the paragraph will lack unity and further confuse the reader. Remember, the responsibility for communicating the intended purpose is solely the writer's, not the reader's. Effective editing will help to bridge any gaps between what is intended and what is understood.

Points to Remember

- *Body paragraphs are miniature compositions.* They contain a central, controlling idea (i.e., the topic sentence) and subsequent sentences develop only that idea.
- *Repeating keywords or phrases increases a paragraph's coherence.* Judicious repetition of keywords or a phrase from the topic sentence helps create consistency and a logical progression of ideas.
- *Creating parallel structures also increases a paragraph's coherence.* Parallel structure refers to words and phrases that contain the same grammatical construction. For example, items in a series should be parallel, with each item beginning with the same part of speech. (See Unit Two for a more detailed explanation.)
- *The point of view and verb tenses should be consistent.* Avoid shifting the document's point of view. For example, addressing your reader as "you," then shifting to a more formal "one," changes the tone and reduces the coherence of the document. Also, avoid shifts in verb tense from past to present, or present to past, unless necessary to describe events that clearly occurred at different times.
- *Transitions between sentences and paragraphs emphasize the relationships between ideas.* Using words and phrases such as *again, furthermore, first, second, although, above, afterward,* etc. connect the previous ideas to the next idea, adding coherence and cohesion to what is being said. (See Figure 15.1.)

Type of link	Purpose	Examples
Logical	Creates logical relationships between ideas.	*accordingly, as a result, if, furthermore, because, in addition, in effect, therefore, consequently*
Qualification	Establishes further evidence for a line of reasoning.	*in addition, this is evidenced by, moreover, subsequently, this is further supported by*
Example	Provides a case in point to clarify a previous idea, theory, argument, etc.	*for example, in fact, that is, for instance*
Summary	Restates the main idea for clarity, to remind the reader what is being argued.	*to conclude, to summarize, in short, in other words, that is*
Time	Establishes time relationships.	*afterward, as long as, before, earlier, subsequently, until, when, while*
Place	Gives a sense of direction.	*above, below, nearby, to the left, beyond*
Contraposition	Demonstrates opposing or qualifying viewpoints.	*however, but, on the other hand, to the contrary, alternately, yet, although, in contrast, nevertheless*

FIGURE 15.1
Examples of Words and Phrases That Create Coherence.

PARAGRAPHS THAT SPECIALIZE: INTRODUCTIONS AND CONCLUSIONS

Because body paragraphs contain the principal content of any document, their importance cannot be overstated. Unfortunately, introductory and concluding paragraphs rarely get enough respect from writers and editors. You probably have read articles and reports where the introductions and conclusions seemed like an afterthought, leaving the impression that the writer was in a hurry to finish the job.

Introductions: Grand Openings

Too often, introductions to technical documents (and other types of writing) provide only background information and little more. Background information is only one aspect; introductions need to orient the reader to the document as well. To fulfill the readers' expectations, good introductions create a good first impression by saying *why* the document was written, for *whom* it was written, its *purpose*, and the general *contents* that will be presented. Introductions can be several paragraphs in length, depending on the reader's expectations and need to know.

Good introductions share common elements (see Figure 15.2). We recommend that when editing a lengthy document, you consider whether the introduction effectively orients the reader by providing the

- *topic* early in the introduction;
- *purpose* as to why it was written;
- *audience* for whom it was intended (e.g., users of Microhard X20C software);

The audience is identified: SUV or car owners who want to save on gas.	No doubt you have recently pulled up to a service station and were shocked by the high price of gas. Not too long ago $25 filled your tank, but now the pump spins past $30, $40, or even $50 for the same fill-up. As you watched the dollars click by, you may have considered trading in your vehicle for a more gas-efficient one. Or you may have paused to think about buying a "green" car, not to mention you'd have the coolest car around.
The topic and purpose of the article are stated early.	Fortunately, modern technology and the auto industry have arrived at a solution for all these needs—the **hybrid car**. In the past few years, many hybrids have taken to the road. Hybrids have been in the news a lot lately. Some auto manufacturers have come out with their own SUV hybrid versions that substantially save gas.
Background information is minimal.	So, how does this new technology work? What have the manufacturers done to increase gas mileage by as much as 40%? How do hybrids reduce the amount of greenhouse gases while reducing global warming? The new technology is truly amazing. In the following pages we will discuss how the different types of hybrids work, their efficiency ratings, and give you advice on why you might want take the plunge and buy a hybrid.
The final paragraph of the introduction provides an overview of the content to be covered.	

FIGURE 15.2
Example of a Good Introduction.

- *background* on the history, theory, motivations that initiated the document;
- *reasons* why the document was written.

    Good introductions usually do not

- apologize for the writer's lack of knowledge, experience, or writing skills;
- announce the writer's intention with phrases such as "The purpose of this report is to . . .";
- mention anything unrelated to the topic;
- provide details other than an overview.

If these elements are integrated skillfully, not only will the reader be better prepared for understanding the contents, but the writing will seem more credible. Also, technical documents

This conclusion summarizes the researchers' purpose and findings. Notice that scientists often write in the passive voice.

We had begun our study with the hypothesis that Randy's vocalizations would progressively improve and begin to mimic speech patterns after implantation of the Werner box in his cochlea. His hearing improved by 25 dB. During the next two months, his speech amplified to exceed the gains found in normal developing children his age. The Werner box, therefore, seems to have substantially improved the quality of Randy's hearing resulting in a marked improvement in his language acquisition. However, before recommending the Werner box be generally accepted for implantation, we suggest additional studies be made on implant recipients.

FIGURE 15.3
Example of a Good Conclusion.

may require more than one introduction. Most technical documents are subdivided into sections, which may be improved with a brief paragraph indicating the new topic and contents. You'll have to use your editorial judgment here, but usually sectional introductions sustain the flow of the information and add coherence.

Conclusions: Grand Endings

Concluding paragraphs should be a thoughtful response to what has previously been written; however, much like introductory paragraphs, concluding paragraphs get little respect. Too often they are brief regurgitations of the thesis. Remember, while introductions create the reader's first impression of the document, concluding paragraphs create the last. They should satisfy the reader's expectation that the topic has been fully covered. (See Figure 15.3.) Although you may think that, by nature, technical documents do not allow for creativity, concluding paragraphs not only provide a brief summary but also may

- call for action on the part of the reader;
- provide speculation how the future will be affected based on what has been discussed;
- offer solutions or suggestions for further study;
- borrow a quote that stimulates the reader;
- ask thoughtful questions;
- reemphasize why the discussion in the document is useful to the reader.

Good conclusions usually do not

- introduce new information or ideas that should have been discussed in the body;
- restate the thesis;
- apologize for any failings on the writers' part;
- preach or get emotional;
- consist of a single sentence.

COHESION: MAKING IT ALL STICK TOGETHER

You're probably familiar with the Segway, a personal transportation device that transfers an individual over a short distance from point A to point B. Have you ever noticed that people use the word *segue* (pronounced "segway") as a term meaning to make a transition from one topic of discussion to the next? "Now that we have discussed Mary's job performance, I'd like to *segue* to the next topic on the agenda." "Since my divorce is final, I can *segue* back into the dating scene."

The term *segue* and its uses may not be very important to you, but the concept of a *transition* is. A transition is like the clutch for a manually shifted car. If you don't depress the clutch while shifting, you will grind the gears and hear the mashing of metal. The clutch, then, makes possible a smooth transition between gears.

Writing without using transitions is similar to shifting without a clutch. Rather than grinding metal gears, though, a document without transitions "grinds" the mental gears of the reader. It's critical for any piece of writing, whether or not technical in nature, to move from one idea to the next without losing the reader. In short, transitions help create coherence and a sense of unity as the reader moves from one idea to the next. Furthermore, transitions link ideas logically. Certainly you have experienced this yourself when someone is talking about one subject and then suddenly jumps to another without your following their thought process. Effective editing calls for the editor to analyze documents for cohesiveness; often, editors must recommend a transitional word or phrase to clarify the writer's meaning.

Repetition of Keywords

One of the most obvious but overlooked methods of creating and maintaining cohesiveness is to repeat the keywords found in the topic sentence. Cognitive studies tell us that from a psychological standpoint, occasionally reminding readers what is being discussed makes a paragraph's central idea more emphatic and keeps the reader conscious of the topic. Examine the following paragraph for repetition of keywords.

The topic sentence initiates this paragraph; the keywords are *conquering, mountains,* and *peaks.* Notice how the writer provides examples of how mountains conquered man.	Before man had ever dreamed of *conquering mountains, mountains* had *conquered* man. Early man believed spirits or gods lived on summits, making them a forbidden place for mere mortals. Lighting storms circulating among the *peaks* were certainly the rumblings of the gods. The Himalayas were idolized by those lived below in its shadow. Mt. Meru, another *mountain* the Hindus thought lay beyond Mt. Everest, was thought to be 84,000 miles high, a so-called *mountain*-above-*mountains* where the gods congregated. Seven concentric rings emanated from Mt. Meru's *peak* where the sun, moon, and planets revolved.

FIGURE 15.4
Example of Good Keyword Repetition.

Transitional Cues

Another method of linking ideas together involves using words and phrases that refer to previous ideas. The most common are pronoun substitutions such as *this, these, that, such*, and *those.* These pronouns replace the nouns they reference and serve two essential functions: referring back to a previously stated idea and avoiding the monotony of overusing one term.

Earlier in this chapter, we provided a partial list of transitional words and phrases that add coherence. Examine the following paragraph for the use of transitions.

The paragraph begins with a topic sentence (under-lined). Subsequent sentences develop the keywords. (**Boldface** words repeat key ideas. *Italicized* words create cohesion through transitional links.)	<u>One of Isaac's greatest **fears** in college was enrolling in a freshman English course that required extensive writing.</u> *His* **fears** were well founded because *previously he* had done poorly on writing assignments in high school. *Nevertheless,* to overcome *his* **trepidations,** Isaac decided to first take a developmental studies class in basic English. *Because* he was highly motivated to succeed, Isaac soon met with success and was writing above-average essays. *In fact,* one expository essay *he* wrote on the unusual mating habits of hummingbirds received the highest grade in the class. *As a result* of his dedication and hard work in the developmental studies class, Isaac enrolled in freshman English and received an A in the course. *In short,* Isaac's **fear** of writing had now turned into confidence.

FIGURE 15.5
Example of Good Transitional Cues.

EXERCISE 15.1

Locate three paragraphs from any textbook, journal article, government document, or memo that you consider poorly written. Find the topic sentence. Revise those paragraphs using more transitions and keywords. Read your revised paragraph out loud after you have finished. How does it *sound* compared to the original?

EXERCISE 15.2

Revise the passage below for better logic, flow, organization, and document design. What is the article trying to tell the readers? Does the article contain more than one message? Is more information needed in places? Check your revision against Jean Hollis Weber's, available at http://www.jeanweber.com/newsite/?page_id=43.

Connecting to OurNetwork

OurNetwork is an Australia wide network with regional offices or agencies in all capital cities. Batch facilities such as printing and plotting are provided in these places as well as access for interactive terminals.

Interactive terminals may be connected with dedicated Telecom lines or used in dial up mode. Dial up requires that the user provide a suitable modem and terminal; the dedicated (leased) line comes with a suitable modem but it is considerably more expensive.

Basic modem types are:

V22bis   2400/1200 bps (240/120 chars per sec) synchronous or asynchronous
V22   1200 bps asynchronous (Full duplex, not V23 1200/75)
V21   300 bps asynchronous

Asynchronous modems are used with asynchronous (dumb, ascii, VT100 style, line by line) terminals and synchronous modems with 3270 Bisync terminals.

The following table indicates the terminal and modem types needed for particular databases. Where a number of choices exists, these are explained later.

Database	Terminal	Modem
Database 1	Sync	V22bis
	Async	V22bis V22 V21
Database 2	Async	V22bis V22 V21
Database 3	Async	V22bis V22 V21

If two terminal types are listed against the database, the first is the preferred method but the other will perform satisfactorily. Similarly, where a number of modems are listed, the first will give the highest (fastest) performance.

Some examples of modems and terminals are:

Modem

V22 bis	Case	Datacraft	Scitec
V22	Case	Datacraft	
V21	Case		

Terminal

Synchronous   3270 Bisync from IBM, Beehive, Fujitsu, ITT, Telex or appropriate software on a personal computer

Asynchronous   VT100, VT220 (Digital Equipment Corporation), Beehive, TAB, Wyse or appropriate software on a personal computer

Note that these are examples; there are many different machines that can perform these tasks. In general, OurNetwork will be moving towards V22bis services for dial up and for new purchases where access to other networks is not needed, the purchase of V22bis modems is suggested.

EXERCISE 15.3

Go to http://www.jeanweber.com/newsite/?page_id=45 and revise the procedure on that page for better logic, flow, organization, and document design. Check your revision against the one in http://www.jeanweber.com/newsite/?page_id=46.

SUMMARY

Paragraphs contain the same elements found in documents as a whole, only in an abbreviated form. For example, body paragraphs contain a topic sentence that states the central idea, with other sentences developing that idea. Paragraphs that introduce and conclude a document have special functions. The introduction, if written properly, establishes the purpose of the document while providing sufficient background information to orient the reader to what will be discussed. Concluding paragraphs, on the other hand, may ask the reader to take action, offer solutions or suggestions for further study, or reemphasize why the discussion in the document is important. Above all, paragraphs must maintain a sense of cohesion and logic that unites the individual sentences as well as the document as a whole. Good technical editors understand the importance of good paragraphs and how they are held together.

Editing Online Publications

5

Vast amounts of communication and commerce have moved online in the past decade, and more is moving all the time. These changes mean rapidly expanding opportunities—and the necessity—for editors to work with materials intended for use online, whether within a company (on an intranet, for example), in a restricted Internet environment (passworded access for students enrolled in a course, for example), or on the open Internet (for a global audience).

Units Two, Three, and Four discuss editing for correctness, visual readability, and effectiveness. These principles apply as much to online publications as they do to printed publications, but the application of some aspects of readability—in particular, document design—varies between print and online.

This unit focuses mainly on web pages and websites (collections of web pages and their associated graphic, video, audio, and other files). The emphasis is on information sites, not entertainment or sites designed primarily for shopping, although the latter often have a significant information component.

In this unit, you will learn how online publications differ from print publications; how readers use online publications; the editor's role in a team producing online publications; what standards and issues editors must consider when working with online publications; what to look for when editing online publications; and how to mark up publications intended for use online.

Some publications that are intended to be printed are also provided in an electronic form and placed on a CD/DVD, the Internet, or a company's intranet as a convenient means of distribution. Such publications are not primarily intended to be used interactively and typically do not include hypertext links or other features unique to online publications. This unit does not discuss these types of publications.

CHAPTER 16

Online Publications: What Technical Editors Need to Know

At the conclusion of this chapter, you will be able to

- describe the differences between online and printed publications and the significance of these differences;
- discuss the audience requirements for online publications;
- explain the importance of metadata and other technical concepts related to online publications;
- find more information about standards and guidelines for online publications;
- understand basic HTML code.

O nline and printed publications share many characteristics, but they also differ in significant ways. Like printed publications, on-line publications fit into a variety of categories, which serve different purposes and are used and read differently by their audiences. Websites generally include several types of online publications. For example, a website for a typical software development company might include advertising, instructional material, reference material, online help, news items, research reports, order forms, annual reports, and other company information.

Online publications can be delivered to the audience in several ways:

- On interactive CDs or DVDs; for example, demonstrations or training materials
- In information kiosks; for example, at the entrance to a store
- As web pages, either on a company intranet or on the Internet
- As online help, which may be available through a website or in-stalled on the user's computer along with a software program

DIFFERENCES BETWEEN ONLINE AND PRINTED PUBLICATIONS

The major differences between online and printed publications are

- Online publications often do not have an easily determined sequence of subject matter. Unlike turning the pages of a book, links within the text and site navigation allow readers to move through the text in unpredictable ways.
- Online publications may include multimedia components (audio, video, and animated graphics such as Flash) as well as text and static graphics.
- Online publications are generally more difficult to read than printed publications (primarily due to monitor-inflicted eyestrain), so people are likely to skim online material looking for what they want.
- Web pages are more likely to be read (or at least visited) by people from a wide variety of cultures, with a range of ability to understand English.
- Websites may be visited by people with a range of disabilities, including visual disabilities.
- Websites may be visited by people with a range of computer equipment, operating systems, software (including web browsers), and speed of Internet connection, so individual visitors' experiences may differ greatly.
- Readers of web pages can change the size of fonts, hide graphics, disable features, or make even more dramatic changes to the appearance and functionality to suit their own preferences, so what you produce may not be what everyone sees.
- Web pages can be customized for individual readers or groups of readers by pulling information out of a database as the page is assembled and downloaded, such as in response to a reader's request on a previous page.
- Web pages can be updated easily and often (compared to printed publications), so readers expect them to be up to date.
- Websites may be translated into other languages by online services such as Google Translate and Yahoo!'s Babel Fish, as well as (or instead of) being translated by trained professionals.

These differences and their effects on the role of the editor are discussed in detail in this chapter and in Chapters 17, 18, and 19.

HYPERTEXT

The main characteristic that differentiates online publications from printed publications is *hypertext*: text linked together in a complex, nonsequential web of associations in which the user can move through related topics.

Hypertext publications are connected by *hyperlinks* (also called simply *links*). Links can be from one place to another in the same publication, or they can be from one publication to another on the same website or different websites. Figure 16.1 shows some links among publications within one website and between publications on several websites.

Printed publications such as this book often contain cross-references to other parts of the publication (such as figures, tables, or other chapters) and to other publications. Readers must find the referenced material in some way before they can access the information in the reference. In a hypertext publication, readers simply click on the cross-reference (link) and they are taken directly to the referenced item, as shown in Figure 16.2.

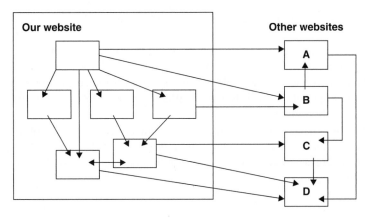

FIGURE 16.1
Links within a Website and between Websites.

FIGURE 16.2
Links within a Document.

For most people, the most familiar form of hypertext publication is a web page, but many other forms of electronic publication, including spreadsheets, presentations, and word processor documents, can also include links. Links can be from text or graphics, as shown in Figure 16.3. The only difference between the two is their appearance; the code that makes them work is the same.

Links can lead to various types of files, including text, graphics, video, and audio files. Indeed, many "hypertext" publications are more correctly called *hypermedia* because they link to several media types in addition to text. *Graphic links* direct you to another web page when you click on the graphic; *text links* also direct you to another page, but here the clickable link is a word or a line of text. The code that makes links work is the same in both cases.

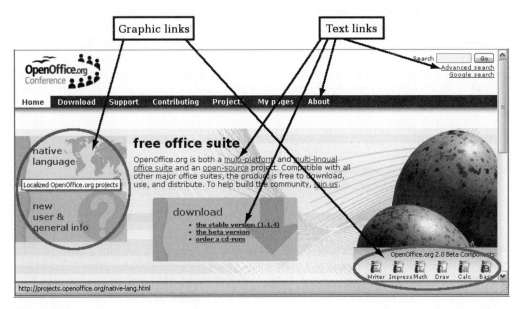

FIGURE 16.3
Page Showing Both Text Links and Graphic Links.
Source: OpenOffice.org.

AUDIENCES

As with any editing, before starting to work you need some background information about the audience and the purpose of the publication. Much of the information you need is the same for readers of both printed and online publications, but you also need to consider some other factors. For example, people read differently online than they do from printed pages (see "How People Read Online Publications").

In addition, the audience for many online publications includes a wide range of other characteristics, as mentioned in "Differences between Online and Printed Publications" at the beginning of this chapter. There are also questions specific to audience analysis for online publications (see Chapter 17).

TIP FOR TECHNICAL EDITORS

Globalize, localize, or translate? Many organizations are multinational and hence multi-cultural. Whether publications are used internally or published for wider distribution, their audience is often likely to include a wide range of people. Some publications may be designed for use by the entire audience, while others may be localized or translated for use in specific countries or cultures. Writers, editors, and publishers must therefore consider the cultural factors discussed in Unit Six.

HOW PEOPLE READ ONLINE PUBLICATIONS

Research in the 1980s, reported by Jakob Nielsen in *Designing Web Usability,* found that reading from a computer monitor is 25% slower than reading from a printed page. This difference may have diminished since then as high-resolution monitors have become common and people have become more accustomed to reading from a screen. However, most people find large blocks of text (paragraphs of more than a few lines) and long lines of text more difficult to read online than in print. Online readers are also likely to be in a hurry to find what they want and leave again quickly, so they are likely to skim a publication, looking at headings to decide whether to read the text.

Many web pages have a great deal of information (including navigation aids and advertisements) surrounding the main content of each page, so readers develop strategies for ignoring the surrounding material until they need it. (Help systems for computer software are less likely to include as many distractions.)

Eye-tracking studies have shown how people orient themselves on a page and focus quickly on the part that interests them. Will Schroeder found that "users typically looked first in the center, then to the left, then to the right. . . . New and experienced web users scanned essentially the same way." Schroeder's study included only one user who was new to the Web. "At first, the new user scanned pages from left to right, as if reading a book. But he quickly changed to the center-left-right sequence." Figure 16.4 illustrates these differences.

More recent research on eye-tracking, reported by Nielsen and Pernice, shows that users often read web pages in an F-shaped pattern: two horizontal stripes followed by a vertical stripe. This research emphasizes that users don't read text thoroughly. Thus the first two paragraphs must state the most important information, and subheads, paragraphs, and bulleted

How a printed page is read

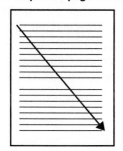

How a screen layout is read

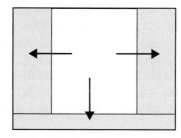

FIGURE 16.4
Differences in the Way People Read Printed Pages and Screen Layouts.

points need to be started with information-carrying words that users will notice when scanning down the left side of the content. Reading in the F pattern is now considered to be more common than reading in the Z pattern, although the Z pattern is still used to teach speed reading.

Readers often look for links to more specific information, jumping from one page to another until they either find what they want, become lost, or give up. They may begin by searching the site or the page, using various strategies and tools (see Chapter 18).

Some people use assistive technology to read web pages. For example, people with some visual disabilities use screen readers, software that reads the underlying code of the page and renders it as speech. See Chapter 18 for information on accessibility and assistive technology.

STANDARDS AND GUIDELINES

In Unit Two, you learned about the importance of standards and guidelines for grammar, punctuation, and mechanical consistency. Standards are equally important for online materials and cover a range of technical issues in addition to those for the content itself. Without standards, web browser software could not display pages consistently. Indeed, a major complaint for many years has been the differences in the way software browsers adhere to standards. Technical editors are not expected to be experts in all of the standards, but you do need to know what they are and why they are important. If you know enough about them to ask intelligent questions and understand the answers, you'll earn more respect from programmers and web designers and you'll be able to debate knowledgeably when someone says "that's not important" or "that can't be done."

The main standards body for web display is the World Wide Web Consortium (W3C), http://www.w3.org/. It develops specifications, guidelines, software, and tools and "is a forum for information, commerce, communication, and collective understanding." Some of the specifications and guidelines developed or maintained by W3C are

- Markup languages, including XML (http://www.w3.org/XML/), HTML (http://www.w3.org/html/), and XHTML (http://www.w3.org/MarkUp/), as well as CSS (http://www.w3.org/Style/CSS/).

- The Web Accessibility Initiative (http://www.w3.org/WAI/), which develops guidelines widely regarded as the international standard for web accessibility, support material to help understand and implement web accessibility, and other resources.
- The Web Content Accessibility Guidelines (WCAG), at http://www.w3.org/WAI/intro/wcag.php, which explain how to make web content accessible to people with disabilities. WCAG is primarily intended for web content developers (page authors, site designers, etc.), web authoring tool developers, and web accessibility evaluation tool developers, as well as policymakers, managers, and others.

A related standard is the Darwin Information Typing Architecture (DITA) from the Organization for the Advancement of Structured Information Standards (OASIS). DITA "defines an XML architecture for designing, writing, managing, and publishing many kinds of information in print and on the Web. DITA consists of a set of design principles for creating 'information-typed' modules at a topic level. DITA enables organizations to deliver content as closely as possible to the point-of-use, making it ideal for applications such as integrated help systems, web sites, and how-to instruction pages. DITA's topic-oriented content can be used to exploit new features or delivery channels as they become available." For an introduction to DITA, see http://xml.coverpages.org/dita.html and http://www.ibm.com/developerworks/xml/library/x-dita1/.

Common usability guidelines include books, articles, tutorials, and courses from Jakob Nielson at Usable Information Technology, http://www.useit.com/.

Among many books and online tutorials on web accessibility, Jim Thatcher (coauthor of *Web Accessibility: Web Standards and Regulatory Compliance*) has a particularly good introduction to standards and practices at http://www.jimthatcher.com/webcourse1.htm. See also Shawn Lawton Henry's *Just Ask: Integrating Accessibility Throughout Design*, free online at http://www.uiaccess.com/JustAsk/ (also available in printed editions), and W3C's *Essential Components of Web Accessibility* (http://www.w3.org/WAI/intro/ components).

METADATA

Metadata is information that describes another item or collection of information. For example, the copyright page of a book lists the title, author, publisher, copyright date, ISBN, topic categories, and other information used for cataloging the book. Metadata is used to describe more than books or other text documents; audio files, photographs, videos, and other material all contain metadata.

Files on your computer (or on the Web) typically contain metadata, such as the time and date last modified, the person who made the modification, the program used to create the file, and so on. Some of this metadata is automatically generated by the program, but other information (such as the title of the document, a description, and subject keywords) is manually entered.

This information goes into databases from which it can be retrieved using a variety of search mechanisms. When you search for a file on your computer, you can specify a range of dates, the file type, and other details. When you search for something on the Web, one of the items used by search engines is the metadata that someone has defined for the file. (For information on defining metadata tags for web pages, see Chapter 19.)

To make metadata as useful as possible, people have developed hierarchical arrangements known as *ontologies* or *schemas*. By using terms defined in these schemas, people can help ensure that documents are suitably classified and likely to be retrieved in a search. Obviously, editors should be involved in checking metadata as well as other contents of a document.

Here are some metadata standards that you should be aware of, even if you aren't familiar with their technical details:

- *The Dublin Core* (http://dublincore.org/) is a metadata standard for describing digital objects, including web pages. The Dublin Core Metadata Element Set, part of the Dublin Core Metadata Initiative (DCMI) Recommendation, consists of sixteen optional metadata elements, any of which could be repeated or omitted: Title, Creator, Subject, Description, Publisher, Contributor, Date, Type, Format, Identifier, Source, Language, Relation, Coverage, Rights, and Audience. Several elements have schemas or a controlled vocabulary; for example, the Type element has twelve recommended terms: collection, dataset, event, image, interactive resource, service, software, sound, text, physical object, still image, and moving image.

- *Exchangeable image file format (Exif)* is a specification for the image file format used by digital cameras. Depending on the camera, the metadata may include date, time, and camera settings (model, make, aperture, shutter speed, focal length, etc.). Other information, such as location, descriptions, and copyright details can be added to the metadata manually if the camera does not support automatic inclusion of this information.

- *ID3* is a tagging format for MP3 files. It allows metadata such as the title, artist, album, track number, etc. to be added to the MP3 file.

THE SEMANTIC WEB

The adjective *semantic* means "of or relating to meaning in language or logic." The Semantic Web project (http://www.w3.org/2001/sw/) aims to create a universal medium for the exchange of data, where data can be shared and processed by automated tools as well as by people. To do this, documents are marked up with machine-readable information (metadata) about the human-readable content of the document (such as the creator, title, description, and so on) or metadata representing a set of facts (such as resources and services elsewhere in the site).

Common metadata vocabularies (*ontologies*) tell document creators how to mark up their documents so that software (known as *agents*) can use the information in the supplied metadata to perform tasks for users.

An early example of the Semantic Web in action is the Creative Commons initiative (http://creativecommons.org/), which gives content publishers a simple way of clarifying how their content may be used by others. Publishers place a Creative Commons license on their website by adding a piece of code. This code links to pages at the Creative Commons site; each of those pages defines a particular element of the license, such as the content's redistribution policy. When search engines and other automated tools pick up that website, they can access these pages and "understand" the site's copyright policy.

TOOLS AND TECHNOLOGY

Editors of online publications need to know enough about common tools and technology to be able to discuss them with technical staff and, more importantly, understand the answers. If you know enough to solve problems on your own, so much the better.

A *web browser* is a software application that enables a user to display and interact with HTML documents. Popular browsers available for personal computers include Internet

Explorer, Firefox, Opera, and Safari. Associated tools include media players for audio and video files.

Assistive technology tools include screen readers (such as JAWS), alternative keyboards and pointing devices, switches, and scanning software.

Although HTML pages can be written in a text editor such as Notepad or gEdit, visual *HTML editors* (also known as web authoring tools) such as Adobe Dreamweaver or Microsoft Expression Web are popular because they allow the creation of web pages to be treated much like word processor documents by hiding the underlying code. Visual editors have a "code inspection" option whereby you can view and edit the HTML code if you want to.

New methods of writing, editing, and managing web pages online using a web browser are being developed so rapidly that any list of them is likely to be out of date before it is published. Some examples are blogging software such as WordPress, which handles ordinary web pages as well as blogs; wikis (such as Mediawiki, used by Wikipedia); and other content management systems (CMS) including Plone, Drupal, and Aukyla.

Java is a programming language used to write browser-based applications. *JavaScript* is a scripting language used by HTML authors to write functions for use on web pages, for example to make an image change to another image when you move the mouse pointer over it. Scripts also automate the sort of task that a user might otherwise do interactively at the keyboard; this use is similar to a *macro,* the term commonly used in word processing and spreadsheet applications. Macros and scripts are often called *extensions* because they extend the capabilities of a browser, word processor, or other application.

An *applet* is a Java-compliant software component that runs in a web browser, a plug-in, or a variety of other applications, including mobile devices. Unlike a program, an applet cannot run independently. It characteristically performs a very narrow function that has no independent use.

On the Internet, nobody knows you're a dog.
—*Peter Steiner*

MARKUP FOR STRUCTURE AND LAYOUT

The two basic types of markup are *change indicators* and *structural or layout indicators.* (Change indicators are discussed in Chapter 3.) In this section, we look at *markup languages*: sets of codes that can be inserted into an electronic document to indicate the structure or appearance of the document.

In most cases, writers and editors won't see the actual codes because the software used to produce documents hides the codes. For example, when you "tag" a paragraph as "Heading 2" in a word processor or website editor, or select the paragraph and manually format it, that paragraph displays the characteristics (typeface, typesize, spacing, and other attributes) that you selected, but you don't see the underlying code. Some word processors and most website editors provide ways for you to look at the code if you want to.

Professional writers and editors should use styles, not manual formatting. The advantages of styles are that they

- Are faster and easier to apply than manual formatting
- Ensure consistency of formatting
- Make formatting changes easy. For example, if you need to change the font, size, or alignment of a book element such as chapter titles, level-three headings, or quotations, you only need to change the style; you do not need to find and change every paragraph of that type

- Greatly assist in producing output in more than one form (printed, web page) or when moving a file from one program to another (for example, when moving a manuscript draft from a word processor to a desktop publishing program for layout and final production)

Editors need to be familiar with many terms and how the markup languages and technologies they describe are used in the production of online documents.

SGML (Standard Generalized Markup Language) is a metalanguage in which one can define markup languages for documents. SGML is a descendant of IBM's Generalized Markup Language (GML), developed in the 1960s. SGML was originally designed to enable the sharing of machine-readable documents in large projects in government and the aerospace industry. It has also been used extensively in the printing and publishing industries, but its complexity has prevented its widespread application for small-scale general-purpose use.

HTML (HyperText Markup Language) is designed for the creation of web pages and other information viewable in a browser. It can be used to define the semantics of a document through metadata and provide structural information such as headings, paragraphs, lists, and information that allows the document to be linked to other documents to form a hypertext web. HTML is a text-based format that is designed to be both readable and editable by humans using a text editor.

XHTML (Extensible HyperText Markup Language) does much the same job as HTML but has a stricter syntax. Whereas HTML was an application of SGML, a very flexible markup language, XHTML is an application of XML, a more restrictive subset of SGML.

XML (Extensible Markup Language) is derived from SGML. XML is a specific subset of SGML, designed to be simpler to process than full SGML for general-purpose applications, such as the Semantic Web.

DocBook, another markup language originally created as an application of SGML, is designed for authoring technical documentation. It is now also available as an XML application. In the past, DocBook was used mainly in the open-source community, including the Linux Documentation Project. In recent years, however, its use has become more widespread, and several commercial documentation tools based on or supporting DocBook XML have become available.

CSS (Cascading Style Sheets) is a stylesheet language used to describe the presentation of a document written in a markup language. Its most common use is to define colors, fonts, layout, and other aspects of document presentation on web pages. CSS is designed primarily to enable the separation of document structure (written in HTML or a similar markup language) from document presentation (written in CSS). It can also control the document's style separately in alternative rendering methods, such as onscreen, in print, by voice (when read out by a speech-based browser or screen reader), and on Braille-based devices. These style sheets are called "cascading" because they specify a priority scheme to determine which style rules apply if more than one rule applies to a particular element. A series of style sheets can be defined, each controlling a different style characteristic: colors, font sizes, page layout (placement of elements on a page), and alternative output methods.

Many website editors work with *source files*: files containing visible HTML and CSS markup. Even if you never need to edit HTML or CSS yourself, you need some basic knowledge of the codes so you can talk with the people who do work with the source files.

If you have used a word processor like Microsoft Word or OpenOffice.org Writer, you should be familiar with the built-in styles for headings, paragraphs, lists, and other structural elements of a document such as tables. For web pages, the separation of structural markup and formatting markup is done through a combination of HTML tags and CSS definitions.

TIP FOR TECHNICAL EDITORS

Although in many circumstances today, HTML and CSS are not case sensitive, it's good practice to use only lowercase because XML requires it, and most documents on the Web will be coded in XML soon, if they aren't already. For the same reason, be sure to always include ending tags.

INTRODUCTION TO HTML MARKUP

This section introduces the most basic HTML structural elements (also known as *tags*): those for headings, paragraphs, lists, images, and tables. Elements have *attributes,* many of which specify formatting in a manner similar to manual formatting of word processor text.

TIP FOR TECHNICAL EDITORS

Formatting attributes (used in earlier versions of HTML) have been replaced by styles in HTML 4. Various HTML tags and their attributes are changing in HTML 5, so if you need to work on code, be sure to check the latest information online at http://www.w3.org/webdesign/.

Markup for Headings, Paragraphs, and Lists

A number of tags are used to signal the beginning and ending of a page, elements related to page content, etc. Although most website editing can be done without viewing the code (analogous to the way you can add and delete text in Microsoft Word's Track Changes mode while hiding the markup), sometimes viewing the code helps you quickly identify what's causing a formatting problem.

HTML provides six levels of headings: H1 (most important), H2, and so on down to H6. Each heading starts with a tag like <h1> and ends with one like </h1>.

Each paragraph starts with a <p> tag and ends with </p>.

HTML supports three types of lists: *ordered* (numbered), *unordered* (bulleted), and *definition*. (Definition lists are a bit different from the other two, so they are covered separately below.) Ordered lists start with <ol> and end with </ol>; unordered lists start with <ul> and end with </ul>. Each item in an ordered or unordered list starts with <li> (for "list item") and ends with </li>. Lists can be nested one within another.

Figure 16.5 shows an example of markup for headings, paragraphs, and lists and shows the result as seen in a browser. The appearance of the result varies slightly from one browser to another.

Definition lists are used for a variety of other purposes, such as a series of headlines with short descriptions underneath. (Chapter 18 has more about the use of definition lists.) They start with a <dl> tag and end with </dl>. Each term starts with a <dt> tag and ends with </dt>, and each definition starts with a <dd> and ends with </dd>. Figure 16.6 shows an example.

Markup	Result
`<h1>Chapter heading (name)</h1>`	**Chapter heading (name)**
`<p>Introductory paragraph.</p>`	Introductory paragraph.
`<h2>Subheading (first level)</h2>`	**Subheading (first level)**

`<p>Following is a numbered list.</p>`
`<ol>`
`  <li>First list item.</li>`
`  <li>Second list item.</li>`
`  <li>Third list item.</li>`
`</ol>`
	Following is a numbered list.
	
	1. First list item.
	2. Second list item.
	3. Third list item.

`<p>Following is a bulleted list.</p>`
`<ul>`
`  <li>First bulleted list item</li>`
`  <li>Second bulleted list item</li>`
`  <li>Third bulleted list item</li>`
`</ul>`
	Following is a bulleted list.
	
	• First bulleted list item
	• Second bulleted list item
	• Third bulleted list item

`<p>Following is a bulleted list nested inside a numbered list.</p>`
`<ol>`
`  <li>First list item.</li>`
`  <li>Second list item.`
`    <ul>`
`      <li>First nested item</li>`
`      <li>Second nested item</li>`
`    </ul>`
`  </li>`
`  <li>Third list item.</li>`
`</ol>`
	Following is a bulleted list nested inside a numbered list.
	
	1. First list item.
	2. Second list item.
	  • First nested item
	  • Second nested item
	3. Third list item.

FIGURE 16.5
Sample HTML Markup and Browser Display.

Markup for Images

To add an image, use the <img> tag. This tag has several attributes that should be included, as shown in this example:

```
<img src="jean.jpg" width="200" height="150" alt="Photo of
  Jean" />
```

The *src* (source) attribute tells the browser the name of the image file (and where it is located, if not in the same folder on the website). The *width* and *height* attributes help to speed the display of the page. For accessibility, use the *alt* attribute to add a description that people can read if they can't see the image. Notice the / before the > at the end of the <img>

Markup	Result
```<dl>```	
``` <dt>First term</dt>```	**First term**
``` <dd>Its definition</dd>```	Its definition
``` <dt>Second term</dt>```	**Second term**
``` <dd>Its definition</dd>```	Its definition
``` <dt>Third term</dt>```	**Third term**
``` <dd>Its definition</dd>```	Its definition
```</dl>```	

FIGURE 16.6
Sample HTML Markup and Browser Display for a Definition List.

tag. Because the <img> tag does not surround anything, it is closed by the / within the tag it-self, instead of by a separate closing tag.

Alternate Text for Images, Forms, and Objects

To meet accessibility guidelines, the *alt* attribute must be specified on two HTML elements: *img* (for images) and *area* (used in image maps). *Alt* is optional on some other elements not covered here: *input, applet,* and *object*.

Why is the *alt* attribute important? Software for screen readers and Braille output reads the *alt* attributes on images, to give some information to people who cannot see the images themselves. In addition, some people on slow modem Internet access choose to turn off images in their browsers to speed up page loading; they will see the *alt* information instead.

The World Wide Web Consortium's recommendations on the use of *alt* include

- Do not specify irrelevant alternate text when including images intended to *format* a page, for instance, `alt="red ball"` would be inappropriate for an image that adds a red ball for decorating a heading or paragraph. In such cases, the alternate text should be `alt=""`.
- Do not specify meaningless alternate text (e.g., "dummy text"). Not only will this frustrate users, it will slow down software that converts text to speech or braille output.

Markup for Links

Links are defined with the <a> tag. To define a link to the page "jean.htm" on the same website:

```
This is a link to <a href="jean.htm">Jean's page</a>.
```

The text between the <a> and the </a> appears on the web page. Links are typically displayed underlined; for example:

This is a link to <u>Jean's page</u>.

To link to a page on another website, you need to give the full address (commonly called a URL, for Uniform Resource Locator). For example, to link to Jean's website, you need to write

```
This is a link to <a href="http://www.jeanweber.com/">Jean's
   website</a>.
```

You can turn an image into a hypertext link. For example, the following allows you to click on the photo of Jean to get to her website:

```
<a href="http://www.jeanweber.com/"><img src="jean.jpg"
width="200" height="150" alt="Photo of Jean" /></a>
```

Markup for Tables

Tables are used for layout as well as for tables of information, but in most cases you should not use them for layout; use CSS (Cascading Style Sheets) instead. When you need a real table for data, you'll need to use the table tags. The basic table tags are <table>, <tr> (row), <th> (heading cell), and <td> (data cell), each with its own end tag. Each tag can have several attributes, most of which should be handled by CSS instead. The exceptions are described in "Making Your Tables Accessible" below.

Figure 16.7 shows a simple example of table markup. The <table> element acts as the container for the table. The <tr> element acts as a container for each table row. The <th> and <td> elements act as containers for heading and data cells respectively.

Making Your Tables Accessible for People with Disabilities

If a visitor to the web page is unable to see the table, it can be quite difficult to understand what the table is about. Marking up a table for accessibility can become quite complex, but a few simple concepts can get you started.

Use the table element's *summary* attribute to describe the structure of the table for people who can't see it. Add a description of the purpose and structure of the table using the <caption> element, which should appear before the <tr> element for the first row.

Markup	*Result*
```<table>``` ```<tr><th>Year</th><th>Sales</th></tr>``` ```<tr><td>2000</td><td>$18M</td></tr>``` ```<tr><td>2001</td><td>$25M</td></tr>``` ```<tr><td>2002</td><td>$36M</td></tr>``` ```</table>```	Year / Sales; 2000 / $18M; 2001 / $25M; 2002 / $36M

FIGURE 16.7
Table Markup Using Only the Most Basic Elements.

```
<table summary="The first column gives the year and the second,
the revenue for that year">
<caption>Projected sales revenue by year</caption>
<tr>
  <th scope="col">Year</th>
  <th scope="col">Sales</th>
</tr>
<tr><td>2000</td><td>$18M</td></tr>
<tr><td>2001</td><td>$25M</td></tr>
</table>
```

FIGURE 16.8
Table Markup Including Accessibility Information.

When a table is rendered to audio or to Braille, it is useful to be able to identify which headers describe each cell. For instance, using an audio browser you can move up and down or left and right from cell to cell, with the appropriate headers being spoken before each cell. The simplest way to provide this information is to add the *scope* attribute to table header (<th>) cells.

Figure 16.8 shows the HTML for the table in Figure 16.7 after all of the extra features described above have been included.

INTRODUCTION TO CSS MARKUP

HTML tags define the structure of a document or web page, but they do not contain formatting information like font attributes, positioning (left, right, center), indentation, and so on. You can use attributes to define many of these things, but that is like manual formatting for a printed document. Instead, it's much better to use CSS (Cascading Style Sheets) to define styles for each element.

CSS is a markup language: a collection of design properties and the values for those properties. CSS is extremely powerful and flexible. For an impressive set of examples of the use of CSS, see the website Zen Garden (http://www.csszengarden.com/). Writing CSS specifications is a specialist skill. Using CSS is fairly easy once you have learned a few basics. This section will get you started.

A simple example of CSS markup is shown in Figure 16.9. This markup defines the attributes or style of a Heading 1. This style will be applied to all paragraphs surrounded by <h1> . . . </h1> tags, just like tagging a paragraph in a word processor as H1 applies the H1 style. If you want to change the appearance of all the level-one headings on the website, you simply change it once—in the CSS.

Similarly, you can define the characteristics of ordinary paragraphs, list items, and other structural elements of your document. Editors should check style code for some common accessibility errors.

- Font size should be expressed as percentages of a base font size (which can be defined or left blank); this assists in correct display when a user chooses to enlarge or decrease the font size. Do not specify font sizes in specific units.
- Use <strong> . . . </strong> instead of <b> . . . </b> (bold) and <em> . . . </em> (emphasis) instead of <i> . . . </i> (italics), so that screen readers and other assistive

```
h1 { color: #641A64;
        background-color: transparent;
        font-family: Arial, Helvetica, sans-serif;
        font-style: normal;
        font-size: 200%;
        text-align: left;
        line-height: normal;
        margin-left: 10px;
    }
```

FIGURE 16.9
Example of CSS Markup.

devices will render the page correctly. Better still, don't use these tags at all; define the equivalent font attributes in CSS and use <span> . . . </span> tags to mark up content.

• Code table dimensions in percentages, not a specific number of pixels or other measurement units, to ensure that tables will resize correctly for different screen resolutions, window sizes, and font sizes.

The Class Selector

What if you want to have some paragraphs look different from others? A common requirement is for the first paragraph under a heading not to have a first-line indentation, but all other paragraphs to be indented. Simple! You define the basic paragraph style to include the attribute *text-indent,* then you define a *class* for the first paragraphs (with no indent) and mark the first paragraphs to be of that class. A typical set of paragraphs and class definitions might look like this (note the period preceding the definition). The names of classes can be anything you want.

```
p { ...
text-indent: 30px; }
.firstpar { text-indent: 0px; }
```

To apply a class to a tag, use this markup:

```
<p class="firstpar">This is the first paragraph after a head-
  ing.</p>
```

The Span Selector

Sometimes, you'll want a selector to apply to only part of a paragraph: a word or phrase. A common use is to make a word italic, bold, or a different color. For this purpose, you'll use the *span* selector. For example, you could define a dark purple color (using numbers from a color chart):

```
dkpurple {color: #641a64; }
```

Then apply it to some words within a paragraph:

```
<p>Here are a few words in <span class="dkpurple">dark
  purple</span>.</p>
```

The Div Element

The <div> element is used to define parts (divisions) of a page, which are usually containers for something: paragraphs of text or images, for example. Divisions can be used to define the header area, footer area, and main page content area, but they can also be used for things like floating menus.

The ID Selector

Use the *id* (identification) selector to identify a unique element in a HTML document. A common usage is with the <div> element, to identify the header area of a page; another *id* would identify the footer area, and a third *id* the main body of the page. For example, the CSS would contain

```
div#header {
position: absolute;
left: 0px;
top: 0px;
vertical-align: top;
background: #ffcc35;
width:100%;
}
div#content {
position: absolute;
left: 0px;
top: 120px;
vertical-align: top;
margin-left: 0px;
width:85%;
}
```

These selectors are then applied in the document like this:

```
<div id="header">Page header</div>
<div id="footer">Page footer</div>
```

EXERCISE 16.1

Look at your university's website and answer the following questions:

- What audiences is it aimed at (for example, students, future students, parents, faculty and staff, alumni, visitors, others)?
- What categories of information does it publish (for example, news, tutorial material, reference material such as course details, guides to services on campus, other)?
- Find examples of different types of hypertext links (text, graphics, pulldown menus). Can you tell which links look like text but are in fact graphics? (Turn off images in your browser and reload the page to be sure.)
- Find examples of links to other websites, links to other pages on the same website, and links to other parts of the same page. (If all the pages are short, you may not find any examples of the last type of link.)
- What search facilities does the website provide?

SUMMARY

This chapter noted significant differences between online and print publication. These differences are crucial as they affect the quality of the document as well as its accessibility and usability. The chapter introduced two specialized concepts (metadata and markup), some common technical terms, and some of the standards and guidelines related to online publications. These concepts, standards, and guidelines are subjects of entire books, so they have not been fully developed here. Having a general idea of these topics is essential if you are going to become a successful technical editor of online documents; specializing in any of them can enhance your career prospects.

CHAPTER 17

Editing Online Documents: Players and Processes

At the conclusion of this chapter, you will be able to

- describe the roles of the editor and other members of the website production team;
- conduct an audience analysis for online publications;
- develop specifications for a website.

This chapter looks mainly at websites, but the principles apply equally to the production of online help, computer-based training (CBT), and interactive CDs and DVDs.

THE EDITOR'S ROLE IN THE WEBSITE PRODUCTION TEAM

Producing a website, online help, CBT, or interactive CD/DVD requires people to fill many roles. In a large company, the roles may be filled by several different people including an editor. In a small company or organization, the editor may fill many roles such as planning, writing, and usability testing in addition to editing.

The roles required vary with the project. In a large, established organization, the editor may be responsible only for editing content, working with templates designed by other people. This scenario is much like working for a large publishing company where book designers determine the appearance of the final product. In a team developing a new website, the editor may have far more input into the design of the product.

This section describes typical possibilities for the roles required in an online publication team, as illustrated in Figure 17.1.

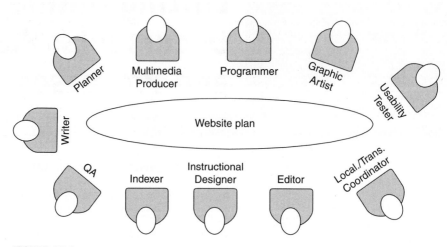

FIGURE 17.1
The Website Production Team.

Project Manager/Planner/Producer

A project manager oversees and coordinates the project, acquires and allocates resources, tracks progress against the plan, and negotiates and makes decisions as necessary to manage changes in the plan or to resolve conflicts.

A project manager is like the conductor of an orchestra. . . . The conductor assumes that all the musicians are proficient in playing their individual instruments and know their individual parts, but need leadership to work together most effectively . . . the conductor envisions how all the individual sounds must fit together to result in a pleasing whole that delights the audience.

—*JoAnn Hackos*

Writer

The writer's role can be quite complex. Depending on the size of the team, writers may be responsible for editing, graphic design, layout, programming, and testing in addition to research and actual writing. Writers need to work closely with other team members to produce and link illustrations, audio, video, and other content as needed. They may organize reviewing, editing, and testing of written materials, and review, edit, and rewrite all copy as necessary in cooperation with the editor, reviewers, and other writers.

Editor

As in any publishing project, the editor should do far more than simply check the material for grammar and spelling errors after the writing is done. Editors should interpret and adapt company standards to the needs of a particular project; review and enforce company standards for user interface design; determine the suitability of material for the target audience; and review, test, edit, and rewrite all copy as necessary in cooperation with writers.

Graphic Artist or Illustrator

Graphic artists should suggest additional or alternative ways to use images to present information; create, crop, edit, annotate, and otherwise clean up images; provide advice to writers on color choices, resolution, file types, file size, and other technical matters related to images; create animations as needed; and design packaging and covers as needed.

Instructional Designer

Websites, online help, CDs, and DVDs often fill a training need. Instructional designers are experienced at breaking complex tasks into chunks and providing each chunk at an appropriate level of explanation for the audience. Instructional designers should suggest ways to improve the presentation of instructional information, assist writers in breaking complex tasks into chunks suitable for different audiences, and advise on ways to link this material.

Programmer

Many projects require programming to make them work. For example, web pages might be dynamic (assembled from chunks of content in a database, or displaying different content in different situations) rather than static. Programmers should set up any required programming and test the website to ensure that it works as intended.

Multimedia Producer

If the project includes audio, video, animations, or interactivity, a multimedia producer can help; indeed, such a specialist may be essential. Multimedia producers should advise on suitable media, considering the audience, the product, and the time and money available to do the work; produce or oversee production of materials; and work closely with writers and programmers to incorporate multimedia material into the project.

Indexer

Website indexing is different from book indexing because it involves keywords and other metadata used for a variety of search purposes (both within the site and to assist people in finding the site online). Some websites include a "back of the book"–style index of terms as well, but many do not. Indexers should either create the index or work with the writer to produce the index. The writer might create the initial entries and the indexer amend or add to them as necessary. (See Chapter 19 for more on optimizing terms for searching.)

Localization and Translation Coordinator

Will the publication be localized or translated? If so, will the work be done in-house or contracted to a specialist? Localization and translation coordinators should work with writers, illustrators, and others to ensure that materials are produced in ways that make localization or translation easier and faster; coordinate the work of localizing or translating the content; and organize testing of the resulting materials. See Chapters 18 and 21 for more about writing to optimize localization and translation.

Usability Tester

Usability testing is essential for producing good online publications because it determines whether the material is suitable for the audience. Usability testers should produce a usability testing plan that clearly states exactly what the test covers. Testing preferably starts with an early prototype so that changes can be made as early as possible in the writing cycle. The testers make recommendations on ways to solve any problems identified during usability testing and follow up by usability testing the revised materials. See Chapters 19 and 23 for more about usability testing.

Quality Assurance (QA) Person

Quality assurance is an area where responsibilities vary greatly from one company to another. Depending on how your company defines "quality," and whether QA covers websites and other publications, QA may include usability testing as well as other activities. Quality assurance people should produce a quality assurance plan that clearly states exactly what type of testing or other work the QA person will do, perform whatever activities are identified for this role, make recommendations on ways to solve any problems identified during QA activities, and follow up by reviewing the revised materials.

PLANNING THE WEBSITE

Before starting work on a website or other electronic publication, you need to gather information and develop a plan. Required information includes both the company's requirements and the audience's requirements for the website. If you will be editing work for an existing website, this information should be readily available. If you are participating in a team developing a new website, members of the development team may need to do some research to answer the questions.

Keep clearly in mind that the audience's purposes for visiting the website may be quite different from the organization's purposes. For example, many companies, universities, government departments, and other organizations want their websites to reflect their internal management structure: its departments, divisions, and subdivisions. But visitors often do not know—or care—what part of the company is responsible for providing a service, product, or piece of information; they just want to find what they are looking for.

Krug (*Don't Make Me Think*) and Nielsen and Tahir (*Homepage Usability*) provide numerous examples of sites that fail to meet the audience's needs because they were designed to put the company's purposes first. It is usually much better to structure the site from the audience's point of view, not the company's.

Figure 17.2 shows the sequence of steps to take when planning and developing a website. These steps are described in the remainder of this chapter.

TIP FOR TECHNICAL EDITORS

Be sure to get agreement on how—and by whom—the website will be reviewed, edited, and tested, and record the agreement in the website plan.

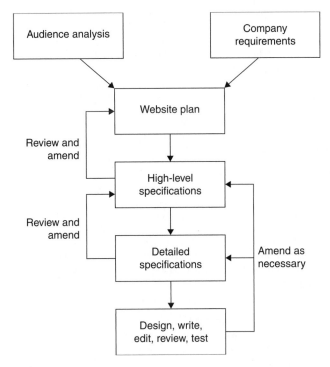

FIGURE 17.2
Steps in Planning and Developing a Website.

ANALYZING THE AUDIENCE

Audience analysis is vitally important, especially when large sums of money are involved (for example, in setting up a shopping website). Many books, articles, and websites are devoted to describing techniques designed to get the most accurate picture of the audience and put this information to use. You can find out a lot of basic audience information fairly quickly, though, without needing to become an expert in analytical techniques.

Figure 17.3 contains a checklist of information you should gather about the audience and what they are likely to be looking for on the website. The answers to these questions will determine how you look at the site when editing it and what the usability testers should test for.

Some ways to analyze the audience include observing how typical members of the audience perform tasks and asking them what they want to do (in interviews, focus groups, or using questionnaires). Profiles of typical users are often created by the marketing department, business analysts, trainers, or others in the company. If you're working on a project that doesn't already have user profiles, you may want to create your own.

A popular variation on user profiles is the *user persona*. While a "profile" is a list of characteristics of a typical user in a specific user category, a "persona" is a fictional person who has those characteristics.

TIP FOR TECHNICAL EDITORS

Avoid stereotyping audiences or assuming that audiences for a new or revised product or service are the same as those for existing or previous products or services.

Who are the member of audience?
- ❑ Employees or members of the organization (internal site).
- ❑ Existing customers, clients, or suppliers of the organization.
- ❑ Prospective customers, clients, or suppliers.
- ❑ Casual browsers or anyone interested in the topics on the site.

How diverse is the audience?
- ❑ International or including many people with English as a second language.
- ❑ English-speaking but from diverse cultural backgrounds.
- ❑ Located in several countries.

How will the audience access the page or site?
- ❑ Through the organization's high-speed intranet connection.
- ❑ Over the Internet, mostly using a high-speed connection.
- ❑ Over the Internet, often using a dial-up connection.

What is the audience's purpose in visiting the page or site?
- ❑ To gather information about the company, its products, or its services.
- ❑ To look up other information provided on the site (for example, research papers).
- ❑ To be entertained.
- ❑ To make a purchase or to conduct other business.

What is the audience's attitude toward the site?
- ❑ Hostile (I don't want to be here, but my boss told me I had to).
- ❑ In a hurry (I need this information to complete an urgent task).
- ❑ Interested (I want to learn about this topic, and I'm willing to spend time).

What is the audience's knowledge level about the topics on the website?
- ❑ Low.
- ❑ Some familiarity.
- ❑ High.

How will the audience use the material on the page or site?
- ❑ Read and interact online.
- ❑ Download to read or print later or keep for reference.
- ❑ Print from the website.

FIGURE 17.3
Checklist for Audience Analysis.

To create user personas, start with an audience analysis and describe one or more typical users in more detail. Tell a story about each of them, as if you were writing about a real person. Be specific about personal details such as job title, age, gender, disabilities, residence, attitudes (toward computers in general and your website in particular), learning style, knowledge, and experience. Include anything else that's relevant to the story you are developing. Sometimes a mock photograph will help you focus on the details.

Example: Two Personas

The following personas describe two typical users of a web-based outliner and organizer program.

Tim A. is a 57-year-old freelance journalist. He has stacks of handwritten notes on ideas for articles and research supporting those ideas: websites, email and newsgroup postings, pages in books, and contact names and phone numbers. He also has photocopies and printouts of material, as well as files saved on his computer. He tries to keep the paper organized in manila folders and ring binders, but some things are relevant to more than one topic. Cross-referencing them is a problem, and keeping track of the hard copies and the computer files gets confusing. Tim hates wasting time chasing down all the information he wants for an article when he needs it; he knows he's got it there somewhere, but it's hard to find.

Tim has a broadband Internet connection and an older desktop computer with out-of-date software (Microsoft Windows XP and Office 2000) that serves his needs quite well. He knows how to use email, a browser, a word processor, and other tools of his trade, but he doesn't know much about their technical aspects and is completely uninterested in learning. If he has a problem with the hardware or software, he calls his son or a friend for help. Tim doesn't want to upgrade his software or install new programs because he doesn't want to spend time learning how to use them. On the other hand, he wants to organize his notes better, so he's willing to try something new, as long as it's easy to use and the help or manual tells him exactly what to do; he doesn't want to have to figure it out for himself.

Emma B. is a 22-year-old university student. Her parents can't help her financially, so she's on a tight budget. Most of her research is done on the university's machines, but she has a secondhand laptop running up-to-date versions of Linux and other open-source software for personal use and for essay writing at home. Her needs for organizing a lot of information are similar to Tim's, but her attitude toward software is very different. She's used to working with databases, she likes to know what's going on behind the scenes with the software she uses, and she likes to be able to customize her experience as much as possible.

ANALYZING THE COMPANY'S REQUIREMENTS

Figure 17.4 shows a worksheet that you could use to gather information about the organization's purposes for the website; it includes data from the audience analysis.

Please answer the following questions. At the end of this planning phase, we will have enough information to begin the high-level design.

What is the purpose of your website?

What are your immediate (short-term) goals for the site?

What are your long-term goals for the site?

FIGURE 17.4
Website Planning Worksheet—Part 1: Company Requirements.

How will you measure the success of the site?

_____

_____

Who are your audience members? What are their characteristics?

_____

_____

What expectations or preferences might your audience have regarding your website?

_____

_____

Where will the website be hosted?

_____

Do you plan to provide any financial transactions through the site?
❏ Yes   ❏ No

Will the website be connected to a database?
❏ Yes   ❏ No

Will user logins (passworded) be required for any section of the site?
❏ Yes   ❏ No   ❏ Not now, but maybe later

Who will design and develop the site?

_____

Who will manage the design and development of the site?

_____

Who will maintain the site?

_____

How often do you plan to update the site?

_____

Will pages be ❏ static or ❏ dynamic?

What tools will be used in the design and implementation of the site?

_____

_____

What existing material do you want to incorporate into the site, or use as source material?

_____

_____

What features do you want to include on the website?
After reviewing this list, you may wish to fill in the table of features and the time frame to implement them (now, mid-term, long-term). Consider the implications of the time required and whether the necessary hardware, software, and skills are available to you.

Some possibilities for things to include in a website:

  ❏ Information about your company and its products and services
  ❏ Blog, newsletter, or other frequently updated content
  ❏ Dynamic content (pages compiled "on the fly" from a database)
  ❏ Shopping cart
  ❏ Search facility, index, site map

FIGURE 17.4 (_continued_)   Website Planning Worksheet—Part 1: Company Requirements.

❏ Comment form or other way for people to give you feedback on the site

❏ Links to outside information sources

❏ Downloadable files (for example, PDFs)

❏ Passworded sections for specific user groups

❏ Mailing list, forum, or other user-interactive pages

❏ Audio or video content

❏ Other? Specify _____

Time frame for implementing features
Fill in the following table to help estimate how long the initial site development will take.

Feature	Now	Mid-term	Long-term

FIGURE 17.4 (*continued*) Website Planning Worksheet—Part 1: Company Requirements.

DEVELOPING HIGH-LEVEL SPECIFICATIONS

If you are hired to work on an established website, you probably won't be involved in developing specifications, but you will need to follow the stated specifications closely where they apply to your work. The main role for editors, as in print publishing, is in the development and maintenance of style guides and templates for content. For a new project, the website design team needs to develop specifications so that everyone on the project has the same expectations.

High-level specifications should include information types, media types to be used, navigation aids, accessibility and localization criteria, and the tools to be used in developing the project.

Figure 17.5 shows the second part of a website planning worksheet: site specifications.

Website Features, Page Types, and Information Types

The page and information types appropriate for a website depend on the purpose and type of the website, the features offered on the site, and on the knowledge, skill levels, and other characteristics of the intended audience. For example, a shopping site (where users perform tasks more complex than simply searching for information and then reading or printing it) is different from a news or other informational site, which is again different from an online learning site where students and instructors interact in various ways, or a support site for a company's software or hardware products. You can probably think of websites that are quite different from any of these examples.

Typical website features are listed in Figure 17.4. A major choice that the planning team must make at an early stage is whether the site's pages will be stored as static pages or in a database from which the pages are compiled when someone visits them. Dynamic pages can be highly customized for individual visitors, often based on choices that the visitors themselves make.

> When I took office, only high energy physicists had ever heard of what is called the Worldwide Web. . . . Now even my cat has its own page.
> —*Bill Clinton*

Use this section to specify the presentation and navigation for the site.

1. **Design principles**
   - Design for quick page loading and display—especially the first page.
   - Do not use frames.
   - Design for ease of use.
   - Design for display on a variety of screen sizes.
2. **Site presentation**
   - Describe the color scheme for the website.
   - Describe the design (layout); include a sample (mockup) or storyboard.
3. **Site structure and navigation**
   - Describe how the site will be structured, from the users' point of view; include a diagram.
   - List the navigation aids to be used.
4. **Website features, page types, and information types**
   - List and describe the features, page types, and information types to be included.
5. **Media types**
   - List the types of media to be included.
6. **Localization or translation criteria**
   - List your requirements for localization or translation.
7. **Accessibility criteria**
   - Use CSS (Cascading Style Sheets).
   - Design for accessibility; meet WAI (Web Accessibility Initiative) accessibility guidelines.
8. **Features**
   - List each feature selected in part 1, and specify how each will be implemented.
9. **Tools and technology**
   - List the selected tools and technology.

FIGURE 17.5
Website Planning Worksheet—Part 2: High-Level Specifications.

As a technical editor, you could work on any type of website, but you likely will work on one that deals with technical information in some form: product sales, support, and training. In such a context, the information types you are most likely to deal with are those categorized according to the DITA (Darwin Information Typing Architecture) standard, described in the "Standards and Guidelines" section of Chapter 16. These information types are concepts, tasks (procedures), and reference material.

- *Concept:* a description or explanation of something—a definition, a fact, a picture, a process description (how something works), and so on.
- *Task (procedure):* a set of steps that someone takes to achieve a specific result (for example, solving a problem with the use of a piece of equipment, or using graphics software to produce a 3-D drawing).
- *References:* lists or tables of data (ranging from statistical, such as rainfall and temperature, to such things as U.S. presidents and dates in office or Academy Award winners by year), specifications, rules and guidelines, and so on.

Other websites might have additional or different information types. Blogs, for example, might be classified as *news,* although individual blog entries often fit into one of the three DITA information types above.

Media Types

Will the documents be mainly text or will they include graphics, animated graphics (such as Flash), video, or audio? Here are some issues to consider.

- What's best for the users? Will they benefit from the inclusion of multimedia? Is it essential, a nice-to-have feature, or perhaps unnecessary?
- What sort of equipment do your users have? Do they have desktop computers with large monitors, large hard disks, DVD drives, and/or high-speed Internet connections? Or are they using laptops in the field, with no DVD drive or no Internet access? Or perhaps they will access your website on the their cell phones. (Think of an engineer or scientist working outdoors.)

Localization Criteria

Publications intended for an international market will probably be translated into other languages. In some countries, for example Canada, more than one language is a legal requirement on publicly accessed signs and websites. Even if the publication is provided in only one language, some localization may be necessary if it's being used in more than one country. For example, websites for accounting software distributed in the UK, Canada, the United States, Australia, and New Zealand may need to consider different accounting terminology or tax regimes.

Localization of products and the accompanying documentation may involve two things: translation into other languages and changes to account for regional differences, including weights and measurements, date formats, temperature scales (Celsius or Fahrenheit), paper size (letter or A4), currencies and formats, legal requirements, local contact information, and cultural differences. (See Chapter 21.)

TIP FOR TECHNICAL EDITORS

Think smart; think simple. Concentrate on

- Information flow
- Intuitive navigation
- Well-designed sites
- Crisp, clear, easy-to-read pages

Accessibility Criteria

Site specifications should include criteria for ensuring that the site is accessible. *Accessibility* is the extent to which people with disabilities can use a help system, website, or other software as effectively as people without disabilities.

Here are the main reasons for making online materials accessible:

1. Website accessibility may be required by law, for example Section 508 of the U.S. Rehabilitation Act; other countries have similar requirements.

2. Increased accessibility is good business, by providing for a wider potential customer base and enabling more people to find answers for themselves instead of calling customer support.
3. Anyone can become temporarily or permanently disabled (for example, by breaking an arm), so everyone potentially benefits.
4. Older members of the population have changing abilities, including difficulty in reading small type sizes.
5. More choices help and empower everyone, not just the disabled.

TIP FOR TECHNICAL EDITORS

http://www.w3.org/WAI/intro/components is a good summary of the essential components of web accessibility.

http://www.w3.org/WAI/quicktips/Overview.php is a great set of quick tips for making accessible websites.

"VISUALLY IMPAIRED" DOES NOT EQUAL "BLIND"

Although website designers are usually aware of some accessibility issues, often the only requirements they think of are making sure the site can be navigated and used by blind people using screen readers, and that the colors are okay for color-blind people. These are important considerations, but there are many other vision-related aspects to think about.

For example, does the audience for the website include people over 50? The two most common vision-related issues for people over 50 are increasing difficulty in reading small print and a decrease in contrast vision. Of course, people in any age group can have these problems. You may think, "Why don't they get new glasses?" but for many people, glasses do not solve the problem. Besides, why should someone need to buy new glasses just to read your company's website? They are more likely to go somewhere else that doesn't make life difficult for them.

Figures 17.6 and 17.7 summarize some common accessibility problems and ways to avoid or solve these problems. Clearly, not all solutions are relevant to all websites; the main thing is to avoid doing things that make sites inaccessible.

Navigation Aids

The site specifications should include what navigation aids to use. When choosing and designing navigation aids:

• Consider your users' expectations about the use of color, icons, topic types, navigation aids, and the placement of information on the screen.
• Be consistent in the placement of navigation aids so that users know where to look for them, for example, at the top or bottom of a page or on the left-hand or right-hand side.

Category	Examples	What to do (examples)
Vision	Partial or complete blindness	Ensure pages or files can be read by a screen reader. Use CSS, not tables, for page layout. Use tables only for tables of data. Provide audio as a choice. Do not have all information in graphics only.
	Color-blind	Choose foreground and background colors carefully. Do not use color as only clue to meaning (e.g., links).
	Focus and field of vision	Allow fonts to be enlarged and column or page width to be changed.
	Difficulty seeing contrast	Use strong contrast, not pale grey; ensure user can impose own color and contrast preferences.
Hearing	Partial or complete deafness	Provide alternatives to audio (don't have audio as the only source of information).
	Noise distraction	Don't have music or voice come up automatically; allow user to choose.
Other physical factors	Arthritis, tremors, muscle deterioration, poor coordination, all causing difficulty using mouse or keyboard	Provide alternative ways to navigate without using mouse or other pointing devices; make sure keyboard navigation works as expected (for example, use of Tab and Enter keys). Avoid necessity for two-handed input.
	Paralysis, partial or complete. Amputation of hand or arm	Provide alternative ways to navigate. Avoid necessity for two-handed input.
Mental, cognitive	Flashing/blinking/moving text or images cause headaches, confusion, distraction	Avoid or make optional.
	Dyslexia	Write clearly; avoid cryptic instructions. Provide error checking of input. Provide alternative ways of input, such as voice.
	Poor memory, inability to concentrate	Make interface easy to understand and use. Embed instructions for use in the interface. Provide easy ways to get help.
	Unfamiliarity with online conventions	Make interface easy to understand; avoid jargon and programmers' terms.

FIGURE 17.6
Accessibility Issues Related to Human Disabilities.

Category	Examples	What to do (examples)
Hardware	Older, slower computers	Avoid requiring latest browser software. Avoid requirement for audio files or special plug-ins.
	Slow modem or telephone limitations (mobile phone access, frequent line dropouts, etc.)	Reduce size of graphic files and use height and width attributes on IMG tags to reduce page loading time. Avoid large downloads unless essential and make them optional if possible. Warn user before large download. Avoid requirement for audio files or special plug-ins. Provide ways for people to do as much offline as possible.
	Small monitors and/or low screen resolution or few colors	Test at different resolutions and color settings. Design so resizing window does not cause horizontal scrolling
	May not have sound enabled	Provide alternatives to audio.
Software	Operating systems other than Windows—may not have Internet Explorer	Don't use features specific to one operating system or browser, or provide alternative ways to do things.
	Older versions of common programs, e.g., Acrobat Reader (especially if non-Windows)	Ensure PDFs work in older versions of Acrobat Reader; don't require users to upgrade their software.
	Not everyone has Microsoft Office	Provide downloads in other file types; RTF is one possibility, if plain text does not suffice.
Personal preferences	Security: may have cookies or JavaScript disabled	Provide alternatives to JavaScript, for example in links. (JavaScript is often used for links where a simple HTML link would suffice.)
	Speed: may have graphics off	Provide clues to help users decide whether to turn graphics on. Don't have all information in graphics. Use *alt* attributes on graphics.
	Browsers other than Internet Explorer	Ensure site works in Opera, Firefox, Safari.

FIGURE 17.7
Accessibility Issues Related to Technology Factors.

Common Navigation Aids

- Menus at top or side
- Table of contents
- Index
- Search facility
- Site map
- Visual aids such as icons
- Next and Previous buttons
- Cross-references within text
- "Related topics" or "See also" links
- Breadcrumbs (path to a topic)

TIP FOR TECHNICAL EDITORS

Breadcrumbs are an orienting technique frequently seen on web pages at the top of a page; they show the path to the topic on that page. Here is an example:

Creating graphics > Creating a bar chart > Setting up a drawing page

Tools and Technology for Creating the Website

Here are some categories of tools that you may need for the project in addition to a suitable computer and monitor. The tools, particularly software, change rapidly, so this chapter does not mention any specific products.

Hardware

- Audio recorder
- Video camera
- Digital camera for still photos
- A modem and dial-up line for testing (in addition to a broadband connection)
- Different sizes and resolutions of monitors, including laptops and cell phones, for testing

Software

- Editing and reviewing
- Change tracking
- Graphics editing
- Content management
- A range of browsers for testing purposes
- If relevant, different operating systems for testing purposes
- HTML and CSS editing

DEVELOPING DETAILED SPECIFICATIONS

Detailed specifications expand upon the principles defined in the high-level specifications. Detailed specifications should include the following topics, most of which are covered in detail in Chapters 18 and 19:

- Related documents (primary sources): documents containing relevant information that is not repeated in the specifications, for example company style guide, preferred dictionary, and preferred writing guide
- Features, page types, and information types
- Media types
- Localization criteria
- Accessibility criteria
- Search engine optimization (SEO) criteria
- Writing conventions
- Terminology (the terminology list can also form the basis of a glossary and the definitions provided to the translators)
- Design and layout
- Navigation scheme
- Content of topic types (templates)
- Project-specific style guide

TIP FOR TECHNICAL EDITORS

If you already have documents covering writing, terminology, design, etc.—for example, the project style guide—refer to them in the specifications. Don't repeat the information.

EXERCISE 17.1

1. Develop user personas for three people who might use your university's website: a student, a parent, and an instructor. What are their personal characteristics? What information might they be looking for? How much computer experience might they have? How much do they know about the courses and services provided by the university?
2. Choose a small business or volunteer organization in your community that does not have a website or that has only a very basic website. Analyze the organization's needs and the audience's needs, and fill out the two Website Planning Worksheets (Figures 17.4 and 17.5).
3. Set up a small team of three or four people to block out the website in item 2 and list their respective duties.

SUMMARY

The technical editor should be familiar with the roles of the editor and the other members of the production team for a website or other online document. To effectively edit online documents, the editor should be involved with conducting a specific audience analysis and developing specifications for the website and should have a basic understanding of the HTML code used in designing and formatting web pages.

CHAPTER 18

Editing Online Content: Tools and Techniques

At the conclusion of this chapter, you will be able to

- edit for organization and readability;
- edit links to make online content more useful;
- choose the best page length for an online publication;
- organize a set of online help pages;
- describe several methods for editing online content electronically.

Editing content (text, images, other media) for web pages and websites involves two distinct processes, which are covered in this chapter: What do you look for? What tools and techniques do you use?

Janice Redish describes three basic types of web pages: the home page (identifies the site and gives readers an overview of what they can find there), pathway pages (menu pages after the home page), and information pages (where the real content is).

This chapter looks at the third type of page, which may provide instruction, reference information, procedures, essays, research reports, news, and other articles, as these are the types of pages and topics that technical editors are most likely to be responsible for editing.

WHAT DO YOU LOOK FOR?

Often, the editor's only job is to edit the content, not the presentation. Just as in print publishing, editing is often done on material before it goes to the people responsible for layout, or the layout is determined by someone else. You might be able to influence some elements of presentation, such as the length of pages or the use of lists, and you might proofread or do production editing on the final version before publication, but in general you may have no input into the design and layout.

This chapter covers what to look for when editing page length, writing style, graphics and other media, and links.

TIP FOR TECHNICAL EDITORS

Two in-depth but easy-to-read books suitable for aspiring technical editors of websites (and their managers) are Steve Krug's *Don't Make Me Think* and Janice Redish's *Letting Go of the Words*. Both books include numerous examples and case studies to supplement the brief overview in this book. Lynch and Horton's *Web Style Guide* is another excellent reference for beginners.

EDITING FOR ORGANIZATION AND ORIENTATION

The principles of editing for effectiveness covered in Unit Four apply to online publications as much as to printed ones. A few points are particularly important online, where readers are even more likely to skim, scan, and skip information:

- Organize information into short, logical chunks.
- Use lots of meaningful headings.
- Break up long paragraphs into shorter ones, or turn paragraphs into lists.
- Write in newspaper style (put the most important information first, summarize the main points, and link to details).

Chunk Information

Break up long stretches of information into logical chunks. Depending on the material, you might put these chunks on separate web pages, or you might keep them on one page but add a contents list with hyperlinks at the top of the page, as we will see in the section on choosing a page length below.

TIP FOR TECHNICAL EDITORS

Even if you don't provide a table of contents in the published version of an online document, compile one for your own use. You might be surprised what you learn about the document's structure and its possible problems.

Add Headings

Add headings and subheadings to the chunked information. Schultz and Spyridakis (2004) reported that moderately frequent headings led to higher comprehension levels by the sample audience; their article includes a good literature review on the topic.

TIP FOR TECHNICAL EDITORS

Think in questions and answers even if you don't write headings as questions.

EDITING FOR ONLINE READABILITY

Guidelines provided by Hackos and Stevens, Nielsen, Kilian, Price and Price, Redish, and many others suggest that writing for the Web is different from writing for print, but Gregory points out that good practices for writing web pages apply equally to print publications. Gregory suggests that editors should

- Cut out unnecessary words, as long as you don't sacrifice meaning.
- Use short paragraphs and short sentences, to assist scanning.
- Use strong verbs and direct writing.
- Use alternatives to paragraphs (for example, lists).

EDITING LINKS

Links can be text, graphics, or part of a graphic or image map. Following are some common linking problems.

Poor Choice of Words

If the link text is part of a sentence, be sure to choose words that establish the context so that the reader can get a good idea of what the link leads to. Don't make the words "click here" into a link because screen readers for the blind will be unable to provide the context to the user.

Links Too Close Together

In text, if adjacent words (for example, in a list) are separate links, the user may be unable to tell that they are not all part of one link, especially if the links are not underlined.

In image maps, if the links do not correspond with discrete elements in the graphic, users may be unable to tell that more than one link is present. (One way to overcome this is by using JavaScript to change the color of the hotspot when the mouse cursor is on it; this works well on maps, for example.)

Links Too Large or Too Small

In text, avoid links that run over more than one line, or links that are very short, perhaps only one or two characters long. In graphics, avoid links that are only a few pixels in height or width. Many people, especially these using cell phones or PDAs, have difficulty clicking on a small area, even if they can see it's there.

Links Not Easy to Distinguish from Other Text

Don't get too creative with your text links. It's often best to leave them in the default color and style, which readers can usually change to suit their preferences, or add an icon that is used consistently throughout your pages to indicate a link. Because the default link indicator is usually blue or green with underlining, don't use underlining for emphasis or to indicate headings in other parts of the text.

Too Many Colors

Readers won't understand or remember a complex color scheme for links, and some people won't be able to distinguish the colors anyway, so there's no point in cluttering your files with irritating and confusing color differences within the text.

Using a Definition-List Style for Links

Links with descriptions are often presented in a "definition list" format, which is recommended for web pages but can be useful in many other situations as well. Compare the following lists of links.

Example of a link list without descriptions

See also:

Chapter 6
Chapter 8
Chapter 9

Example of a link list with brief descriptions

See also:

Chapter 6, Producing the Table of Contents and Index
Chapter 8, Meeting the Needs of Novices to Experts
Chapter 9, Linking from Application to Help

Example of links with longer descriptions presented as a definition list

See also:

Chapter 6, Producing the Table of Contents and Index
    Diagnosing problems, what to look for, how to fix problems, and examples of good and bad tables of contents and indexes.
Chapter 8, Meeting the Needs of Novices to Experts
    Some common problems, with suggestions for how to overcome them; also, how to try to provide information for everyone at the right level of detail.

Chapter 9, Linking from Application to Help
How to avoid linking problems by agreeing with programmers on how linking will be done, and some suggestions for overcoming problems if these agreements are not followed.

CHOOSING A PAGE LENGTH

Studies by Nielsen, Krug, Redish, and others show that most of the time, users seeking information online "read to do" rather than "read to learn." But some people, especially younger people (under 25), often use the Web as their primary or only source of information, and many of them *do* read to learn. For example, editing students might learn about writing HTML or CSS by reading an online tutorial.

But even those who read to learn want to find information quickly, so they skim, scan, and search, as well as use a table of contents, site map, or (rarely) an index. The challenge for an editor is to find solutions that balance the needs of various readers.

Advantages of short pages:

Short pages load faster.
Readers get only the information they want (if the pages are well written).

Advantages of long pages:

Long pages are easier to skim, print, and search.
It's often faster to scroll and follow internal links than to retrieve another page.

The best solution depends on the nature of the material. "Scan and get information" pages can be of several lengths: long, medium, or short. Each length has its advantages and disadvantages and is most suitable for specific types of content.

Long Pages

Long pages require scrolling vertically more than two or three times to see all of the content. They are good for articles, essays, and tutorials that develop a complex topic. To be most usable, long pages need to include ways for readers to quickly jump to specific parts of the page if they want to.

Figure 18.1 shows the beginning of a long article. Readers can quickly skim the summary and contents list to see if the article is of interest. The items in the list of contents are hyperlinked to headings farther down the page so that a reader can easily go to a particular topic.

Many sites provide a medium-length to long page of abstracts of articles, with links to the full articles. Figure 18.2 shows an example.

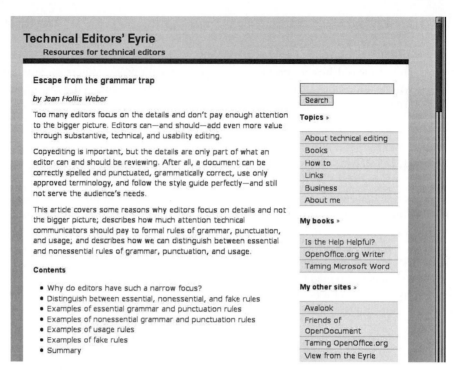

FIGURE 18.1
Top of Long Article with Hyperlinked Contents List.

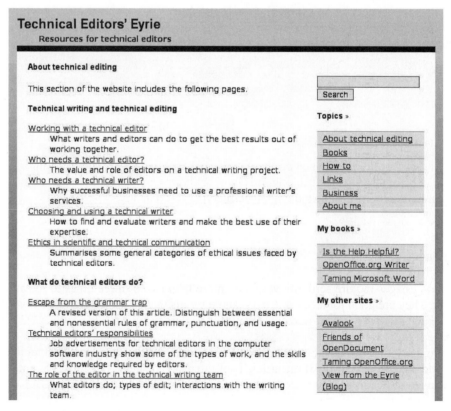

FIGURE 18.2
Top of Page of Abstracts of Articles with Links to the Full Articles.

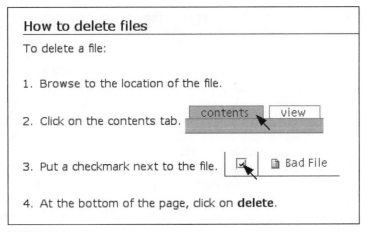

FIGURE 18.3
Example of Short Page of Information.

Short Pages

Short pages require no scrolling when viewed in a typical browser window and at a typical font size. They are best for short items or subsets of longer topics. Online help topics are best done as short pages when possible because readers of online help want to find a quick answer to a question, usually of the "how do I?" variety. Figure 18.3 shows an example.

If you break up a long topic into a series of short topics, you'll need to look at what type of navigation to use to help readers remain oriented and move from one subtopic page to another. Solutions to navigation issues are discussed in Chapter 19.

TIP FOR TECHNICAL EDITORS

Provide an option for the reader to display, download, or print in full any article that has been subdivided into shorter pages, in addition to printing the pages individually.

Medium-Length Pages

Medium-length pages require readers to scroll once or twice to see all of the content. They are typically used for content that is too long for a short page but cannot logically be broken into more than one page. Many procedural topics are in this category.

Medium-length pages are also used for shorter articles, essays, and tutorials, and for many news items. Medium-length articles are less likely than long pages to need a hyperlinked list of contents. There is little difference between many medium-length pages and either short or long pages in terms of content; some topics simply require more length to be developed thoroughly but don't need to be divided into short pages.

EXERCISE 18.1

Figure 18.4 shows the top of a CSS tutorial from the W3C website. This page is less useful than the article shown in Figure 18.1 because the list of topics is not hyperlinked to headings farther down the page; readers must scroll to find specific topics of interest.

Some editors would divide this tutorial into separate pages, with each of the main headings starting on a new page. If you look at this page online, you'll see that although the topics are grouped under main headings, the headings are not shown in the list of topics. Skim the full article, note the headings used, and decide whether the user would be better served by keeping this tutorial in one long page (but improving the contents list) or dividing it into subtopics on separate web pages.

FIGURE 18.4
Tutorial on W3C Website.
Source: *http://www.w3.org/MarkUp/Guide/Style*.

Special Case: An Online Book

Many publishers are providing online copies of books that can also be purchased in printed form. Reference manuals in online form are particularly popular. Their advantages to readers are the ability to search for specific topics and the convenience of not having to drag a thick book around with them when they are not at their usual workplace.

Some online books are provided as PDFs, or as a series of PDFs (perhaps one for each chapter of the book); others are provided in HTML format. We'll look mainly at HTML, but the principles are much the same for a series of PDFs because they can include hyperlinks to other documents.

The question for an editor is how much should a book be subdivided? At the chapter (Heading 1) level? At the next level (Heading 2)? At a lower level (Heading 3 or lower)? Consider the online book shown in Figure 18.5. This book is subdivided at the Heading 3 level. The contents page is fully hyperlinked to separate pages. Clicking on a chapter title leads to a page with some introductory comments and a contents list for that chapter (see Figure 18.6); these pages may be short or medium-length. Any section page with subsections includes a contents list for the section. Sections without subsections are typically medium-length, but a few are long.

WHAT TOOLS AND TECHNIQUES DO YOU USE?

Now that you know what to look for when editing a website, online help, or other online documentation, how are you going to do the work? Your choice depends on a variety of factors, including the available technology, the stage of development of the website, the location of the writers and editors, the skill levels of the writers and editors (both in using the tool and

---

Self-publishing using OpenOffice.org Writer

Contents

Preface

- Why use OpenOffice.org?

Part I: Essentials

1. Introduction to Writer
    - Features of Writer
    - The Writer workspace
    - Using menus and toolbars
    - Using the Navigator
2. Setting up Writer
    - Choosing options to suit the way you work
    - Language settings
    - Preparing to check spelling
    - Controlling automatic functions
3. Designing your book
    - Templates and styles
    - Planning the book design
    - Basic page layout using page styles
    - Defining paragraph styles
    - Headers and footers
    - Page numbering
    - Creating a template

FIGURE 18.5
Contents Page of an Online Book.
Source: *http://taming-openoffice-org.com/*.

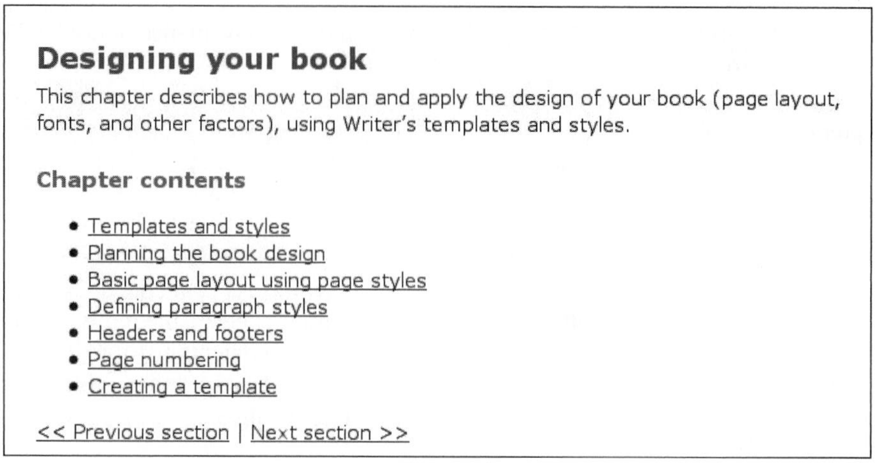

Designing your book

This chapter describes how to plan and apply the design of your book (page layout, fonts, and other factors), using Writer's templates and styles.

Chapter contents

- Templates and styles
- Planning the book design
- Basic page layout using page styles
- Defining paragraph styles
- Headers and footers
- Page numbering
- Creating a template

<< Previous section | Next section >>

FIGURE 18.6

First Page of a Chapter in the Online Book Shown in Figure 18.5.

Source: *http://taming-openoffice-org.com/*.

in writing or editing), the working environment and its pecking order, and the requirements of managers. This section describes some common methods for editing online publications.

Before Layout as Web Pages

If you are editing content before layout, the text may be in Microsoft Word or some other word processor, so you can edit it using the same program, just as you would for print publication. You can use the change tracking feature to advantage here. If content is created, maintained, or updated from a database, spreadsheet, or other files, you can edit those files using the appropriate software. That software may or may not have a change tracking feature.

Pages Already in Web Layout

If the text is already in a format for online display, you have several choices, each of which has good and bad points. Figure 18.7 summarizes these choices.

Printouts (Hard Copy)

Editors print individual web pages and write on the paper. Someone else then transfers the changes to the electronic copy.

Separate Comments Files

Editors type their comments into a separate file or email message and then send the file to someone else to transfer the changes to the electronic copy. This method is especially useful for more general comments or things that need to be discussed with the writer. (See the two case studies later in this chapter for some examples.)

Editing method	Is markup easily related to text and topic?	Is markup easy to do?	Does editor need copy of web publishing tool?	Is double handling required?
Printouts (hard copy)	Yes	Yes	No	Yes
Separate comments files	No	No	No	Yes
RTF, Microsoft Word, or other editable files	Yes	Yes	No, but may need other software	Maybe
HTML files with text editor	Yes	Maybe	No	No
PDF files	Maybe	No	No, but may need Adobe Acrobat	Yes
Web publishing program	Yes	Yes	Yes	No
Web-based collaborative writing and editing tools	Yes	Yes	No	No

FIGURE 18.7
Advantages and Disadvantages of Common Methods of Editing Online Documents.

RTF (Rich Text Format), Microsoft Word, or Other Editable Files

The online document is exported in RTF, Word, or another format. Editors insert changes and comments into the file using change tracking. Someone else then transfers the changes to the electronic copy. A variation is to copy text from the online document and paste it into a word processor file for editing.

HTML Files with Text Editor

Use a text editor or word processor to edit a copy of the pages. In this case, you'll probably see the HTML codes and be able to edit them as well.

PDF Files

Use Adobe Acrobat to capture the page as a PDF file, and then either print and hand-annotate it, or use Acrobat's tools to electronically annotate it (as described in Chapter 3).

Screen Captures

Use a screen capture program (in Windows, you can simply press the Print Screen key on your keyboard) to take a picture of problems that show on screen but not when you create a PDF or print a web page from a browser. You can print these out and mark changes on them by hand, or you can use a graphics program (Microsoft Paint will do) to mark changes and then send the file electronically.

Web Publishing Programs

Editors can edit a copy of the pages using the same web publishing program as is used for the site. If necessary, you can use various techniques (such as colors) to mark your changes or insert editorial comments online.

Web-Based Collaborative Writing and Editing Tools

Editors and writers can use a collaborative writing tool that enables everyone to work directly on the website or a copy of the files. New tools are appearing all the time. Two popular tools at the time of writing are Google Documents and wikis. With Google Documents, you can upload a variety of document types (including HTML, Word, OpenDocument Text, and RTF); several people can edit them within their web browsers, and the files can be exported to the same range of formats.

Wikis are web applications that enable documents to be written and edited in a simple markup language using a web browser. Although many wikis can be edited by anyone, they can be password-protected to restrict access to writers and editors.

EDITING ONLINE HELP

Online help may be displayed in a web browser or in a special help viewer. In each case, the principles of good organization and navigation are much the same.

Checklists are very useful when editing any document, as a reminder of the type of questions to ask (beyond the obvious ones of correct grammar and usage). Online documents have special requirements. As an example, use the checklist in Figure 18.8 to guide you when editing online help, whether it is provided in a web browser or in a specialized help viewer.

If any answers are "no" or "has weaknesses," identify specific examples and either correct them (if you can) or offer suggestions to the writer or website designer, especially if you don't know what might be the best way to resolve the problems.

Writing

❑ Is the help system well written?
❑ Is the writing style appropriate for the audience and the topics being covered?
❑ Is the language consistent throughout the help system?
❑ Is the language appropriate to the subject?
❑ Are procedures (if used) presented in clear sequential steps?

Content design

❑ Do the titles and headings clearly identify the information that follows?
❑ Are lists, tables, and graphics used effectively?
❑ Does the entry provide signposts to orient the user?

Overall quality

❑ Do all of the navigational elements, such as hyperlinks, behave in the expected manner without error?
❑ Is the help navigation (e.g., hyperlinks) error-free?
❑ Is the content consistent and appropriate for the audience?
❑ Is the interface consistent, easy to use, and reliable?

FIGURE 18.8
Checklist for Editing Online Help.

Organization/integration

❏ Is the help system well organized and is the organization appropriate for the audience?

❏ Is the organization obvious?

❏ Is the information organized into appropriate topics?

❏ Is the information organized into appropriate subtopics?

❏ Can you easily navigate between topics?

❏ Are there direct links or text references to relevant external documents, topics, or subtopics?

Table of contents (or navigational equivalent)

❏ Are the contents clearly identified in a table or navigational equivalent?

❏ Is the table of contents or navigational equivalent complete and comprehensive?

❏ Does the table of contents or other navigational equivalent provide an easy way to access the contents or move through branches of information?

Search

❏ Does the help system have an effective search mechanism?

❏ If there is a full text search capability, is it easy to use? Does it support wildcards, case sensitivity, and word variation?

Navigation

❏ Is it easy to find specific information, to navigate through information, and to return to where you started?

❏ Are navigational aids present and are they used consistently throughout the entry?

Usability

❏ Is the interface intuitive, easily interpreted, and consistent?

❏ Is information provided to assist the user, such as help, help-on-help, action cues, etc.?

FIGURE 18.8 (*continued*) Checklist for Editing Online Help.

CASE STUDIES

See the following case studies for some examples of conversations between editors and writers/designers as they work together to solve problems.

CASE STUDY: EDITING ONLINE HELP

Figure 18.9 shows the main contents page of a help file for an outliner or PIM (Personal Information Manager), which can store almost any kind of information and organize it in any way the user chooses. The audience for this program could include people from any age or background, but it would be especially useful for students and researchers.

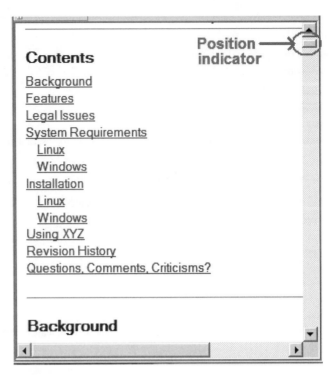

FIGURE 18.9
Contents List for Sample Help File.

Notice the location indicator on the vertical scroll bar. The help for this program is in one long file. Clicking on the "Using XYZ" link takes you a bit farther down in the file to another list of contents, as shown in Figure 18.10. As is often the case, this help was written by the programmer.

Put on your editorial hat and look at these contents lists from the audience's point of view. What do you notice first? The contents lists (especially the second one) are *function-oriented* instead of *task-oriented*—that is, they look at the various parts (functions) of the program and what those functions can be used for, instead of looking at what a user might want to do and then describing how to go about accomplishing those tasks. *Function-oriented writing* is suitable for reference material, but *task-oriented writing* is generally more appropriate for telling people how to use a program.

If you scrolled through some of the help, you would probably conclude (as the editor in our case study did) that some of it could be turned into task-oriented help if you reworded some of the headings and text and did some reorganizing. Most of the existing text could be retained as reference material. Separating the "technical" reference parts from the "how-to" parts will make the help much more useful. This is good to know because the time and money available to edit the help is very limited, so the editor won't be able to make major changes. You also want to break up the big file into a series of linked pages instead of one very long page.

But first, you need to get approval to make major changes to the file because your original task was to "do a quick copyedit." You produce some sample pages, do a cost-benefit analysis of the proposed changes, and prepare a proposal to take to your

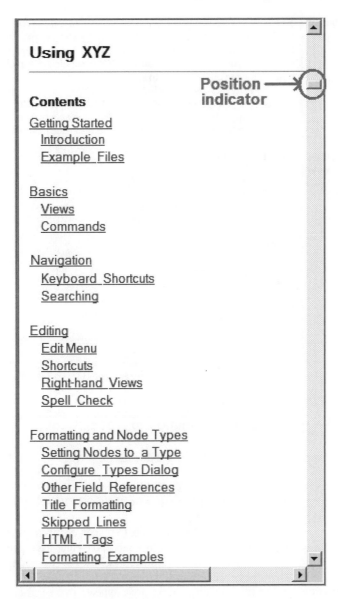

FIGURE 18.10
Contents List for a Subsection of the Help File.

manager and the programmer. Figures 18.11 and 18.12 show two examples of help topics to include in the proposal. A meeting with the manager and programmer will probably be necessary. At that meeting, you can explain more fully your reasons for suggesting the reorganization, and you can point out that although it looks like a lot of work lies ahead, the changes you suggest are actually relatively small, yet they will make a huge difference to the usability of the help file.

Because you have researched your proposal thoroughly and presented your case convincingly, your manager and the programmer agree to allow you to do the work.

Title of help topic	**Welcome to XYZ Help**
Brief description of process or purpose	This help file has two parts, the "how-to" section and the reference section. The "how-to" section contains step-by-step instructions for how to accomplish certain things with XYZ. It does not explain everything you could possibly do with XYZ, but clearly explains basic tasks so you can get up to speed quickly.
	The reference section includes a screen-by-screen description of the program and a detailed explanation of the program's functions.
	The two parts are cross-referenced, so you can move easily between the "how-to" explanations and the reference material.
Links to related information	**How to use XYZ**
	• A tour of the XYZ main window
	• Starting a new document
	• Adding information to a document
	• Getting information out of a document
	Reference
	• Formatting and Node Types
	• Field Types
	• File Import and Export
	• Output
	• Tree Data Manipulation

FIGURE 18.11
Sample Overview Topic.

Title of help topic	**Adding information to a document**
Purpose	To add information to a document:
Steps	1. Click Edit > Add Child on the menu bar. A text input box (showing the word "New") appears in the left-hand pane.
	2. Type a title for this node. (You can change it later.)
	3. Click on the Data Editor tab in the right-hand pane.
	4. Type the information to be included in this node.
What happens now?	The title (heading) of the node appears in the left-hand pane. To add subheadings, create additional "child nodes" under this node.
	You can now rearrange the order of nodes, add nodes, format the information, manipulate the information in a variety of ways, and output the information.
Related topics	**Related topics**
	What is a node?
	Editing information in a document
	Formatting information in a document
	Outputting information from a document

FIGURE 18.12
Sample Procedural Topic.

CASE STUDY: DO I HAVE TO LOG IN?

This case study shows how an editor can contribute to a project by identifying a problem and helping others find a solution. Editors don't need to know all the answers, and often there isn't any one "right" answer.

The scenario: A volunteer organization is developing a new website. The contents will be visible to the general public, but only registered users will be able to add content to the website or update existing content. The site therefore needs to have some way for people to register (create a new account) or log in to their existing account.

The first draft of the website design looked like this:

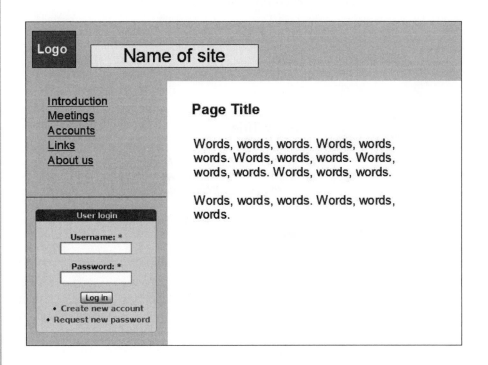

One editor's first reaction on seeing this page: "Do I have to log in?" Here is some email correspondence between the editor and the site's designer.

Editor: Can the user login section be made less prominent or put in another location?

Designer: "Less prominent" is very vague. We can edit the CSS. "Another location" is very vague. Where do you want it?

Editor: I'm just wanting to know what the possibilities are. The way the login area looks now, visitors—especially first-time visitors—might think they need to log in or create an account before they can read what is on the site. If they start clicking on the menu items, they should quickly discover that they don't need to log in, but some might just go away without doing that. Some people react very badly when faced with a login box!

So my thought is to make the heading "User login" in a smaller font, or add a few words—in the login box, not in the text of the page—to say that login is not required to read the site, or do something else—I don't know what—to counter the impression that login is required. I don't know how best to handle this, I'm just noting it as a possible problem. Perhaps just "User login (optional)" would do?

Designer: Our choices for making the login box less prominent are limited. We will need to get creative. I think that this is a legitimate concern, but I'll need time to think of a solution. Nothing comes to mind right now.

[Next day] I removed the login box entirely and put two links at the bottom that say Login and Register. These take you to a separate page. What do you think?

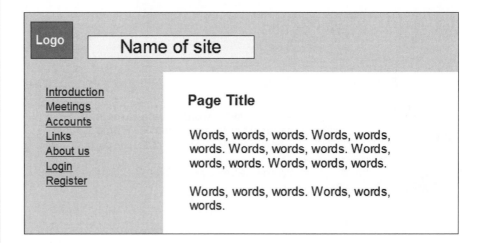

Editor: Excellent! Lots of sites have links like that. They are unobtrusive but easy to find and do not carry any implication that login or registration are necessary. I think it's a perfect solution.

CASE STUDY: WHAT DO I DO NOW?

This case study shows how editors and writers/designers may need to go through several steps to arrive at a good result. Each step corrects one problem, which then makes it easier to see other problems and possible improvements. Notice how the editor reinforces a cooperative approach by providing positive feedback at appropriate times.

The scenario: An education and training company wants to improve its website. In one section of the site, teachers can record the progress of their pupils in achieving various levels of competence (marked by three awards: Bronze, Silver, and Gold) and print award certificates.

The existing site works, but many teachers find it confusing, especially at first, so they ask a lot of questions, or they make mistakes. The company's website designer asks an editor to help reword (and if necessary, reorganize) the problem pages.

Designer: On the Pupils page, here is what I would like you to do: add one pupil, then add a class of three pupils using a CSV file, give all the pupils Bronze awards, and print one or more award certificates. Please comment on how easy or hard those tasks were, where you got stuck, how you figured out a solution, etc.

Pupils   Add a new Pupil | Add a class using a CSV file

Registration: Any time ▾   Select/unselect all ☐

Name	Registered	Award	Award date	Action	Print
Maya Toitovna	05 Apr 2005	Bronze	03 Aug 2005	🖼 ❗ [Add award]	☐
Saxifrage Russell	02 May 2006	Bronze	02 May 2006	🖼 ❗ [Add award]	☐
Frank Chalmers	02 May 2006	Bronze	02 May 2006	🖼 ❗ [Add award]	☐

[Print awards]

Before printing, please read our **printing instructions**

Editor: I had no problem adding one pupil.

Add Pupil

Add Pupil
Name: [＿＿＿＿＿＿＿＿]

[Add Pupil] [Cancel]

When I returned to the Pupils page, I could see that I needed to click the "Add award link" in the Action column. However, I wondered what the two icons in that column are for, why the pupils' names and the items in the Award column are links, and why most of the column headings are links. Without more instructions, I have no clue, and I'm nervous about just clicking around to see what happens, in case I mess something up.

Moving on, my first reaction to "Add a class using a CSV file" was: how the heck should I do that? I didn't want to click the link because I didn't have a CSV file (I assumed I would have to create one at some point), and I was not sure what should be in it. After clicking the link, I found some instructions.

Add a class of students

The text file should have a single column of student names. A "Comma Separated Value" export from a spreadsheet will produce the correct format.

Choose the award for your class, then upload your text file and press "Add class".

Add Class
Award: [Choose award ▾]
CSV File: [＿＿＿＿＿] [Browse...]

[Add class] [Cancel]

I opened a spreadsheet program, put in three names, saved the file as CSV, and returned to the website, thinking, "What a nuisance for only three names, but doing one name at a time would be a nuisance too."

I noticed that the "Add a class" page uses the word "student" but the other pages all use "pupil." This makes me wonder if there is a difference that I'm not aware of.

I also noticed that when adding a class, I must choose an award at the same time. This is different from adding a pupil, where it is done in two separate steps. I tried to add a class without choosing an award, but I got an error message. I guess if some of the pupils are getting a higher award, I can change it later. I chose Bronze award, browsed for the CSV file, clicked the "Add class" button. All seemed okay, and the three pupils I added are now listed on the Pupils page.

Why can I not choose an award when adding a single pupil? If I had done a class first and then added a single one later, I would have had an expectation of being able to do both in one step, and I think I would have been annoyed.

Oh good, there is a link to printing instructions. Let's see, I guess I mark the checkboxes in the Print column and click "Print awards." Yes, that works. I'm glad to see a "Select/unselect all" checkbox as well; that makes it easier to give the same award to everyone.

Recommendations: Provide more explanation and instructions (perhaps a Help link), use the same method to add an award for one pupil or a group of them, give some clue of what the Action icons are for, and use the same term (*pupils* or *students*) throughout.

Designer: I plan to remove the part where you assign an award at the same time as adding a class because that functionality creates all sorts of problems. For example, say you have thirty kids and give them all Bronze. A while later, they get Silver. If the CSV file is seen as a way to give out awards, then you'll use the CSV file again. Which means that all thirty pupils will be duplicated. They will look like thirty *new* pupils instead of a new award for the old thirty. This is made worse by the fact that the old site made it very inconvenient to assign many awards to existing pupils.

Editor: I had not thought about that problem, but now that you mention it, I completely agree that your suggested change would be much better, for the reasons you gave.

Designer: Here is the next iteration. Please do the same tests and tell me what you think.

Pupils   Add a new Pupil | Add a class using a CSV file | **Help**

Registration: Any time ▾   Select/unselect all ☐

Name	Registered	Award	Award date	Action	Select
Maya Toitovna	05 Apr 2005	Bronze	03 Aug 2005	📋 ❗	☐
Saxifrage Russell	02 May 2006	Bronze	02 May 2006	📋 ❗	☐
Frank Chalmers	02 May 2006	Bronze	02 May 2006	📋 ❗	☐

Choose award ▾ [Add awards] [Print awards]

Before printing, please **adjust your print settings**

Editor: Yay! A Help link. Oh good, lots of instructions, including ones for creating a CSV file. Hmmm . . . still no clue what those icons in the Action column are for. Add pupil: No problem. Add a class: Oh good! A link to "Learn how . . ." The instructions are good and clear; no confusion here. I feel confident. Add class: No problem. Giving awards: No problem. I hardly had to think about what to do; it was really obvious to me: select names, choose award, click "Add awards."

I think the new site is a great improvement in layout, wording of instructions, and consistency in the method of adding one or several pupils, without specifying an award at the same time. Having only the award level (Bronze, Silver) in the Award column, not the redundant word "award," is good too. The Pupils page is much less cluttered now.

Designer: I'm glad that you noticed all the things I changed. That's very encouraging. I've been thinking of other changes, and your test encourages me that they are good ones. Tell me what you think:

Remove the Action column entirely. Those icons are completely meaningless and people may be afraid to click them because the change might not be reversible. Instead, I'll have buttons at the bottom that say "Edit" and "Delete."

Replace the two links for adding pupils by a single one that says "Add pupils." The new page will have a text field where you can type in one or many names (one per line) or upload a CSV file.

Editor: Yes!! on both ideas. I especially like your solution to the one-or-many pupils problem. If I have only a short list of names, using a CSV file is a nuisance, and so is entering them one at a time. The ability to type several names in a text field is, in my opinion, a perfect solution to the short-list issue. And retaining the CSV file for a longer class list is good, too. Gives everyone options that suit them best.

Designer: I made a new test site with the changes I told you about. I was hoping you could take a quick look and verify that it works, and it's better.

Pupils Add new pupils | **Help**

Display pupils registered: Any time ▾   Select/unselect all ☐

Name	Registered	Award	Award date ↓	Select
Maya Toitovna	05 Apr 2005	Bronze	03 Aug 2005	☐
Saxifrage Russell	13 Jun 2006	Bronze	13 Jun 2006	☐
Frank Chalmers	13 Jun 2006	Bronze	13 Jun 2006	☐
Ann Claybourne	13 Jun 2006	No awards yet		☐

Choose award ▾ [Add awards] [Edit] [Delete] [Print awards]
Before printing, please **adjust your print settings**

Editor: Yes, it works, and it's better. A great improvement, actually, and a great example of how small changes can make a big difference in the user's experience. The interface is cleaner and less cluttered, the additions to the Help are good, and I especially like the ability to add a list of pupils by hand.

Add Pupils

Please enter the pupil names, one name per line.

Or upload a CSV file (how do I make a CSV file?).

[ ] [Browse...]

[Add pupils] [Cancel]

Being able to sort the list in different ways by clicking on the column headings is a nice touch that I forgot to mention in my previous review.

Another thing I forgot to mention earlier (as it wasn't relevant to the set of tasks) is the Registration drop-down confused me until I realized it was a filter.

Lastly, I spotted a few trivial copyediting errors in the Help file, but thought I would wait until the UI is finished and the Help is stable before I wrote them up and sent them along. Let me know when you want them.

Designer: On the drop-down, I replaced the word "Registration" with "Display pupils registered." Do you think that's better? Now is a good time to send the copyediting corrections.

Editor: "Display pupils registered" is a definite improvement. I agree that sorting by column should be useful to teachers. Lastly, I've attached the copyedits. Congratulations on a great job!

Designer: People never realize how much work goes into making very usable interfaces. Because it looks simple, they figure it was easy. It's taken many iterations to reach this point. Some previous versions were more complicated but added needed functionality. The main problem was that teachers could easily have 200 pupils, and I needed ways to reduce the time spent searching for the one they want. So I had a lot of filters. Then I replaced those filters by the ability to sort by column and just the one registration date filter. Sorting by column could be very useful. At least, I hope so. For example, you could sort by date to find the most recent pupils. Or sort by award to find the ones that don't have an award yet.

EXERCISE 18.2

1. Find a particularly weak website. Using the technical editing strategies in this chapter, analyze the website and write up your analysis for your instructor. Then, using your own advice, revise the website so that it is more effective.

2. Have your instructor find a website that the class should revise. Or the instructor could find several poor examples and have the class split up into groups.

SUMMARY

This chapter covers several components of editing online documents. As a technical editor, you will need to know how to edit for organization and readability, including editing text links and choosing the best page length, in order to make the material more helpful and effective for users. This chapter also addresses several methods for editing online content electronically. We believe that these software skills will be some of the most crucial that you will encounter as a technical editor. Case studies demonstrate how editors can interact with writers and designers to improve the usability of websites and online help.

CHAPTER 19

Production Editing of Websites

At the conclusion of this chapter, you will be able to

- explain twelve common complaints about web pages and websites and their causes;
- edit a website for layout and presentation;
- edit a website for accessibility;
- edit a website's navigation;
- edit a website's keywords and text to improve searchability.

In this chapter, we'll look at some common problems with websites and suggest ways that you as a technical editor can look for them and make recommendations to the site design team.

COMMON COMPLAINTS ABOUT WEBSITES

1. The page takes too long to download or display. *Cause*: This is almost always related to the images: too many, too large, no height/width specified, or being loaded from a different website.
2. The page is an image with no information content, no alternative text, and no obvious way to bypass it and get to something useful.
3. The page has one or more images that show essential information, but there is no alternative text, so I don't see the information if graphics are turned off or if I'm using a screen reader.
4. The page downloads a video or audio file without asking me first if I want it.
5. The text is too small to read comfortably, and I can't easily enlarge it. *Cause*: The type size has been specified as a fixed size instead of a variable one, or (worse) the text is actually a graphic.
6. The colors are difficult to read onscreen or the background obscures the text (not enough contrast between the background and text colors).
7. I have to scroll sideways to see the whole page. *Cause*: The page width has been set to a fixed size suitable for a higher-resolution monitor, or something (usually a graphic or a table) on the page is forcing the minimum page width to be large.

8. When I print the page, some of it is cut off on the right-hand side. *Cause*: The page or column width is fixed, so it won't adjust to fit within the margins of a printed page, and no alternative style sheet has been defined for printing the page.

9. The page doesn't display (or work) properly on my browser. *Cause*: The page was produced using nonstandard features, or it uses standard features that work only with the latest browsers, and no provision has been made for older browsers or for browsers that don't support the standards.

10. I can't find what I'm looking for. *Cause*: Poor navigation aids, poor or no search or index facility, and site design not logical from users' point of view.

11. I have to click through too many pages to get to what I want.

12. I can't bookmark a page so I can return to it easily. *Cause*: The site uses frames or some other programming construction that does not give users a direct link to specific page content.

How much can you as a technical editor do about these problems? Most of these complaints are about design, organization, and presentation of the site, not content, so you may not have much opportunity for input; but some complaints do have a content component. If you are restricted to editing content, try to keep in mind how that content might be presented to the audience. If you're working with new material for an established site, you should have a good idea of what the final product will look like and a style guide to follow so that you can edit appropriately.

THE TEN-MINUTE LAYOUT EDIT

If you get a chance to view the pages online, you can check for a variety of other problems. An easy ten-minute check of some sample pages can reveal some of the more glaring potential problems, which you can then bring to someone's attention. Here is what to do:

1. Change the font size to something large. Do some parts of the page overlap other parts, or does some text disappear (as shown in Figures 19.2, 19.5, and 19.6)?
2. Using the normal font size, make the browser window narrower. Do some parts of the page overlap other parts, or does some text disappear?
3. Use several browsers (preferably including one older version, one on a different operating system, and one on a cell phone), or use an online service like Browsercam.com.

EDITING FOR LAYOUT AND PRESENTATION

Some things that an editor should check must be done on the finished product, just as you would check the page proofs for a book. At the production editing stage, you should not be looking for problems with the words; you should be looking for problems with layout and presentation:

- Some headings are at the wrong level.
- One or more images are in the wrong place.

- Columns in tables don't line up.
- Indentation or spacing of numbered or bulleted lists is inconsistent or incorrect.
- The text is hard to read on a patterned background.
- Links are difficult to distinguish from other text.

TIP FOR TECHNICAL EDITORS

Make sure the page title (as shown in the browser's title bar) is correct and meaningful. How many times have you seen a page title that says "Page Title" or a website where every page has exactly the same title, typically the name of the site or the company?

Checking for layout and presentation problems is best done on a monitor instead of on paper. Any printout other than a screen capture is likely to be just a bit different from what shows on a monitor; unless you use a monitor, you can't see what happens if you change the size of the font or the browser window.

TIP FOR TECHNICAL EDITORS

Don't obsess over minor differences in appearance of a design in different web browsers or on different operating systems. Fonts, layouts, and even colors may vary slightly or greatly. The important thing is that the page is usable—that is, visitors can read the text and use the navigation or other controls.

In addition to the problems described earlier, here are two specific problems that editors should look for when reviewing the content of a web page or website online:

- Text or images that overlap other text or images. See Example 1 (Figure 19.2) and Example 3 (Figure 19.6). Note: If done well, overlapping is okay; see Example 2 (Figure 19.4) for overlapping that works well.
- Some text that is not visible because the space available does not expand when needed. See Example 3 (Figure 19.5).

Example 1: Text Overlaps Other Text (Bad)

Even when pages are designed using CSS positioning (as they should be), enlarging fonts may cause one column to become wider and overwrite the next column to the right, as demonstrated in Figures 19.1 and 19.2. As an editor, you should bring the problem to the attention of the site designer, who can fix it easily by changing some positioning attributes in the CSS.

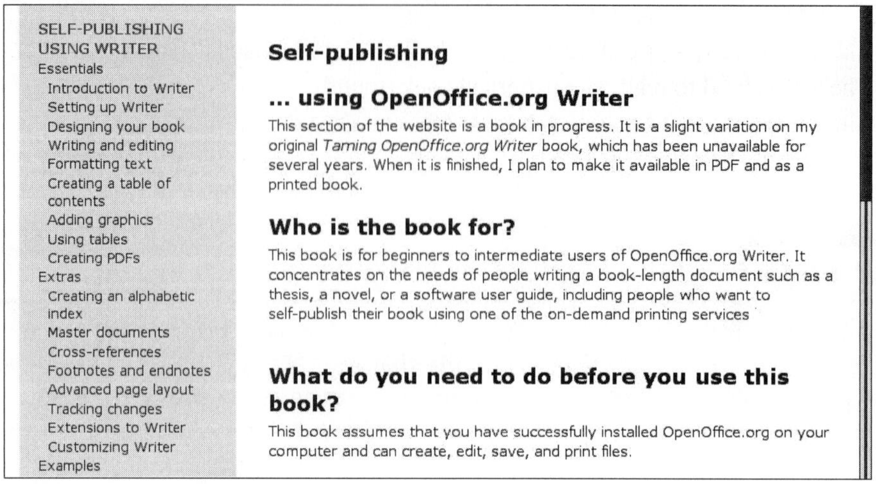

FIGURE 19.1
A Two-Column Web Page with Optimum Font Size.

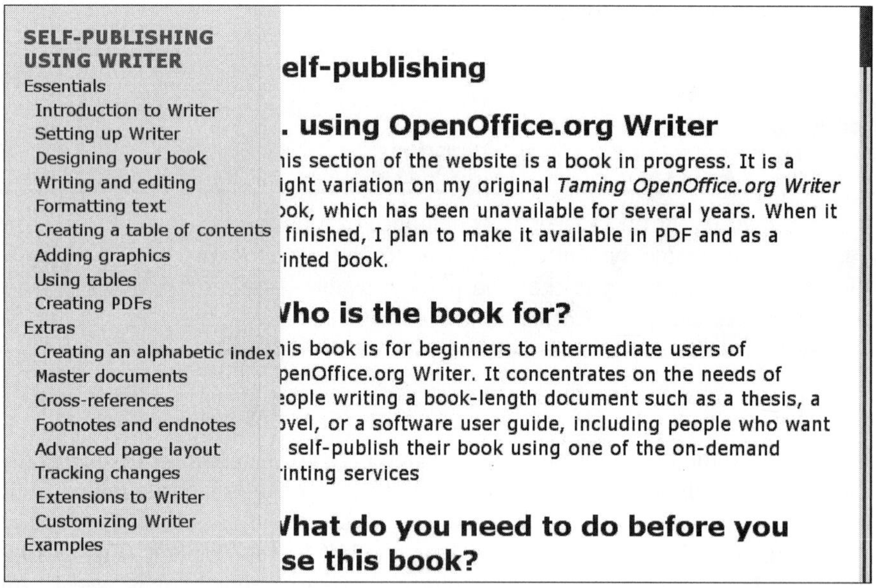

FIGURE 19.2
The Same Page with Enlarged Font; the First Column Overlaps Text in the Second Column.

Example 2: Essential Text Overlaps Nonessential Images (Good)

Figures 19.3 and 19.4 show one way that designers have dealt with the problem of narrow browser windows. Notice how some of the text adjusts to fit the available space, and another block of essential information slides on top of a nonessential image when the browser window is narrow. This technique ensures that the pages work even though some of their layout changes.

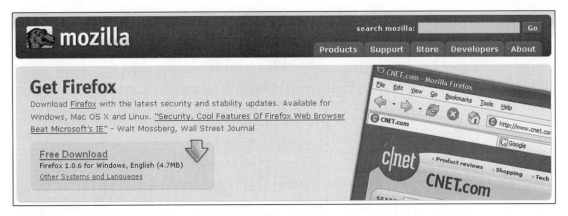

FIGURE 19.3
Page Displayed at Nearly Full Width of a High-Resolution Monitor.
Source: Mozilla®.

FIGURE 19.4
Same Page Displayed in a Narrow Browser Window.
Source: Mozilla®.

TIP FOR TECHNICAL EDITORS

The tabs on the navigation bar on the Mozilla page are not graphics; they are text links. You can verify this by viewing the page with the display of graphics disabled; the words on the tabs are still visible, even though the "tab" background isn't.

Example 3: Text Disappears or Overwrites Other Text (Bad)

When text within a fixed-height area is enlarged, some of it may disappear, as shown in Figure 19.5. This type of problem is commonly found in sidebars, including ones on websites targeted at senior citizens, an audience group that contains many people who need to enlarge the fonts in order to read them comfortably.

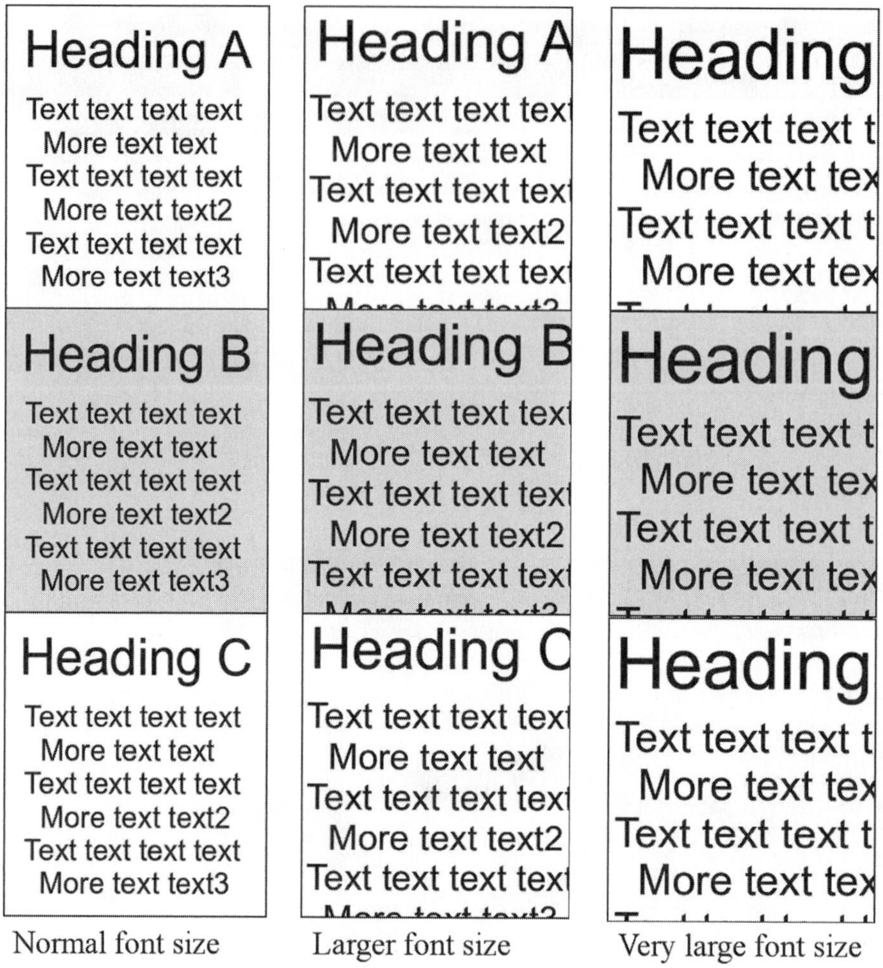

Heading A	Heading A	Heading

Normal font size   Larger font size   Very large font size

FIGURE 19.5
Text in Fixed-Height Boxes May Disappear when Fonts Are Enlarged.

In other cases, enlarged text within a fixed-height area may overwrite text in other areas. This effect (shown in Figure 19.6) is different from the one shown in Figure 19.2. The site designer can fix both problems by changing some attributes in the CSS.

EDITING FOR ACCESSIBILITY

Website accessibility is a hot topic these days, but most of the same issues apply to any electronically delivered documentation including PDFs. Audiences include blind and vision-impaired users and users with hearing, physical movement, and cognitive disabilities.

When people talk about website accessibility, they are usually thinking about visual impairments that require visitors to use a screen reader. But there are many other disabilities, both permanent and temporary, that designers should consider when making a site accessible. In addition, even when users are not disabled, technology issues may affect the accessibility of websites they visit. The "Accessibility Criteria" section in Chapter 17 summarizes reasons

Increased accessibility for users with disabilities almost always leads directly to improved usability for *all* users.
—*Jakob Nielsen*

Heading A	Heading A	Heading A
Text text text text	Text text text text	Text text text text
More text text	More text text	More text text
Text text text text	Text text text text	Text text text text
More text text2	More text text2	More text text2
Text text text text	Text text text text	Text text text text
More text text3	More text text3	More text text

Heading B text (overlapping in the two enlarged columns)

Heading B	Heading B	Heading B
Text text text text	Text text text text	Text text text text
More text text	More text text	More text text
Text text text text	Text text text text	Text text text text
More text text2	More text text2	More text text2
Text text text text	Text text text text	Text text text text
More text text3	More text text3	More text text3

Normal font size   Larger font size   Very large font size

FIGURE 19.6
Text in Fixed-Height Boxes May Overwrite other Text when Fonts Are Enlarged.

for making online materials accessible and lists common accessibility problems and how to avoid or solve them.

To edit a website quickly for some accessibility problems, do these tests in addition to the items given in "The Ten-Minute Layout Edit" earlier in the chapter:

1. Turn off graphics. Is all the essential information still available? Can you still navigate? Figure 19.8 shows a typical result of turning off graphics. The easiest way for an editor to find missing or poorly written (uninformative) *alt* information is to turn off the images. For more about the *alt* (alternative text) attribute, see Chapter 16.

2. Test the site at modem speed to check the download time and whether the text loads before the graphics.

3. Turn off the sound and check whether audio content is still available through text equivalents.

4. If you can, change the display color to grayscale (or print out the page in grayscale or black and white) and observe whether the color contrast is adequate.

5. Without using the mouse, use the keyboard to navigate through the links on a page (for example, using the Tab key), making sure that you can access all links and that the links clearly indicate what they lead to.

Other methods to use, if you have the time and tools, are summarized on the W3C's page "Preliminary Review of Web Sites for Accessibility" (http://www.w3.org/WAI/eval/preliminary.html), and the linked pages of automated tools for evaluating web accessibility (http://www.w3.org/WAI/ER/tools/).

TIP FOR TECHNICAL EDITORS

The method of turning off images depends on your browser, but it is generally found in Options or Preferences. Some browsers provide a clickable icon to turn images off or on quickly.

Example 4: Images Disabled—Good Use of Alternative Text

Figures 19.7 and 19.8 show a web page with images enabled and disabled respectively. Notice the alternative text visible in the no-images view; this text can be read by a screen reader.

EDITING FOR USABILITY

Usability is the extent to which a product can be used by specified users to achieve specified goals with effectiveness, efficiency, and satisfaction (see Chapter 23 for a more in-depth discussion of usability).

The usability of printed publications depends on factors including

- Design elements like typefaces, type sizes, and the use of white space, as discussed in Chapter 10.
- The logical organization of ideas, as discussed in Chapter 14.
- Navigation aids such as the table of contents, index, headings, and page headers and footers.

Photos left and right: some of the siltstone "sculptures" on Edeline Island. We continued northeast past an area known as "The Graveyard", where many pearl divers perished due to treacherous currents or the "bends" (nitrogen narcosis). Then on to the Whirlpool Passage, a channel that funnels the tides into whirlpools. Tides along this coast are extreme, up to 9 metres, and can get quite treacherous, especially in narrow areas.

Along the way we saw some very interesting tilting of ancient sandstone rock, and a group of yellowfin tuna leaping through a school of baitfish.

I took some photos of the Whirlpool Passage, but they don't convey the sensation of being whirled around by the water as we were. (You had to be there!) I would not enjoy doing that in a sailboat, although it was fun in the Explorer.

FIGURE 19.7
A Web Page with Images Enabled.

```
     Siltstone sculpture on        Photos left and right: some      Siltstone "sculpture" on
     Edeline Island                of the siltstone "sculptures"    Edeline Island
                                   on Edeline Island. We
                                   continued northeast past
                                   an area known as "The
                                   Graveyard", where many
                                   pearl divers perished due to
                                   treacherous currents or the
     "bends" (nitrogen narcosis). Then on to the Whirlpool Passage, a channel that
     funnels the tides into whirlpools. Tides along this coast are extreme, up to 9 metres,
     and can get quite treacherous, especially in narrow areas.

     Tilted sandstone,             Along the way we saw some        Yellowfin tuna
     Strickland Bay area           very interesting tilting of
                                   ancient sandstone rock, and a
                                   group of yellowfin tuna
                                   leaping through a school of
                                   baitfish.

                                   I took some photos of the
     Whirlpool Passage, but they don't convey the sensation of
     being whirled around by the water as we were. (You had to
     be there!) I would not enjoy doing that in a sailboat,
     although it was fun in the Explorer.
```

FIGURE 19.8
The Same Page with Images Disabled.

Online publications often need somewhat different design elements, and they organize ideas nonsequentially using hypertext links and use very different navigation aids. Jim Thatcher lists five elements of website usability:

- *Learnability:* Can visitors use the website effectively the first time they visit it, without becoming frustrated?
- *Memorability:* Will visitors remember how to use the website the next time?
- *Effectiveness:* Can visitors easily navigate through the website, determine what to do next, and understand the content? Is the design consistent and predictable?
- *Efficiency:* Can visitors find what they need and accomplish their goals in a reasonable amount of time?
- *Satisfaction:* Do visitors have a good feeling about using the website? Will they use it again? Is the content presented effectively?

Numerous books and articles describe studies of website users and the many ways in which they can misunderstand and be confused by the layout and navigation, thus failing to find what they want. For example, Jakob Nielsen and his colleagues, in a series of books including *Designing Web Usability, Homepage Usability, Prioritizing Web Usability*, and *Eyetracking Web Usability*, analyze web pages and offer specific recommendations for improving usability. Steven Krug, in a book cleverly titled *Don't Make Me Think*, provides an analysis of specific pages' problems and shows examples of improved versions of the same pages. Janice Redish's *Letting Go of the Words* is an excellent, in-depth, easy-to-read book containing numerous examples and case studies.

> If customers can't find a product, they won't buy it.
> —*Jakob Nielsen*

Krug's book is a good introduction to the topic because it is shorter and easier to read than Nielsen's books. Use Krug's book if you need to convince your boss that changes should be made to a site's design. Use Redish's and Nielsen's books to go into more depth and extend the basic principles into more specific situations.

Technical editors must understand how people use materials online and edit those materials with these requirements in mind. Editors must also know the usability requirements of websites and individual web pages and ways to test for usability. Do usability testing with real users if at all possible. (See Chapter 23 for more about usability testing.) But before testing, use your editorial skills to evaluate the site. Often relatively minor changes to the design will make major changes to its usability.

SOME USABILITY PRINCIPLES FOR WEBSITES

- Use a standard layout that readers are familiar with—for example, a logo in the upper left-hand corner, company name or other page title at the top to the right of the logo, graphic buttons down the left-hand side, text links below the page content, contact information, and the date of the last update below the text.
- Be consistent in the placement and wording of navigation aids.
- Provide more than one way for users to navigate through the site, but don't confuse them with too many choices that aren't readily identified as different ways to find the same thing.
- Allow mouseovers (change of images or text highlighting) on navigation buttons, as long as the site will still work even if the mouseovers don't work.
- Avoid using navigation icons without associated text because not everyone will know what the icons mean. (The text can be in a *title* attribute, so it pops up when you hover the mouse pointer over the icon, and in an *alt* attribute that specifies text that is given when the visual can't be displayed.)
- Avoid cryptic or ambiguous words or phrases in navigation. Whenever possible, use common terms that your intended audience will recognize.

Editing a Website's Navigation

Navigation is a necessary evil that is now a goal in itself.
—*Jakob Nielsen*

A major aspect of a site's usability is its navigation. Good navigation helps users find what they want quickly and easily. Navigation is not the only way to help users, nor is it always the best way, but poor navigation is almost guaranteed to be a problem.

Questions to Ask When Editing for Navigation

Is the site organized in a way that is meaningful to the audience, not just to the organization that produced it?

If possible, have a list of typical tasks that a user might do when visiting the website. To edit for navigation, perform those tasks, while asking the following questions:

- What are the users' most likely questions?
- Are the answers easy to find?
- Are the navigation labels useful?

Is it always clear to users

- Where they are?
- How to get back to where they came from?
- How to get somewhere else on the site?

Are the navigation aids

- Easy to locate?
- Clearly and meaningfully labeled?

Look at the hyperlinks.

- Are the links helpful, logical, trivial, confusing?
- Is it clear where the links go and what the reader will find there?
- Does summary information tell readers which links they need?

The KISS Principle for Websites

KISS ("Keep it Short and Simple" or "Keep it Simple, Stupid") definitely should apply to websites, or at least to those sites that are intended to provide information or sell products or services. A very helpful exercise is to make some notes when you visit a site that you find particularly helpful or annoying. What are the site's good or bad features? Here are some questions to ask.

- Does the first page have enough useful information to lead me to the part of the site that I want? (Or is it so "simple" that it has little or no information content at all?)
- Is the first page cluttered? If so, how could it be improved? (Do I have difficulty finding the information I want because it's buried in a lot of other stuff?)
- Are the navigation aids helpful? If not, how could they be improved?
- Do I feel that I know where the links lead, or do I have to guess and hope I don't waste too much time finding the link I need?
- Are the color combinations easy or difficult to read?
- Is the site's design consistent from one page to the next, thus assisting me in becoming familiar with what's there?

EDITING FOR SEARCHABILITY

Searchability refers to the ease with which readers can find the information they want. It applies at two levels: finding the website on the Internet and finding specific information on the website.

The audiences for many websites will not need to search for them; they will find out about the site through other means. But if the site you are editing is intended for a wide audience, you may need to consider how the audience will find it.

Search engines and directories can bring people to the main page of a website or to specific pages on the site. Once within the site, visitors should be provided with more than one way to find the specific information they're looking for. These methods can include within-site links, a search facility for the site, a book-style index, or a table of contents.

Finding Pages Using Search Engines

Readers typically look for information on the Web using a search site such as Google. Search engines use several parts of a web page when evaluating it: the page title (as seen in the browser's title bar), the text on the page, and the metadata information placed in the code but not visible in a browser.

Most websites provide a site-specific search function. A good website provides several ways for visitors to find information. A well-designed site helps users find what they want without searching, but many visitors prefer to use a search facility rather than attempt to find information through menus or links (site navigation).

Editors should ensure that several ways are provided for visitors to find information. The methods might include tables of contents, indexes, and site maps as well as a search function. Many online publications include tables of contents or indexes, similar to those found in printed books. The main difference is that the user can click on the heading or index entry and jump directly to the associated topic.

A site map is similar to a table of contents in that it groups topics in a hierarchical manner in order to assist readers in finding the topic they want. Despite the term "map," most site maps are text in outline form, not a graphical representation. One advantage to a site map is that a topic can be listed in several places, with all the links leading to the same page. Figure 19.9 shows a very simple site map, much like a table of contents but with the topics in alphabetical order.

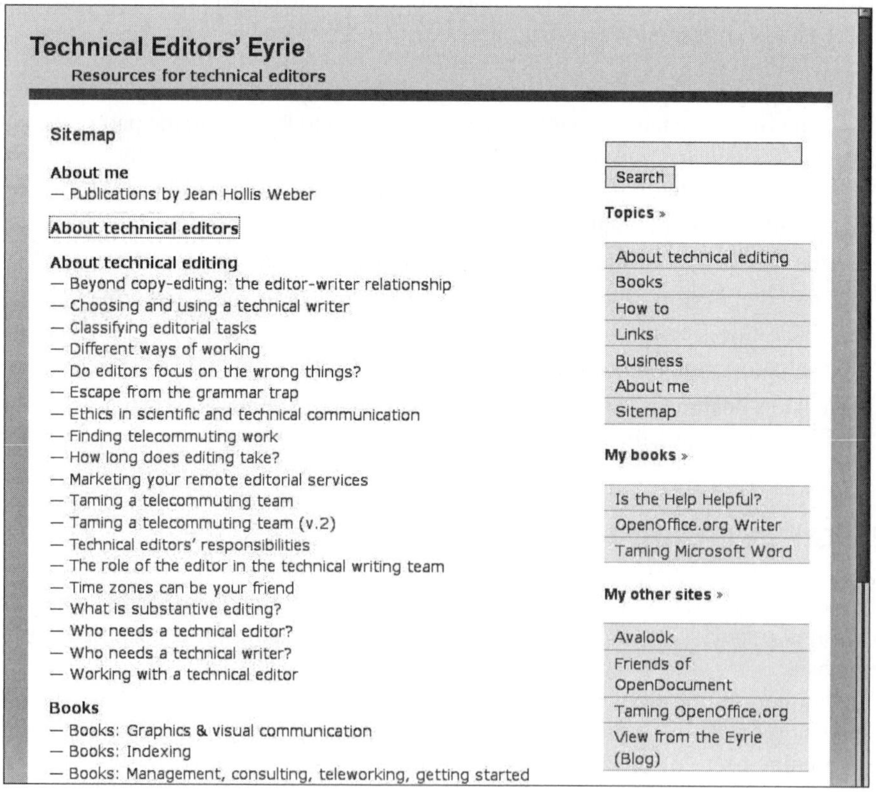

> While it is important to provide information on the web, it is equally important to provide methods for users to retrieve that information.
> —*Glenda Browne and Jonathan Jermey*

Technical Editors' Eyrie
Resources for technical editors

Sitemap

About me
— Publications by Jean Hollis Weber

About technical editors

About technical editing
— Beyond copy-editing: the editor-writer relationship
— Choosing and using a technical writer
— Classifying editorial tasks
— Different ways of working
— Do editors focus on the wrong things?
— Escape from the grammar trap
— Ethics in scientific and technical communication
— Finding telecommuting work
— How long does editing take?
— Marketing your remote editorial services
— Taming a telecommuting team
— Taming a telecommuting team (v.2)
— Technical editors' responsibilities
— The role of the editor in the technical writing team
— Time zones can be your friend
— What is substantive editing?
— Who needs a technical editor?
— Who needs a technical writer?
— Working with a technical editor

Books
— Books: Graphics & visual communication
— Books: Indexing
— Books: Management, consulting, teleworking, getting started

Search

Topics »

About technical editing
Books
How to
Links
Business
About me
Sitemap

My books »

Is the Help Helpful?
OpenOffice.org Writer
Taming Microsoft Word

My other sites »

Avalook
Friends of OpenDocument
Taming OpenOffice.org
View from the Eyrie (Blog)

FIGURE 19.9
Site Map Example.

Browne and Jermey, in their book *Website Indexing*, note that site-specific search engines are useful if the site is too large to browse easily; if users come to the site to search for information; if the site is dynamic, with content changing often; or if related information is scattered throughout the site. Most users are now accustomed to looking for a site-specific search box as one way to locate information.

Search engines use small programs called *robots, bots, spiders*, or *web crawlers* to visit publicly accessible websites. The programs follow all the links they find and bring back information that goes into the search engine's index of web pages. The search engine then uses a set of rules to determine each page's rank in a listing of pages corresponding to particular search terms. The rules give different weights or values to different types of information, and the people who run the search engines change the rules from time to time.

The type of information needed to optimize search results is the same for site-specific searches and for Web-wide searches: carefully chosen keywords that match the search terms the audience uses. One of your jobs as a technical editor should be to assist in choosing keywords and ensuring that they are included where they are needed on web pages and in metadata.

The following methods may help to improve the page's rank in search engines:

- Include synonyms and common misspellings of key concepts in the keywords.
- Include unambiguous keywords in titles and headings within the text.
- Make sure the first 100 words on the page make a good description.
- Use <meta> tags for keywords, title, author, and description.

It shouldn't be too much of a surprise that the Internet has evolved into a force strong enough to reflect the greatest hopes and fears of those who use it. After all, it was designed to withstand nuclear war.
—Denise Caron

Finding Sites Using Subject Directories

In contrast to search sites, a subject directory (sometimes called a *subject gateway* when limited to a specific topic) is a collection of evaluated links to websites. At some point in the process, people evaluate websites to determine whether they appear in the directory. The Social Science Information Gateway (http://www.sosig.ac.uk/) is an example of a subject directory; many disciplines have similar directories.

From the point of view of the user of a subject directory or gateway, the process of searching is the same as for a general search engine. The information you need to put into your web pages so that people can find them is the same in both cases.

The Importance of Metadata and Keywords

The term *keyword* is used for two different but related concepts: (1) words or phrases (also called *search terms*) used when searching for a topic; and (2) metadata terms. The words or phrases may occur in the text itself, or they may be synonyms for concepts or terms in the text.

Metadata Tags

As discussed in Chapter 16, metadata is structured information about information. *Metadata tags* in the <head> section of a web page contain metadata information about the page itself. These tags are not visible in a web browser, but they are visible to search engines or to anyone who examines the source code of the page.

Many of today's generalized search engines pay less attention to keywords in metatags than they once did, but for specialized searching or searching with a site itself, they are still a valuable tool.

You can edit the contents of metatags using a text editor or web authoring tool, or for specialized work, you could use a metadata editor like one of those listed at http://metadata.net. Specialized programs are particularly useful if you are following a recognized schema such as the Dublin Core or the U.S. Library of Congress's Encoded Archival Description (EAD). For example, the Firefox Dublin Core Viewer Extension adds a button to the browser's status bar (and, optionally, the toolbar) to access an overview list of Dublin Core Metadata.

EXAMPLES OF METADATA TAGS

```
<meta name="description" content="Technical editors work in
a variety of fields, editing both printed and electronic
materials.">
<meta name="keywords" content="technical editor editing
electronic">
<meta name="author" content="Jean Hollis Weber">
```

Use the meta keyword tag to list search terms relevant to the page. You can also include common misspellings or other words (synonyms) that may not appear anywhere on the page. A good set of keywords encapsulates the specific topics the page covers.

Use the meta description tag to summarize the contents of each page. Many search engines display this description as part of their results. The description is often the same as the first paragraph of visible text on the page.

To assist in metadata creation, you might use a controlled vocabulary of metadata terms or a thesaurus (a structured list of approved subject headings). These lists are particularly useful for scholarly or other specialized subject areas (science, medicine, law, for example). Refer to the U.S. Library of Congress's Encoded Archival Description (EAD) metadata standard.

For a detailed explanation of metadata creation, controlled vocabularies, and related topics, including many references to resource material, see Browne and Jermey's *Website Indexing*.

Keywords in Text

In addition to keywords in metatags, you need to ensure that many of the same keywords also occur in the text of the web page—especially the headings. Even if someone else deals with the meta keywords, your job as editor should include ensuring that relevant terms appear on the page, preferably in the first 100 words. So you might need to edit the titles or the first few paragraphs of pages or provide summaries or abstracts.

Many search engines use headings to rank a page in relevance for a particular search. They assume that words in headings are more important than words in the text, so the pages are more relevant to that search. Mark these headings with HTML heading tags to make sure the search engines recognize them as headings.

Page Titles

Page titles are the words you see in the browser's title bar and are also used as the label for the page when someone adds the words to a bookmark list. A page title is often the same as the highest-level heading on the page. The title is placed in the <head> section of a web page and looks like this:

```
<title>About technical editing</title>
```

The importance of a good title becomes clear when you consider how the results of a web search are displayed. The list usually contains page titles and some kind of text, either the meta description data, the first few lines of the page, or a summary of the most important text.

SEARCH ENGINE OPTIMIZATION

Search engine optimization (SEO) is the craft of choosing keywords and using them carefully in titles, headings, text, and metatags to improve the position in which a particular page will appear in search engine results.

Links into a Website (Backlinks)

Links into your website, also known as *backlinks*, are similar to citations of a book or research paper by other writers. Search engines may consider the number of links into a site as an indicator of the site's importance or relevance to a topic.

CASE STUDY: WHY IS MY WEBSITE NOT SHOWING UP ON GOOGLE?

A technical writing instructor created a website to advertise the courses she offered. After some months, she asked an editor why her website was not showing up on Google although other websites that had links to her site (but did not offer courses of their own) did show up.

> "What keywords are you searching for?" asked the editor.
> "Technical writing courses Australia" or "technical communication courses Australia" or "documentation courses Australia," answered the instructor.

Knowing that Google and other search engines use a combination of factors when determining page rank, the editor looked at the website and immediately found several problems:

- The page title (the words that show in the browser's title bar) was either the name of the company (XYZ Communications) or the word "Courses." These titles contain either none of the keywords in some of the searches or only one keyword ("communication" or "course") in others.

- Most pages had the same title, instead of a unique title for each page, relevant to the contents of the page.
- The heading on the homepage said "Welcome" (no relevant keywords), and the first paragraph contained no relevant keywords either. The pages describing the courses had relevant headings but no descriptive paragraphs—they were just course outlines, and the topics often did not contain relevant keywords.

Looking at the "page source" (the HTML code), the editor found more problems:

- The headings on each page (for example, the name of the course) were not tagged as headings (such as <h1>). Instead, the headings were surrounded by code specifying the typeface (bold) and the font size. A search engine would not know that there were supposed to be headings.
- There were no metadata tags for keywords or descriptions on any pages.

The editor recommended the following changes:

- Make the heading of each page unique and contain relevant keywords. For example, change the heading "Welcome" on the homepage to "Technical Writing and Documentation Courses" (or even "Technical Writing and Documentation Courses in Australia").
- Make the title of each page unique, and be sure to include keywords relevant to that page. Put the company's (or university's) name at the end of the title, not at the beginning, or remove it completely. The title does not need to match the heading for the page, but often the best and easiest solution is to make them the same. For example, change the title for the home page (which now has a heading of "Technical Writing and Documentation Courses") from <title>XYZ Communications</title> to <title>Technical Writing and Documentation Courses in Australia—XYZ Communications</title>. On the course page for "Business Writing," use <title>Business Writing Course in Australia—XYZ Communications</title>.
- Change the tags on each headings from <p><font face="Arial Black" size= "5"> . . . </font> to <h1> . . . </h1> and put the font information into CSS.
- Write or reword an introductory paragraph for each page that contains keywords relevant to that page.

Here are some of the changes that were made.

Original homepage:

Title:	XYZ Communications
Heading:	Welcome [not tagged as a heading]
First two paragraphs:	If you are looking for expertise in technical communication, you have come to the right place. This website details the activities and accomplishments of Sondra Gatlin of XYZ Communications in Sydney.
	Sondra Gatlin presents many technical communication courses and contracts for a variety of documentation projects. She also teaches for several universities.

Revised homepage:

Title: Technical Writing and Documentation Courses in Australia—XYZ Communications

Heading 1:   Technical Writing and Documentation Courses [tagged as <h1>]

Heading 2:   Expertise in technical communication [tagged as <h2>]

First two paragraphs: Sondra Gatlin presents many technical communication courses and contracts for a variety of documentation projects. She also teaches for several universities.

                        If you are looking for expertise in technical communication, you have come to the right place. This website details the activities and accomplishments of Sondra Gatlin of XYZ Communications in Sydney.

As a first step, these paragraphs were put in a different order, but the page could be much improved by rewriting them with a different emphasis.

An original course description page:

Title: Courses [all course pages had the same title]

Heading 1:   Technical Writing Skills for Non-writers—2 days [not tagged as a heading]

First paragraph:

                        Communication and the Writing Process
                        Introduction to communication
                        Communication in the workplace
                        The communication triangle
                        Technical writing defined
                        Writing models

Revised course description page:

Title: Technical Writing Skills for Non-writers—2 days

Heading 1:   Technical Writing Skills for Non-writers—2 days [tagged as <h2>]

First paragraph: [This was left unchanged in the first revision, but it would be better to have an overview first paragraph (containing more keywords) preceding the course outline.]

U.S. CREDIT CARDS ONLY!

If people outside the United States are looking to purchase goods or services from a U.S. website, they want to know immediately upon entering the site whether they can do this. Otherwise, they are wasting their time, even if the site offers information they can use when looking to purchase the same product elsewhere (perhaps through a distributor in their own country).

    Similar problems can arise with a website in any country. Suppose you had visited Australia and heard a musician whose music you especially liked. When you get home, you decide to buy a CD of that person's music. You find the musician's website, and the CD is for sale there. "Great!" you think, so you start through the purchasing process. But imagine your reaction when you discover—after filling out the purchase form—that the website will not accept your credit card. Would you be merely disappointed or would you also be annoyed or angry? Would you decide not to buy the CD even if you could find it elsewhere?

Now imagine how you would feel if the website had told you, right up front, "We're sorry, we can only accept U.S. credit cards. We're trying to fix this problem, but for now the only thing we can do is suggest you try buying through ThisOtherStore.com." No doubt you would still be disappointed, but probably you would not be angry.

What lesson should we learn from this?

If a website sells goods or services online, editors should check whether any statement about sales restrictions is easy to find. A lack of restrictions can be a selling point, too "We ship anywhere, and accept credit cards from around the world" is very welcoming. If the site accepts only *some* cards, it should say that too. "We accept Visa and Master-Card from around the world, but we cannot accept Diners Club or American Express" may not be ideal, but at least it's honest and does not leave visitors to find out halfway through the purchasing process that their card isn't accepted.

SENDING COMMENTS TO DESIGNERS

As a technical editor, you probably won't be able to directly change anything in the design and layout of the website; instead, you'll need to tell whoever is responsible what the problem is and your recommendation for fixing it.

Often the easiest way to describe the problem is to provide a picture of the web page. You can try printing the page directly and marking on it by hand. If the problem doesn't show on the printout or if you're working with someone remotely and need to send the report by email, then you could make a screen capture, put that into a word processor or graphics program, use the program's tools to draw a circle around the problem area, and type some explanatory text.

EXERCISE 19.1

Go to http://www.webpagesthatsuck.com/. Find out why these websites are considered inferior, and talk about with your classmates how the criteria for "sucky" websites matches up with the advice in this chapter. Don't forget to check out the link to "The Daily Sucker: Current Examples of Bad Design," which features examples of websites that contain accessibility and usability problems in addition to poor document design issues.

SUMMARY

Just about anyone can create a website. Unfortunately, this results in many poorly designed web pages. As a successful technical editor, you will be able to identify and solve the twelve common complaints about web pages and websites so that your author, client, or company doesn't make any of these mistakes when they publish on the Web. You will want to make sure that the websites you edit are effective in terms of navigation and searchability; the latter is dependent on your ability to choose appropriate keywords that your users are likely to choose. Finally, this chapter covers the often ignored issue of website accessibility. We need to make sure that websites are available and easy to use for everyone, including those users with disabilities. It is crucial for you as a technical editor to understand that a website must be created so that any of your target audience can access it, and you need to know the basics of how to achieve this goal.

Trends and Issues in Technical Editing

6

This unit covers many technical editing issues not included in the previous units. Units Two through Four cover the three-part process of editing approach: editing for correctness (mostly grammar and mechanics), editing for visual readability (document design and graphics), and editing for effectiveness (style, organization, and logic of texts). However, to be a successful technical editor, you need to develop several skills beyond just editing a document. Just as Unit One and Unit Five discuss important proficiencies like personal communication skills and much-needed software knowledge, this unit provides you with an overview of some expertise you should have of editing situations "beyond the document itself." This knowledge includes ethics and editing, editing for global issues, understanding the document production cycle, and performing usability tests. In short, the chapters covered in this final unit should make you well-rounded and knowledgeable across a range of technical editing contexts.

Ethical and Legal Issues in Technical Editing

At the conclusion of this chapter, you will be able to

- define trademark and patent;
- identify plagiarized information;
- apply the commonsense approach to ethics;
- recognize public domain of intellectual property;
- distinguish between legal and moral ethics.

In Unit One, we discuss how the role of technical editors continues to expand and be redefined. No longer are technical editors simply responsible for grammatical and mechanical correctness; they must possess people skills and knowledge of graphic design, computers, software, and web design. Despite these new demands, one very important quality is often overlooked: technical editors must possess strong ethical principles, have a commitment to honesty and fairness, and be willing to uphold the ideals of the technical communication community. Knowing the current laws relevant to technical communication is also essential.

Perhaps the above seems a bit dramatic, but veteran editors acknowledge that too often they must confront situations that may be illegal, unethical, or both. For example, a recent website carried the photo of a soldier who purportedly had been taken hostage. Upon closer inspection, the "soldier" turned out to be a toy action figure. On May 1, 2003, a *New York Times* reporter confessed to plagiarizing numerous other sources. CBS News, the *Daily Mirror*, *USA Today*, the Associated Press, the *Arizona Republic,* and other publications and organizations have discovered reporters either distorting information or just making it up.

Although we may not hear about it as often, distortion or misrepresentation of the truth occurs in technical documents. Sometimes, the truth becomes altered unknowingly, but often it is a victim of someone's deliberate attempt to deceive others. For example, suppose your boss tells you to add the following sentence to an instruction manual you are copyediting: *This widget has been found to be safe when used as directed*. During the first three months of being on the market, no consumer has reported being injured using this new device. Therefore, your boss reasons, consumers have tested this device through usage and should be able to conclude that the product is completely safe.

Adding such a misleading statement is both illegal and unethical. Why? Because consumers who read this statement assume that the product has actually been tested for safety. Yes, you could argue that the boss's statement never makes such a claim. However, the law provides for the consumer to make certain assumptions. Accordingly, the phrase *has been found to be safe* suggests that some type of empirical testing has been conducted. What would you do if your boss insisted on including this misleading statement?

How you confront legal and ethical issues as a member of the technical communications community is no small matter. Consider how you might react if you believed your client had broken the law by misrepresenting the performance of a product, or your client accused you of overcharging or not living up to the contractual agreement. How would you react after discovering a senior colleague was passing confidential, business-sensitive information to a rival company. Would you confront the client or colleague, take the issue to your boss, or just keep quiet? Suppose you are accused of illegal or unethical conduct by a client? Will you know what to do and how to be prepared to confront these types of issues?

A COMMON SENSE APPROACH TO ETHICS

Ethical behavior in the workplace is generally a matter of common sense derived from rational, ethical principles. It involves doing what we know is right despite the temptations to do otherwise. The student who cheats on an exam, the merchant who fails to give an honest measure, the mechanic who does not make promised repairs, and the accountant who underreports income to the IRS all exhibit unethical behavior.

Always tell the truth seems obvious enough advice. However, misrepresenting the truth by falsifying data, distorting facts, or leading others to make false conclusions is unethical (and, depending on the situation, possibly illegal). Consider the following examples. Did you ever wonder how one television commercial claims nine out of ten people surveyed prefer Brand X toothpaste, whereas Brand Y makes the same claim about its toothpaste? Or have you ever eaten at a fast-food restaurant and received a hamburger that only remotely resembled the burger in the advertisement? Or do you really believe you can develop six-pack abs in only four weeks with the fitness guru's new abdominal device? Perhaps the law has not been broken technically in any of these examples, but clearly the advertisements misrepresent the truth by attempting to lead the consumer to form erroneous conclusions.

These types of distortions are usually obvious and easy to identify, yet apparently they work to deceive the public. However, any distortions in technical documents that rely heavily on facts, data, or statistics may not be as obvious and clear-cut to the reader. A good technical editor will always check the writer's facts, will never assume that subject matter experts (SMEs) do not make mistakes, and will make certain that readers are not likely to make invalid inferences. For example, based on a comparison of ACT scores during 2005, some politicians in Alabama demanded a substantial increase in spending on education. The composite ACT scores for Alabama high school graduates was 20.2; the national average was 20.9. However, without further explanation of the statistics, invalid conclusions may be drawn. First, the difference between 20.2 and 20.9 is not statistically significant outside the margin of error. Second, some states require all of their students to take either the SAT or the ACT, whereas others do not. Obviously, states where all students take the test, not just the college-bound students, will score lower than those that do not maintain this requirement. Other demographic factors influence the scores as well, such as the decision to include or exclude special education students. Good technical writing and editing includes sufficient information and explanations to prevent these types of misinterpretations.

Unfortunately, making an ethical decision may not always be a simple matter of common sense. For centuries, philosophers have tried to create an absolutely irrefutable proof for determining whether a given decision is indeed ethical. Although the scope of this text does not include all modern ethical theory, we do offer a philosophical principle for governing our actions as technical communicators: ***Act only on that maxim through which you can at the same time will that it should become universal law***. This principle, Immanuel Kant's *categorical imperative*, sounds very similar to the Golden Rule. It simply suggests that whatever courses of action we choose to take, we should desire everyone else to act the same. Being honest, for example, is an ethical principle we can assert should be universally adopted. Yet we cannot rationally argue that lying, stealing, or cheating should become universal, so these actions are not ethical. The implications for the field of technical communication have been summarized into ethical guidelines by several technical communication organizations. We encourage you to read these guidelines thoughtfully and abide by them. An example is the Society for Technical Communication's guidelines, shown in Figure 20.1.

Society for Technical Communication Ethical Principles for Technical Communicators

As technical communicators, we observe the following ethical principles in our professional activities.

Legality
We observe the laws and regulations governing our profession. We meet the terms of contracts we undertake. We ensure that all terms are consistent with laws and regulations locally and globally, as applicable, and with STC ethical principles.

Honesty
We seek to promote the public good in our activities. To the best of our ability, we provide truthful and accurate communications. We also dedicate ourselves to conciseness, clarity, coherence, and creativity, striving to meet the needs of those who use our products and services. We alert our clients and employers when we believe that material is ambiguous. Before using another person's work, we obtain permission. We attribute authorship of material and ideas only to those who make an original and substantive contribution. We do not perform work outside our job scope during hours compensated by clients or employers, except with their permission; nor do we use their facilities, equipment, or supplies without their approval. When we advertise our services, we do so truthfully.

Confidentiality
We respect the confidentiality of our clients, employers, and professional organizations. We disclose business-sensitive information only with their consent or when legally required to do so. We obtain releases from clients and employers before including any business-sensitive materials in our portfolios or commercial demonstrations or before using such materials for another client or employer.

Quality
We endeavor to produce excellence in our communication products. We negotiate realistic agreements with clients and employers on schedules, budgets, and deliverables during project planning. Then we strive to fulfill our obligations in a timely, responsible manner.

Fairness
We respect cultural variety and other aspects of diversity in our clients, employers, development teams, and audiences. We serve the business interests of our clients and employers as long as they are consistent with the public good. Whenever possible, we avoid conflicts of interest in fulfilling our professional

FIGURE 20.1
The Society for Technical Communication's Ethical Principles for Technical Communicators.
Source: *http://www.stc.org/about/policy_ethicalPrinciples.asp.*

responsibilities and activities. If we discern a conflict of interest, we disclose it to those concerned and obtain their approval before proceeding.

Professionalism
We evaluate communication products and services constructively and tactfully, and seek definitive assessments of our own professional performance. We advance technical communication through our integrity and excellence in performing each task we undertake. Additionally, we assist other persons in our profession through mentoring, networking, and instruction. We also pursue professional self-improvement, especially through courses and conferences.

Adopted by the STC Board of Directors
September, 1998

FIGURE 20.1 (*continued*) The Society for Technical Communication's Ethical Principles for Technical Communicators.

WHAT CAN YOU OWN?

You own more than you think. It may surprise you to learn that for better or worse, you *own* every email you have ever sent out, and for that matter, every document, painting, graphic, letter, composition, and web page that is distinctly your original work. So if individuals pass along your email without acknowledging you as the original author, technically they are violating the law. (Yes, that is correct, but do not call the email police on them because you are probably just as guilty of the same practice.) Anything you create that is clearly your original work gives you certain rights of ownership. This form of ownership is called a copyright. A copyright is granted by law the moment something is created. Stephen Fishman in *The Copyright Handbook* defines the term *copyright* as "a legal device that provides the creator of a work of art or literature, or a work that conveys information or ideas, the right to control how the work is used."

Despite what we may wish, the law is not that simple. For example, suppose that while under the direct employment of a particular company you invent or create a product. Despite what you may think is unfair, your employer maintains the copyright for the product you created. However, if you are a freelance writer and not directly employed, you may or may not own the articles you write when you sell them. Copyright ownership in this case depends on what is expressly negotiated at the time of the contract. Deciding on ownership of intellectual property is another matter. For example, if a professor develops an online course for a university, who owns that actual course? Can the professor leave and take that online course to another university claiming it's his or her intellectual property, or does the initial university maintain ownership of the course? Perhaps you remember several years ago when late-night television host David Letterman argued that he had the sole rights to the Top Ten List as his intellectual property.

WHAT CAN YOU USE THAT BELONGS TO OTHERS?

Suppose that the use of any material from a journal, magazine, book, news report, or speech required permission from the author. Getting permission for every excerpt or quote you incorporate in your own reports or articles seems a bit silly. More time would be devoted to locating the authors and determining who owned the copyright than writing the material itself.

Fortunately, U.S. copyright laws contain an exemption that allows limited use of commentary, news reporting, and sources intended for educational or noncommercial purposes. This law is called "Fair Use." For example, you can copy several pages from a textbook for your own educational purposes; however, you cannot copy the entire textbook to avoid buying it. You may want to copy a journal article for a research project. The Fair Use laws allow you to copy the article but not the entire journal. Essentially, Fair Use constitutes three criteria for copying: your intent, how much is copied, and whether any commercial repercussions are experienced. Generally, the editor is responsible for verifying that permission has been obtained from the copyright holder. Assume that permission is needed if the document or publication is intended for wide distribution or commercial gain. Certainly, you want to obtain permission in writing, most likely in a standard form used by your organization. This form should briefly state what specifically is to be copied and where it will appear. Frequently, copyright holders charge a fee for reprinting their work. Always protect yourself by keeping records of all emails and forms pertinent to obtaining copyrights. You never know when you may need them to save your organization and yourself from a nasty lawsuit.

EXERCISE 20.1

Copyright and Fair Use laws for online materials are evolving. Please research what you can about laws for online materials. Try to focus on .gov sites as they will probably provide the most credible and up-to-date information. Compare some of the issues with online material laws with laws for print documents. What seem to be the major differences?

The laws regarding ownership of online documents continue to evolve. Below is a brief discussion of print copyright laws.

Public Domain

Under Fair Use laws, you may use small selections or passages from copyrighted works; just make certain you state where you obtained your information. Also, you may use works that exist in the public domain, which means the information belongs to everyone. Ideas, facts, words, names, government documents, and works for which the copyright has expired are considered in the public domain. For example, after a number of years prescribed by Congress, a novel no longer maintains its copyright and belongs in the public domain. You could copy from it freely. However, it is always best to research the copyright on any document because Congress continues to change the length of copyright protection. For example, Martin Luther King Jr.'s "I have a dream" speech is owned by his family, and their permission must be obtained to use it until it becomes public domain.

Public domain laws can be tricky, though. For example, you can copy Homer's *Odyssey* freely; it was written over two thousand years ago and is in the public domain. However, some translations into English from the original Greek may still be copyrighted and would require permission to use. Mathew Brady's photos from the nineteenth century are now in the public domain; however, Ansel Adams's are not.

Original government documents, which are funded by taxpayers, are in the public domain and may be reproduced without anyone's permission. Photos taken of American soldiers fighting in Iraq by an army photographer are in the public domain; similar photos taken by a news correspondent are not. (However, it would be unethical for you to use a data sheet created by the government without acknowledging that it came from a federal agency and is not your original work.)

Basically, the same principles hold true for websites: if you create it, you own it. Remember, though, you cannot "lift" a graphic from another website and use it unless it is in the public domain. Now, here is an unusual twist to the laws: if you create a website or a document on your organization's computer or otherwise use your company's resources, your organization may own the copyright according to the law. Professors who create web-based courses or receive grant money to conduct research probably do not own any product they create. So what do you really own? Most of the time questions of ownership are straightforward, but you only have to look in the phone book to see how many lawyers make their living interpreting them.

Trademarks and Patents

Trademarks (also called *marks*) are distinctive symbols, logos, phrases, graphics, words, or other forms of representation that identify the particular source of a product. They help distinguish that product from others' products and are identified with the symbol ^{TM} or ®. Companies must register their product with the U.S. Patent and Trademark office (PTO) to use these symbols. Nike incorporates its checkmark on all its sports apparel, Apple Inc. uses a stylized apple (with a bite taken out of it) on its products, and CBS utilizes a stylized eyeball to achieve name recognition. All are registered trademarks. And as you might expect, national and international companies jealously guard against other companies who they feel might try to infringe upon their trademark. A few years ago, McDonald's legally prevented a hotel from using the name McSleep. McDonald's argued that the name capitalized on their reputation for fast, standardized, inexpensive service.

Technical editors should also be aware that other types of marks are used to identify a company or their service. A *service mark*, for example, such as Amazon.com, identifies a retail website; a *trade dress* suggests a product's particular packaging, such as Gateway Computer's cowhide-looking boxes or the shape of Coca-Cola's bottle; a *collective mark* indicates a symbol, word, phrase, or other type of mark used by a group of people, such as the Boy Scouts' slogan "Be Prepared"; and a *certification mark* indicates that a product has been approved by a particular organization, such as UL for Underwriters Laboratories.

Plagiarism

Lifting, bootlegging, copying, cheating, stealing, and *kidnapping* are just a few of the synonyms used to describe plagiarism. With the Internet making information easily available, teachers report that plagiarism has dramatically increased in recent years. Some students do not think anything is wrong with lifting entire ideas from a source without giving credit, or even buying entire papers off the Internet. To combat student plagiarism, new software has been developed that can detect plagiarized documents, such as the software sold by Turnitin.com. Actually, most people do not know what plagiarism means, at least to the full extent. Certainly, copying a document word for word without surrounding the quoted material with quotation marks and giving credit to the author is plagiarism. It is both illegal (violates copyright laws) and unethical to pass someone else's work off for your own. So choose not to do it.

If you steal from one author, it's plagiarism. If you steal from two, it's research.
—*Anonymous*

Although you may believe that plagiarism principally resides with student papers, technical editors frequently have to determine if information they are reading follows a source too closely and may constitute plagiarism. For example, plagiarism is not only copying something word for word; you may not "copy" someone else's style either. Have you ever copied what someone said, but to avoid plagiarizing you scrambled some of the words or phrases? This is known as *paraphrasing* and this also constitutes plagiarism when the author and/or source is not cited. Look at the following example closely from "End of the Binge," an article by James Kunstler that appeared in the *American Conservative.*

> **Original:** The global economy is, in fact, nothing more than a transient set of trade and financial relations based on a particular set of transient, special sociopolitical conditions, namely a few decades of relative world peace between the great powers along with substantial, reliable supplies of predictably cheap fossil fuels.

> **Plagiarized version:** The global economy is a set of trade and financial relations based on transient, sociopolitical conditions that include several decades of world peace and substantial, reliable supplies of oil.

> **Acceptable version:** According to James Kunstler, the last few decades of world peace can be attributed to plentiful, inexpensive, and consistent sources of fossil fuels (cite page number here).

It is difficult to pick up a paper or trade journal without reading allegations of someone accusing someone else of plagiarism. Musicians such as the Rolling Stones threatened to sue when other bands such as the Verve "sampled" their songs too closely. Politicians and journalists have also been found guilty of plagiarism. William Hague, an English politician, was accused of paraphrasing a speech of Tony Blair's. Graham Swift and J. K. Rowling both have endured queries into the originality of their work. Plagiarism is a career buster and can lead to serious legal consequences. Recognize plagiarism for what it is and do not listen when someone tries to dismiss it as "drawing inspiration from someone else."

IS IT UNETHICAL OR ILLEGAL?

You cannot really prepare for all the possible ethical or legal scenarios you will encounter as a technical editor because each situation will be different. However, following a few basic principles, such as always being honest, doing what you instinctively know is right, taking personal responsibility for your actions, and putting your client and the safety of the consumers first, will go a long way to guiding your actions. Too often people want to point fingers or say, "That's not my job," or "I was only doing what I was told to do."

Remember, though, that just because something is legal doesn't mean it is ethical. For some people, abortion is unethical even though in many countries it is legal. Some countries allow men to legally abuse their wives. Slavery remains legal in some countries. This ethical/legal dichotomy applies to the workplace as well: what is legal may not be ethical. For example, a few years ago, a major producer of automobiles knew from casualty reports that their gas tanks easily ruptured on side impact. Even if the federal government did not demand a recall on the cars, the manufacturer had an ethical obligation to fix the problem.

As a technical editor, you may not be confronted often by legal issues; however, you definitely will be confronted with ethical issues. Try to anticipate how your audience or client will perceive your comments. Be clear, and be certain to keep detailed and accurate records of your work.

All ethical questions boil down, at some point, to accepting personal responsibility for one's own actions.
—*Jean Hollis Weber*

EXERCISE 20.2

Following are a number of ethical scenarios. Determine the correct course of action for each one.

1. You are a student editor in a writing lab. You are responsible for editing the OWL's (Online Writing Lab's) website exercises for grammar errors only. However, you come across some sexist language in the website. What do you do?

2. How would you need to revise the following sentence from a website and why? "Make sure to have the edited documents fedexed directly to the publisher."

3. Your company's website provides internal links to several websites, but you are fairly sure that your company has not asked any of the other website owners if they would mind having their sites deeplinked to your company's site. You are your company's technical writer, a job that includes being the web editor. What would you do?

4. After agreeing on a price to edit a book manuscript for a technical writing manual, you realize halfway through the process that the job was not as time-consuming as you thought it was going to be, and that the manual is actually well written. You finish a week ahead of the four-week deadline. What do you do with the money that you were paid for the fourth week?

SUMMARY

In this chapter we provide a common sense approach to ethical issues encountered by technical editors. Ethical guidelines for editing have been established, and it's vital that technical editors make certain these guidelines are followed.

Being ethical means being honest and fair. For example, distorting facts and data will cost scientists their jobs and reputations, as well as impugning the entire scientific community. Failure to be honest and fair can result in serious legal repercussions or breach confidentiality. Therefore, writers and editors must be keenly aware of copyright, trademark, and patent laws as well as laws governing public domain usage. If ever in doubt, cite the source of information. You will notice that the authors of this text have cited their sources and received permission for all copyrighted materials.

Finally, one of the worst breaches of ethics for writers is plagiarism. Taking someone else's idea and passing it along as your own is an act of theft. Many careers have been ruined because someone claimed to have written something and was later discovered to have copied from another source. So, although you don't have to be a lawyer to be a technical editor, you do need to know the basic issues. And, because the laws continue to change, you need to stay up-to-date and avoid potential problems.

CHAPTER 21

Global and Cultural Issues in Technical Editing

At the conclusion of this chapter, you will be able to

- prepare technical documents for international readership;
- avoid cultural bias in technical documents;
- choose graphics that will be appropriate in any cultural setting;
- analyze documents for language that might confuse international readers.

As we mention in Chapter 18, at times you will have to edit a document for international audiences. Editing a document to convey technical information accurately to a native-English-speaking audience can present many challenges in itself. But what about editing a document for use by a non-native-speaking audience, or native English speakers in another country? Applying the various levels of edit to ensure correctness, visual readability, and effectiveness is just part of the process. Preparing documents for an international audience also involves paying close attention to word choice, sentence structure, and graphics.

In our age of instant communications by email and Internet, global distribution of texts and images is more common than ever before. Billions of documents—from marketing to government information to equipment instructions—need to be written, localized, or translated for several audiences. Therefore, it is essential that a technical editor be knowledgeable about intercultural and translation issues.

This chapter presents an overview of the wide range of cross-cultural issues that could affect the readability of a technical document by an audience containing people who speak different versions of English or English as a second language, the ease and accuracy of translation into other languages, and the ease and accuracy of localization of documents for other cultures. We look at global issues of culture that affect text that has been translated to or from English, and discuss how you might edit a document that has been translated from English into another language. We cover the problems of choosing, editing, and translating text and illustrations.

How important is it for technical editors to consider these issues in the documents they edit? The lack of communication between NASA and their contractors over the use of Imperial (inches and feet) versus metric (centimeters and meters) units of measurement caused both the catastrophic demise of the $125 million NASA Mars Climate Orbiter and a major malfunction in the Hubble Space Telescope.

Ideally, you as the technical editor would know many languages and would be able to check the accuracy of almost anything translated into English. Clearly this is unrealistic: although you do need to have a lot of knowledge and skills as an editor, knowing all of the languages you will encounter during your career is pretty much impossible! Even a passing familiarity with a few languages will barely get you by when it comes to detail-oriented editing and translating culturally sensitive text. Nevertheless, a few principles can help you with global issues.

> The biggest problem with communication is the illusion that it has been accomplished.
> —*George Bernard Shaw*

THE G.I.L.T. PROCESS

> Most people are as unaware of their own culture as they once were of oxygen, evolution, or gravity.
> —*Henry Steiner*

According to the Localization Industry Standards Association (LISA), cross-cultural issues in technical communication can be broken down into four categories: *Globalization, Internationalization, Localization,* and *Translation* (known collectively as G.I.L.T.).

Globalization

Globalization is the process of preparing your documents for a global audience. Globalizing a document usually means the document will stay in English but be read by an international audience. A globalized document should communicate clearly and should encourage cross-cultural comprehension.

Internationalization

Internationalization is much like globalization, except the document is written and edited with the knowledge that it will later be localized and translated. Many technical documents are translated (before publication) by people who can make informed decisions about the text, but material on websites is often translated by readers using tools such as Google Translate or Yahoo! Babel Fish. The focus of this chapter is on internationalizing documents.

Localization

Localization refers to the process of preparing a document for a specific culture as though it originated from that culture. Localization may include translation but applies equally to documents intended for use in other English-speaking countries. An essential component of localization is the use of graphics, page layouts, units of currency, references to government departments and policies (for example, taxation), and so on that are specific to the location and culture of the intended audience.

Translation

Translation—converting a document from one language into another—is a central part of localization when the target audience speaks a different language.

Now that you know what G.I.L.T. is, how do you make sure a document is ready to go global? You have to develop an "international style." As Apple Inc. states in its *Apple Publications Style Guide,* "Following international style helps readers with limited English proficiency read what you write. It also helps translators, human or machine, localize your writing by minimizing the burdens of cultural and customary language usage." And because our readers and users today are both people and machines (e.g., robot software that scans Internet résumés for keywords), it is important for documents to be translated effectively and appropriately.

THE INTERNATIONAL STYLE

The International Style—also known as the General Writing Style—includes language strategies for writing text in English that has to be translated for global audiences. English is an inherently difficult language. Consider the fact that English contains a dizzying array of *homophones* (words with different meanings and spellings but the same pronunciation, e.g., *sew, so, sow*), *homographs* (words that have the same spelling but different meanings and pronunciations, e.g., *bass, dove, close*), and *idioms*. (For a little fun, read the English poem in Figure 21.1 and the sentences in Figure 21.2.) What if the title of a document were the homograph "Mobile Homes"? Would this be an article about manufactured housing or houses in Mobile, Alabama?

His death, which happen'd in his berth,
At forty-odd befell:
They went and told the sexton, and
The sexton toll'd the bell.
—from "Faithless Sally Brown" by Thomas Hood

FOUR ALL WHO READ AND RIGHT

We'll begin with a box, and the plural is boxes;
but the plural of ox became oxen not oxes.
One fowl is a goose, but two are called geese,
yet the plural of moose should never be meese.
You may find a lone mouse or a nest full of mice;
yet the plural of house is houses, not hice.

If the plural of man is always called men,
why shouldn't the plural of pan be called pen?
If I spoke of my foot and show you my feet,
and I give you a boot, would a pair be called beet?
If one is a tooth and a whole set are teeth,
why shouldn't the plural of booth be called beeth?

Then one may be that, and three would be those,
yet hat in the plural would never be hose,
and the plural of cat is cats, not cose.
We speak of a brother and also of brethren,
but though we say mother, we never say methren.

Then the masculine pronouns are he, his, and him,
but imagine the feminine as being she, shis, and shim.

FIGURE 21.1
Pluralization Poem (Author Unknown).

Reasons Why the English Language Is Hard to Learn

1. The bandage was wound around the wound.
2. The farm was used to produce produce.
3. The dump was so full that it had to refuse more refuse.
4. We must polish the Polish furniture.
5. He could lead if he would get the lead out.
6. The soldier decided to desert his dessert in the desert.
7. Since there is no time like the present, he thought it was time to present the present.
8. A bass was painted on the head of the bass drum.
9. When shot at, the dove dove into the bushes.
10. I did not object to the object.
11. The insurance was invalid for the invalid.
12. There was a row among the oarsmen about how to row.
13. They were too close to the door to close it.
14. The buck does funny things when the does are present.
15. A seamstress and a sewer fell down into a sewer line.
16. To help with planting, the farmer taught his sow to sow.
17. The wind was too strong to wind the sail.
18. After a number of injections my jaw got number.
19. Upon seeing the tear in the painting I shed a tear.
20. I had to subject the subject to a series of tests.
21. How can I intimate this to my most intimate friend?

Screwy pronunciations can mess up your mind! For example, if you have a rough cough, climbing can be tough when going through the bough on a tree!

Let's face it—English is a crazy language. There is no egg in eggplant, nor ham in hamburger; neither apple nor pine in pineapple. English muffins weren't invented in England or French fries in France. Sweetmeats are candies while sweetbreads, which aren't sweet, are meat. We take English for granted. But if we explore its paradoxes, we find that quicksand can work slowly, boxing rings are square, and a guinea pig is neither from Guinea nor is it a pig. And why is it that writers write but fingers don't fing, grocers don't groce and hammers don't ham?

If the plural of tooth is teeth, why isn't the plural of booth beeth? One goose, two geese. So one moose, two meese? One index, two indices? Doesn't it seem crazy that you can make mends but not one amend, that you comb through annals of history but not a single annal? If you have a bunch of odds and ends and get rid of all but one of them, what do you call it? If teachers taught, why didn't preachers praught?

If a vegetarian eats vegetables, what does a humanitarian eat? Sometimes I think all the English speakers could be committed to an asylum for the verbally insane. In what language do people recite at a play and play at a recital? Ship by truck and send cargo by ship? Have noses that run and feet that smell? How can a slim chance and a fat chance be the same, while a wise man and a wise guy are opposites? How can overlook and oversee be opposites, while quite a lot and quite a few are alike? How can the weather be hot as hell one day and cold as hell another? Have you noticed that we talk about certain things only when they are absent? Have you ever seen a horseful carriage or a strapful gown?

Met a sung hero or experienced requited love? Have you ever run into someone who was combobulated, gruntled, ruly, or peccable? And where are all those people who are spring chickens or who would actually hurt a fly? You have to marvel at the unique lunacy of a language in which your house can burn up as it burns down, in which you fill in a form by filling it out and in which an alarm goes off by going on. English was invented by people, not computers, and it reflects the creativity of the human race (which, of course, isn't a race at all). That is why, when the stars are out, they are visible, but when the lights are out, they are invisible. And why, when I wind up my watch, I start it, but when I wind up this essay, I end it.

FIGURE 21.2
Reasons Why the English Language Is Hard to Learn (Author Unknown).

English is by no means a "pure" language. It is instead a mixture of Old English, Latin, Greek, Old Norse (from the Vikings), and French. We have borrowed many words, and our grammar system reflects parts of the grammar of more than one language (which is why even grammar rules can be confusing and have exceptions). These borrowed words have created "problems" such as homophones, homographs, pluralization dilemmas, and idioms; however, because homophones and homographs deal with pronunciation issues, they are not necessarily the major roadblocks to technical editors dealing with translation issues. There are many terms in the English language that we take for granted as being commonly understood. For example, the idiom "kick the bucket" has both a literal meaning (a foot coming in contact with an actual bucket) and a figurative one (to die). Figurative language uses such as idioms should be avoided because they are so culturally specific. Listed below are the items you need to pay close attention to when editing a document using the International Style. John Kohl's *The Global English Style Guide* provides detailed guidelines for writing clear, translatable documentation for a global market.

Sentence Structure

Try to use short, simple sentences. Compound sentences should be rewritten as two separate sentences. Long, elaborate sentences may sound impressive to you but will likely confuse your international readers. Complex sentences also make translating much more difficult.

> In order to facilitate expeditious remuneration of compensation, fill out the form that you have and send it back to us when it is finalized. (complex sentence)
>
> Complete the form to receive payment quickly. Return the form when complete. (revised)

Active Voice

Use the active voice as often as possible. The passive voice is more difficult for non-native English speakers to understand because the agent is subordinated in the sentence, making the subject of the action difficult to find. If possible, rewrite sentences in the active voice.

> The computer <u>was struck</u> by lightning. (passive voice)
>
> Lightning <u>struck</u> the computer. (revised)

Idioms

An idiom is any phrase that doesn't make sense when taken literally, for example "on the cutting edge," "state of the art," or "hit the road." Idioms will confuse a non-native speaker of English and will be very difficult (if not impossible) to translate. You need to identify and replace all idioms.

> You will need to <u>make up your mind</u> which program to install first. (idiom)
>
> You will need to <u>decide</u> which program to install first. (revised)

Contractions

Replace all contractions with their original words. Contractions may seem simple to a native English-speaking audience, but they are confusing to non-native speakers.

<u>Don't</u> forget to remove the <u>computer's</u> hard drive. (contractions)

<u>Do not</u> forget to remove the hard drive <u>from the computer</u>. (revised)

If users break their keyboards after you've instructed them to "hit the Enter key" when you mean "press," who is to blame?
—*techscribe.co.uk*

Synonyms

Writers often use synonyms to add interest to their material. To a non-native speaker or translator, synonyms create a host of problems. Make sure the same term is used to describe an object or process throughout the document.

This program <u>deletes</u> all files. The <u>erased</u> files cannot be retrieved. (synonyms)

This program <u>deletes</u> all files. The <u>deleted</u> files cannot be retrieved. (revised)

Parts of Speech

Look for the same word being used for different parts of speech. As shown in the example below, the same word in English can often be used as a verb, a noun, and an adjective. This creates unnecessary ambiguity in the document.

The assistant <u>filed</u> the new project <u>file</u> in the <u>filing</u> cabinet. (same word used as a verb, noun, and adjective)

The assistant <u>placed</u> the new project <u>file</u> in the cabinet. (revised)

Abbreviations and Acronyms

Technical documents are often full of abbreviations and acronyms. Differences in spelling mean that abbreviations and acronyms do not translate well into other languages. The acronym for the Central Intelligence Agency, CIA, might become ACI (e.g., Agencia Central de Inteligencia in Spanish) when translated into another language. Spell out all abbreviations and acronyms in the document you are editing.

We are updating the <u>CPU</u> to accommodate the new <u>VoIP</u> service. (acronyms)

We are updating the <u>central processing unit</u> (CPU) to accommodate the new <u>voice over Internet protocol</u> (VoIP) service. (revised)

Jargon

Like abbreviations and acronyms, technical jargon can confuse international audiences. If it is impossible to avoid using the term, at least offer a simple explanation of it in brackets or in a subsequent sentence, especially if the target audience are not members of the same discourse community as the writers of the document.

The part specifications indicated that the acceptable hardness values for the part ranged from Rockwell H30 to Rockwell H75. (jargon)

The acceptable hardness values for the part ranged from Rockwell H30 to Rockwell H75. The Rockwell Hardness test measures the net increase in depth of impression as a load is applied. Hardness numbers have no units and are commonly given in the alphabetical scales (R, L, M, E, H, etc.). The higher the number in each scale, the harder the material. (revised)

Culturally Specific References

Identify and replace all culturally specific references. Analogies and references to sports, politics, holidays, religions, seasons, and geographic place names should be avoided. If the document simply refers to "Manchester," is this Manchester, New Hampshire, Manchester, England, or Manchester, Tennessee? Does "summer" refer to July or January?

I will need a <u>ballpark</u> estimate on the project before <u>Christmas</u>. (culturally specific reference)

I will need an <u>approximate</u> estimate on the project by the <u>end of December</u>. (revised)

The new version of our software is scheduled for release <u>next summer</u>. (geographically specific reference)

The new version of our software is scheduled for release <u>in July or August</u>. (revised)

> One example [of companies failing to consider localization] is the perfume "Mist" that was launched in a German market but failed. "Mist" in German means "manure." Enough said.
> —*Sophie Hurst*

Gender-Specific References

The documents you edit should not contain any gender-specific references. Identify and replace all gender-specific references. Chapter 6 discusses this topic in more detail and gives examples.

Humor, Idioms, and Colloquialisms

Many writers attempt to "lighten up" their documents by using humor. Humor, as well as puns, satire, irony, figurative language, and colloquialisms, will not translate into other languages. A good example is "kill two birds with one stone," which becomes "catch two pigeons with one bean" when translated into Italian! Identify and eliminate all occurrences of humor, satire, etc. from your documents.

Collected below are some stories from various sources that can illustrate what happens when companies attempt to internationalize their slogans.

In China, the name Coca-Cola was first rendered as "ke-kou-ke-la." Unfortunately, after thousands of signs had been printed the company discovered that the phrase means "bite the wax tadpole" or "female horse stuffed with wax," depending on the dialect.

One characteristic but apocryphal tale tells of an American military system designed to translate Russian into English, which is said to have rendered the famous Russian saying "The spirit is willing but the flesh is weak" into "The vodka is good but the meat is rotten."

When Parker Pen marketed a ballpoint pen in Mexico, its advertisements were supposed to say, "It won't leak in your pocket and embarrass you." However, the company mistakenly thought the Spanish word *embarazar* means "to embarrass." Instead, the adverts said, "It won't leak in your pocket and make you pregnant."

In Taiwan, the Pepsi slogan "Come alive with the Pepsi Generation" was originally translated as "Pepsi will bring your ancestors back from the dead."

Measurements

Be sure that all measurements, decimals, etc. follow international conventions. In Europe, 1,000.25 would be written as 1.000,25 or 1 000,25, depending on the country. Other countries use variations on these notations. Remember to convert all Imperial measurements to standard metric units; a document might include both units (one in parentheses after the other). Editors should check that conversions have been done correctly and are expressed in the correct units for the intended audience. Remember also that *metre* (not *meter*) is the standard international spelling of the metric unit of measurement in most countries.

The cable is 3,281 feet long. (nonstandard conventions used)

The cable is 1.000 metres long. (revised for some European readers)

The cable is 1,000 metres long. (revised for Australian readers)

Dates and Times

Many cultures write dates differently. In America, 5/12/2005 means May 12, 2005. In Europe, it would mean December 5, 2005, more commonly written as 5 December 2005. Although some dates can't be mistaken (3/25/99 can't be the 3rd day of the 25th month), they can still cause a reader to stumble. For nontechnical audiences, make sure dates spell out part or all of the month, for example 5 Dec. 2005. For technical audiences, consider using the international date format defined by ISO (International Standards Organization). This format presents dates as YYYY-MM-DD, where YYYY is the year, MM is the month, and DD is the day. For example, April 3, 2009, is written 2009-04-03 in this international format.

When writing time, do not use "a.m." or "p.m." Use the 24-hour format, where 17:30 means 5:30 p.m. Some cultures write times with a period or no punctuation separating the hour and minutes, but a time written with a colon will generally be understood. When publicizing contact hours or organizing meetings, include the time expressed in UTC (Coordinated Universal Time) as well as local time. You can look up UTC and other time zone information on http://www.timeanddate.com/.

When localizing a document, remember that many countries do not use the same Gregorian calendar used in the United States. For example, the year 2005 is 5765 in the Jewish calendar and 1426 in the Islamic calendar.

The webcast will be on 6/2/2009 at 8 p.m. (nonstandard conventions used)

The webcast will be on 2 June 2009 at 20:00. (revised)

Telephone and Fax Numbers

To internationalize your telephone and fax numbers, you need to do three things:

1. Use digits only, not letters, or give your phone number both ways. Not all countries have letters marked on their telephone keys, and if they do, the letters might not correspond to the same digits as in your country.
2. Include the country or region code as well as the area code, in this form: +1 617 4948045 or +1.617.4948045, where +1 is the country or region code.
3. If you publicize "free call" (800 or the equivalent) numbers, check whether that number is good for incoming international calls. If not, provide alternative telephone and fax numbers.

ENGLISHES

Even if you know no other language than English, understanding how confusing English can be might be the first step toward understanding the difficulties a non-native speaker might have. In short, you may speak English, but do you speak American English, Australian English, or British English (just to name a few)? See Figure 21.3 for some examples of differences between Englishes. The list includes some terms that technical editors might come across in their work. It does not include many common social terms or slang; numerous lists intended for travelers are available on the Internet.

U.S.	U.K.	Canada	Australia	India
antenna (TV)	aerial	aerial, antenna	aerial, antenna	aerial, antenna
appraiser	valuer	appraiser	valuer	appraiser, valuer
back up (car)	reverse	back up, reverse	reverse	reverse
baseboard	skirting board	baseboard	skirting board	(not used)
bathroom, restroom (public)	lavatory, toilet	washroom, bathroom	toilet	toilet, lavatory, bathroom
bill (money)	note	bill	note	note
bulletin board	notice board	bulletin board, message board	notice board	notice board
braces { }	curly brackets	braces, curly brackets	curly brackets	curly brackets
brackets [ ]	brackets, square brackets	brackets	brackets, square brackets	brackets, square brackets
busy (phone)	engaged	busy	busy, engaged	engaged
calendar (appointment)	diary	diary, planner, calendar	diary, planner	diary
call (on telephone)	ring	call	ring, call	ring

FIGURE 21.3
Words in Five Varieties of English.
Source: *http://jeanweber.com/newsite/?page-id=67.*

U.S.	U.K.	Canada	Australia	India
cart (shopping)	trolley	cart	trolley	trolley, cart
cell phone	mobile phone	cell phone	mobile phone	mobile phone
check (mark)	tick	check	tick	tick
check (money)	cheque	cheque	cheque	cheque
closet	cupboard	closet, cupboard	cupboard	cupboard
cot	camp bed	cot, folding bed	camp bed	cot
counterclockwise	anticlockwise	counterclockwise	anticlockwise	anticlockwise
crib	cot	crib	cot	crib
defroster (on car)	demister	defroster	demister	(not used)
district attorney	public or crown prosecutor	prosecutor (crown, federal, provincial)	prosecutor	public prosecutor
doctor's office	(doctor's) surgery	doctor's office	(doctor's) surgery	clinic
elementary school, grade school	primary school	elementary school, grade school	primary school	primary school
faucet, tap	tap	tap, faucet	tap	tap
fender (car)	mudguard, wing	fender	mudguard	mudguard
fiscal year	financial year	fiscal year	financial year	financial year
flashlight	torch	flashlight	torch	torch
freeway, expressway, interstate	motorway	highway, expressway	motorway	national highway, highway, expressway
gas pump	petrol bowser	gas pump	petrol bowser	petrol pump
gas, gasoline	petrol	gas, gasoline	petrol	petrol
high school	secondary school	high school, secondary school	secondary school, high school	high school, secondary school
hood (of car)	bonnet	hood	bonnet	bonnet
installment plan	hire purchase	installment plan	hire purchase	installment plan, hire purchase
intermission	interval	intermission	interval, intermission	interval
internship	work experience	internship	work experience	internship
laid off	(made) redundant	laid off	(made) redundant	laid off, relieved from duty

FIGURE 21.3 (*continued*)   Words in Five Varieties of English.

U.S.	U.K.	Canada	Australia	India
lawyer, attorney	solicitor, barrister	lawyer	solicitor, barrister, lawyer	lawyer, solicitor, barrister
license plate (car)	number plate	license plate	number plate	number plate
liquor	spirits	liquor	spirits	liquor
mutual fund	unit trust	mutual fund	unit trust	mutual fund
nearsighted (myopic)	short-sighted	nearsighted	short-sighted	short-sighted
outlet, socket (power)	power point, socket	outlet, socket	power point	point
parentheses ( )	brackets	parentheses, brackets	brackets	brackets
parking lot	car park	parking lot	car park	car parking
pass (a car, going in the same direction)	overtake	pass	overtake	overtake
pavement, asphalt, blacktop	bitumen, macadam	pavement, asphalt	bitumen	tar road
period (punctuation)	full stop	period	full stop	full stop
pharmacy, drugstore	chemist	pharmacy, drugstore	chemist, pharmacy	chemist
plexiglass	perspex	plexiglass	perspex	plexiglass, perspex
prenatal	antenatal	prenatal	antenatal	prenatal, antenatal
property taxes	rates	property taxes	rates	property taxes
railroad tie	sleeper	railroad tie, railway tie	sleeper	sleeper
ranch	farm (no direct equivalent)	ranch, farm	station	(no equivalent)
real estate agent	estate agent	real estate agent	estate agent	(real) estate agent
rent (car), *v.*	hire	rent	hire	rent
rent (property)	let, let out	rent	let, rent	rent
reservation	booking	reservation, booking	booking, reservation	reservation
reverse charges (telephone call)	call collect	reverse charges	call collect	call collect
shot (vaccination)	injection	shot, injection, needle	injection, vaccination	injection
sidewalk	pavement	sidewalk	footpath	footpath

FIGURE 21.3 (*continued*)   Words in Five Varieties of English.

U.S.	U.K.	Canada	Australia	India
slaughterhouse	abattoir	slaughterhouse	abattoir	abattoir, slaughterhouse
streetcar, trolley	tram	streetcar	tram	tram
stub (check)	counterfoil	stub	stub, butt	counterfoil
styrofoam	polystyrene	styrofoam	polystyrene, styrofoam	styrofoam
tailpipe	exhaust pipe	exhaust pipe, tailpipe	exhaust pipe	exhaust pipe
trailer (camping), mobile home	caravan	trailer	caravan	trailer
trailer (goods)	trailer	trailer	trailer	trailer
trailer (horse)	horsebox	horse trailer	float	trailer
trunk (of car)	boot	trunk	boot	dicky
(labor) union	trade union	union	trade union	trade union, labour union
vacation	holiday	vacation, holiday	holiday	holiday, vacation
windshield (car)	windscreen	windshield, windscreen	windscreen	windshield
wrench	spanner	wrench	spanner	spanner
yield (street sign)	give way	yield	give way	(no equivalent)

FIGURE 21.3 (*continued*) Words in Five Varieties of English.

If you need to edit a document for use in an English-speaking country other than your own, always get a native speaker of that variety of English to check your work. Many of the terms in Figure 21.3 are not exact equivalents, and their use in a sentence may vary from one country to another.

In most cases, the U.S. English term will be understood in other countries (especially among people under the age of 30) because of the widespread screening of U.S. films and television, and the use of U.S. terms on the Internet. However, a term in common use in another country may not be understood in the United States.

EXERCISE 21.1

Go to a translation site like Google Translate (http://translate.google.com/) or Yahoo! Babel Fish (http://babelfish.yahoo.com/). Type in some text, perhaps an idiom. Now back-translate the same text. What happens? On Yahoo! Babel Fish, we typed in the idiom "caught between a rock and a hard place" and translated from English to Spanish, resulting in "cogido entre una roca y un lugar duro," which literally means something is stuck between a real rock and a very hard location! When we back-translated from Spanish to English, this was the result: "taken between a rock and a hard place." The same phrase put through Google Translate resulted in an exact back-translation. Now you try it with some idioms of your choice.

You don't need to go across continents to see the variety in language. For example, a Southerner may live *catty-corner* from a Midwestern neighbor who says *kitty-corner* (both terms meaning diagonal and variants of the term *cater-cornered*). A Wisconsinite might refer to the turning signal device in an automobile as the *directional*, while others may call it the *clicker*. Elsewhere, some people use the term *clicker* to describe a *remote control*. A grocery cart can be a *shopping cart* or a *buggy*. A water fountain can be a *bubbler* or a *drinking fountain*. Although most of these variances are easy to interpret, imagine the potential confusion for a first-time user of English. Therefore, it is important to be sensitive to cross-cultural factors when editing translated text in order to help the reader as much as possible.

USE OF APPROPRIATE GRAPHICS

One solution that has been proposed to combat the costly translation market is to use graphics instead of words. Technical illustrator Patrick Hoffmann has created several wordless manuals by replacing text with illustrations. This method is highly cost-effective in that companies don't have to pay for translation software or for writers to translate the document into multiple languages. However, this method is not foolproof. Pictures create another opportunity for cultural difference issues to arise. For example, the hand gestures in Figure 21.4 are not acceptable interculturally. The use of inappropriate graphics may insult an international reader.

Because illustrations are an integral part of technical communication, choosing graphics that will be appropriate in any cultural setting is an important part of internationalizing technical documents. The use of an inappropriate graphic in a technical document can humor, confuse, or even insult the international reader. A good example is the "thumbs up" gesture. While this gesture is positive to an American audience, it is considered vulgar in some European countries.

Another example is the "OK" gesture. To Americans the sign for "OK" means they agree with something or they consent to carry on as planned. However, in Brazil this gesture means "screw you" and is considered obscene. Brazilians often joke about how unwary visiting Americans shoot the "OK" sign to signify their pleasure, not knowing they are really sending an offensive message. This is a classic example of the need to know cultural differences.

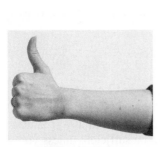

"Thumbs Up"   "O-K"

FIGURE 21.4
Potentially Offensive Hand Gestures.

TIP FOR TECHNICAL EDITORS

Try to use illustrations instead of text when possible. It is often better to illustrate a product or procedure rather than trying to explain it. (Some manuals, such as those from microchip manufacturer Intel, do not contain any text and are composed entirely of graphics.) Good illustrations will aid in the comprehension of the text and help eliminate the need for translation. Because translation is so costly, many companies are working to internationalize their communications by using wordless documents whenever possible. For instance, Hewlett-Packard creates computer equipment for several countries. A text manual to explain how to install an HP printer to a PC might seem simple and straightforward, but if the text has to be translated into sixty different languages, the translation costs can reach into the hundreds of thousands of dollars. Today, Hewlett-Packard saves high translation costs by using color-coded illustrations instead of text, a choice that also saves the company technical support costs because customers confused over any mistranslated text aren't calling in to complain.

Olin Lagon, in his article "Culturally Correct Site Design," relates the following example of how one company attempted to localize a document by using what was thought to be an appropriate graphic.

> A globalization manager with a very large business said his company thought it had the right approach: He translated the American-targeted copy into the local language and changed the photograph, switching to models of the targeted ethnicity. The photograph pictured a man and woman of the correct ethnicity, with the woman demonstrating to the man how to use the company's product. This was an unacceptable breach of accepted gender roles in the target country. Furthermore, even though the models were of the correct ethnicity, they had an American look. The globalization manager quickly learned that his team had made mistakes, as his foreign customer was shocked with the results. With one seemingly innocent picture, his group committed multiple cultural offenses. The content had to be totally redesigned.

People's ethnicity and gender roles are not the only triggers of offense with globally used graphics. Religion also plays a key role in defining cultural expectations and values. Any reference to eating pork is highly offensive in Muslim countries.

In short, it is crucial to study the cultural norms of individual international audiences, as opposed to learning a dozen languages (although learning new languages will certainly not hurt you). Cultural differences are very real, and avoiding offense and misleading information should be your primary goals.

TOOLS AND TECHNOLOGY

One of the most important tools you can invest in as a technical editor is a good dictionary. A recent hardback dictionary should be sufficient unless you're an "e-only" person and want an online dictionary like www.merriam-webster.com. An online encyclopedia might also be useful; try Wikipedia (http://en.wikipedia.org/wiki/Main_Page). Google Scholar (http://scholar.google.com/) also has reliable links to information. As mentioned earlier, translation sites like Google Translate (http://translate.google.com/) or Yahoo! Babel Fish (http://babelfish.yahoo.com/) can help clarify some translation issues. But be sure to back-translate (see Exercise 21.1) to find potential errors.

Although we mentioned that you will not have to be a bi-, tri-, or quadrilingual speaker to be an effective technical editor, you may still find yourself in a work situation where you are being asked to proofread a document that has been produced by your company and then translated into a language you don't understand. What should you do? Roger M. Ribert discusses comparing the translation with the English original and offers the advice shown in Figure 21.5.

- *Count and match everything you see*. If the English source has ten bulleted points in a list, make sure the Portuguese translation matches it.
- *Compare the lengths of sentences and paragraphs*. If the translator changes the size of a text box in a graphic, some of its text may be accidentally cut off.
- *Look for cut-and-paste mistakes in English terms and phrases*. For example, if a translator decides to cut and paste "Acme Software Company" thirty times in a translation, he or she may inadvertently cut existing text once or twice.
- *Trust your instincts*. If a translated paragraph seems somehow incomplete, or if it seems to conflict with text you read three pages earlier, check out the situation.
- *Follow your company's guidelines for presenting trademarked names*. For example, the Acme Company guidelines may require that its name always appear in bold print and caps next to a trademarked product name: "ACME AB400 Software." However, if the text says, "Acme makes software," no bold print or caps would be necessary.
- *Trademark guidelines may also restrict the use of prepositions in a product name*. For example, the French translator might translate "ACME AB400 Software" into "logiciel AB400 de ACME." This use of "de" would violate the guidelines. The correct translation would be "de logiciel AB400 ACME."
- *Watch for a particular language's rules in punctuation and capitalization*. For example:
  - French puts a space between a word and a semicolon.
  - German capitalizes all nouns and words used as nouns.
  - Spanish places an inverted question mark at the beginning of a question.
  - Italian does not capitalize the days of the week or the months of the year.
  - *Look for inconsistencies in punctuation and capitalization*. For example, if a list of bulleted points on page 11 has commas, but a similar list on page 12 does not, you need to find out why. Likewise, if the German word *hinwig* (note) carries a colon on page 20, but not on page 35, you have to find out why.
  - *Check for broken and mislabeled links*. For example, the Acme software manual says, "To see a diagram on page 10, click here." However, in the Italian translation, the diagram is on page 11. The translator may fix the link, but forget to revise its text.
  - *Check all units of measure*. Have all the inches been converted into centimeters, all the Fahrenheit temperatures into centigrade, all the dollars into euros? Are the numbers supposed to be rounded up? Does the translator use centimeters on page 5, but millimeters on page 6 for the same measurement?
  - *Check the headers and footers*. The translator may have forgotten them or assumed they weren't supposed to be translated.
  - *Look over terms that appear both in English and the foreign language*. Does the English term come first or does the Danish translation? If two terms in a list of ten are not translated, is there a reason why?

FIGURE 21.5

How to "Proof" a Translation.

Source: *http://www.stcsig.org/itc/articles/05-proof-transl.htm.*

SUMMARY

Paying close attention to word choice, sentence structure, and graphics will better enable you to edit and prepare documents for an international audience. As trade and communication between cultures increases, so does the need to accurately convey technical information to individuals of different cultures and languages. This chapter provides you, the aspiring technical editor, with skills on how to internationalize words and graphics in a global document.

CHAPTER

Production and Project Management Issues

At the conclusion of this chapter, you will be able to

- practice planning a software user guide project;
- estimate the time required to edit a document;
- prioritize editorial changes;
- explain some editorial management issues;
- use a file naming, tracking, and archiving system;
- describe the document production cycle.

Throughout this book, we have talked about products and documents going through a *production cycle*. This term refers to the sequence of steps (the cycle) required to produce a document or product that is ready for users. Technical editors today often are involved in every aspect of the production cycle for documents and sometimes for the product as well. In addition to working as writers and improvers of text and graphics, having to be familiar with technological issues, and possessing superior soft skills like communicating well with writers, technical editors also need to be ready to function at all levels of the production cycle. This need is not a burden but actually a huge benefit. Why? Technical editors, who are knowledgeable in so many areas, want to be involved because their involvement improves the project at every stage. Although it might seem like an easy job just to edit some manual at the end of the production cycle, the truth is that the more involved technical editors are and the earlier in the production cycle they are working with the design team, the better the product, the happier the user, and the better the documentation that will go with the product.

In this chapter, we first discuss the planning stage, allowing you as a budding technical editor to experience firsthand what it would be like to plan a project. Then we address some project management issues that will help you streamline early, middle, and late stages of the production cycle, such as estimating the time required for the project, organizing a work sequence for user guide development, and tracking your work. To illustrate the steps in the production cycle, we provide the sample project of a software user guide.

PLANNING A SOFTWARE USER GUIDE PROJECT

User guides should be planned and designed as part of the overall software development project, with similar consideration for users' requirements. Planning is particularly important when more than one person is involved in producing the user guide, one person is writing the user guide and another person is writing the online help, or more than one editor is involved, so that all players know their roles and how to ensure that their contributions fit together into a coordinated whole.

Good planning vastly improves your chances of completing a project on time and within budget while producing a usable, helpful, correct, and complete user guide. Lack of planning almost ensures that the project will run into problems.

Therefore, someone (for example, the documentation team leader, the writer, or the editor) needs to write a documentation plan.

The written plan should describe each phase of the user guide development process: task analysis, design, formal reviews, editing, and testing; and the roles of the people involved at each stage of the process. Other people, such as software developers, working on the project can then use the plan to understand how their roles work together with the roles of the writer, editor, and others on the team.

Include a schedule showing how documentation development fits in with user interface development.

Specify all the assumptions, dependencies, and risks involved in producing the user guide. What do you need from others, and when? (See Figure 22.2 later in this chapter.)

Managers, marketing personnel, user interface designers, and software developers should review and approve the plan before major work begins on the user guide. Update the plan when significant decisions are made or whenever the scope or timing of the work significantly changes.

The ideal project would have unlimited time and resources to produce a user guide that meets all the needs of its target audience. It would also be well planned and the development would proceed according to the plan.

In reality, time and resources are always limited and changes often occur at inconvenient times, so we often have to make compromises. The compromises should never include skipping the planning stage of the project, yet this happens all too frequently.

Sometimes, planning is not done because the people involved don't realize how much they can do before learning the details of the user interface for the software product. In fact, you can complete most of the planning steps without this information as long as you know what the product is supposed to do and who the intended audience is.

Let's look at the steps involved in the ideal documentation project, why each step is important, and what might happen if the step is skipped.

1. Analyze the audience, plan the project, and write the plan.
2. Develop specifications.
3. Perform a detailed task analysis.
4. Build and evaluate a prototype user guide.
5. Develop an outline for the guide.
6. Write, index, and edit the guide.
7. Review the guide.
8. Test the guide with the product.
9. Release the guide with the product.
10. Evaluate the guide and plan for improvements.

Step 1: Analyze the Audience, Plan the Project, and Write the Plan

Planning might be done by the documentation team leader, the writer, the editor, or by all of these people contributing their particular expertise. Include all planning decisions in a plan, and have the plan approved by the relevant people.

This step has several substeps, some of which should be done at the same time. If your project is updating an existing user guide, some of this work may have been done previously, but you should check whether any of the factors have changed for this update.

1. Describe the software product and its purpose in terms of its intended users and their tasks.
2. Analyze and profile the intended users, their knowledge and skill levels, and the tasks to be supported by the software product.
3. For products to be sold commercially, study your competitors' products, which may have influenced customers' expectations.
4. Establish the main aims of the user guide.
5. Determine any constraints, such as

   - Will the guide be printed or delivered in electronic form (PDF, HTML, or other), or both?
   - Will the product and the user guide be translated into other languages?
   - Will the product and the user guide be localized—for example, to accommodate different currencies, cultures, or taxation schemes?
   - How much time and money is available for the project?

6. Write down other assumptions and dependencies and the consequences if these assumptions and dependencies are not met; for example

   - Reviewers will return all review comments within X days.
   - Developers will keep to their schedule.
   - Clients will not change the specifications after they are approved.
   - The product's technical specifications are accurate and complete.
   - Subject matter experts (SMEs)will provide technical input when needed.
   - A working version of the product will be available by an agreed time.

7. Estimate the time required to plan, write, edit, review, and test the user guide.
8. Negotiate with developers and others to

   - Determine realistic requirements for the user guide.
   - Specify when the writer can get access to a working version of the software.
   - Specify the type and timing of technical reviews and testing.
   - Specify who will have the final authority to approve the content and what the approval criteria and process will be.

If the planning step is not done, writers, developers, clients, and others may have unrealistic expectations about the complexity of the project and what is needed to complete it. As a result, too much work may have to be completed in the available time, so some steps may be dropped or curtailed, or the project will run over time and over budget. For example, editing may be limited to a cursory check of spelling, grammar, and layout consistency; substantive issues may never be identified. Indexing and testing may be inadequate, contributing

402 Technical Editing in the 21st Century

to usability problems. Problems often cannot be fixed, even when identified, because no time is left in which to make the changes.

All of these problems may be greater if the user guide has been written by someone other than a professional technical communicator, or if it has been written by a team of communicators without overall planning and coordination.

Step 2: Develop Specifications

Specifications should include decisions on the contents and overall design of the book: what topics will be included and in what sequence. As it is a user guide (not a reference manual), the approach should be task-oriented, not function-oriented. The specifications should include writing style, terminology, style sheets, and templates for topic types. Be sure to get management's approval of the specifications.

Developing specifications is one of the two most important planning steps; the other is the task analysis (Step 3). The editor needs to develop a project-specific style guide or style sheet to record specific decisions about the format, presentation, and content of the documentation. Work with the writer and the project leader when making the decisions.

TIP FOR TECHNICAL EDITORS

Task-oriented documentation provides how-to and other information from the point of view of users and their tasks instead of describing how the software works. A user task often does not have a one-to-one correlation with one part of the software, so merely describing what users can do at a particular point is often much less helpful than explaining the sequence of steps and the consequences of choices they need to make along the way.

As a technical editor, you should ensure that terms used in the user guide match those to be used in the user interface and the online help. Create a terminology list, including terms used by your competitors, so you can include relevant synonyms in the index of the help and the printed user documentation. The terminology list can also form the basis of a glossary and the definitions provided to the translators. Ensure that any consistency issues between the help, the interface, and the user guide are resolved. Consider translation and accessibility issues. Even when only one writer is involved, a project leader needs to consider what happens if that writer leaves the project before it is complete and who will maintain the user guide for future releases of the software.

With a detailed plan for the documentation, everyone on the project knows what's expected and doesn't have to take time later in the project to figure out what to do. New members of the team can quickly learn how to fit their writing into the overall look and feel of the documentation.

If specifications are not developed, writers will have to pay more attention to style, terminology, design, and other issues during the writing stage, and they are more likely to forget something. The editor's work will be more complex and will probably include many more consistency changes and negotiations over terminology and other issues. The opportunity for correcting problems later usually is minimal because of time constraints.

Lack of detailed specifications, particularly on a project involving more than one writer, often leads to inconsistency between the user guide, the online help, and the software itself, including style, terminology, and content. Correcting these inconsistencies takes more time later in the project; and if the corrections are not made, users may be confused.

Step 3: Perform a Detailed Task Analysis

Once a plan has been developed, the technical communication team needs to create a detailed task list from the intended audience's point of view. Task analysis may involve use cases (formal descriptions of how the software works), user scenarios (descriptions of how people can use the software to accomplish tasks), and other techniques. The requirements specifications for the product itself should already include much of this information; if not, the documentation team may need to compile it all.

Include your task list in the documentation specifications. In addition, create a list of questions that users might ask. From the task and question lists, you can begin building a list of topics to be covered in the user guide. This list will probably evolve as the project progresses, especially if the application itself is evolving. If this is a revision of an existing product, the user support group is an excellent source of information on where users have problems with the product. Even for a completely new product, you may be able to get good information from beta testers' feedback.

These lists form the basis of the topics in the book and the editorial review of the index, the table of contents, and the overall helpfulness of the book. If task lists are compiled at this stage, the editor can check the writer's work more easily because most questions about what should be included have been resolved. Without task and question lists, the writer may forget to include some information that the users need, and the editor may fail to notice the omissions.

Step 4: Produce and Evaluate a Prototype User Guide

After completing the task analysis, you can produce a prototype user guide. For an early prototype, you don't need to know the details of the software's user interface although you do need the task and question lists developed in Step 3. You can use placeholders such as "Procedure steps go here" and "Insert screenshot here" in an early prototype. Ask yourself if the topics and the planned level of topic breakdown make sense in the context of how the user interface will work. Come back to your prototype after Step 5, build in some more detail, and evaluate it again. Ideally, you will be able to involve real users in the evaluation.

At this stage, the team can identify ideas that don't work well in practice and have a chance to change them before anyone starts writing. A bit of time spent here often saves more time later because you don't have to rewrite material. You can show the prototype to users or customers and get feedback from them about your topic types and planned level of detail; you can then improve the design based on their comments. Sometimes you uncover some serious usability problems in the software itself and can influence the software development too.

If you skip the prototype step, you could find during the writing or testing stages that some or all of the user guide might have to be redesigned or rewritten.

Step 5: Develop an Outline for the Guide

Steps 4 and 5 are often done together, although their focus is a bit different. Develop a detailed outline for the user guide and add it to the specifications. By this stage, if not earlier, the software's user interface design should be available even if the software itself is not working. If this step is not completed, writers, testers, editors, and reviewers may miss some important user tasks or will need to spend more time later on checking to make sure that nothing has been skipped.

Step 6: Write, Index, and Edit the Guide

If design planning was done, terminology agreed on, and templates produced in Steps 1 through 5, the writer can concentrate on content in this step, and the editor can concentrate on checking the writing against the design specifications.

If working portions of the software interface are available, the writer and editor can also check the documentation against the interface. Ask if the user guide accurately describes what the users will see and what happens if they follow the written procedures.

Make sure that the product developers inform the writer and editor about software changes as soon as the decision is made, so any necessary changes can be incorporated in the user guide during this step. Some development processes make this information exchange difficult because changes are frequent and often documented later, if at all. You may need to negotiate ways to ensure that writers are informed, but don't get bogged down in changing details too early in the writing process. Concentrate on the bigger picture and fill in the details as they become available and are stabilized, as late in the writing cycle as possible.

Obviously you can't skip the writing, but indexing and editing are very important, too. If inconsistencies, grammatical errors, inadequate indexes, and unhelpful topics are not detected, they can lead to poor user perceptions and an increase in support calls. If software changes are not communicated to the writer, errors may be detected in the next step, leading to extra work, or they may slip through undetected. Users may see the user guide as unhelpful or unprofessional.

Step 7: Review the Guide

The software developers or other SMEs typically do this step. Usability experts should also be involved. If possible, include your clients and some users and their managers. Writers then make necessary changes. The writing team should use a change-control procedure to track documentation changes arising from changes to the application as well as those arising from reviewers' comments. Developers and others should sign off on the content of the user guide at the end of this step.

Documentation that has been technically reviewed is more likely to be accurate, especially if changes were made to the design of the application after the design documents were prepared. A user guide that has been reviewed by usability experts is more likely to be helpful to the users.

If previous planning steps have been done well, major problems should not persist until this stage. However, you may find that technical reviews are limited to the technical accuracy of the book, with no one assigned to review helpfulness.

Step 8: Test the Guide

Test the user guide with the software. Revise it as needed. Ensure that details of the expected testing process have been included in the approved plan because testers often do not realize what is involved in thorough testing of a user guide (see "What Types of Testing Are Required?" later in this chapter). Ensure that the tester is formally committed to test the book in the way specified in the plan.

Step 9: Release the Guide with the Product

Beta or test versions of software may be released without a user guide or with an early draft of the book, but finished software should include a fully written, indexed, edited, and tested user guide—whether printed, supplied as a PDF, or available on the Web.

Step 10: Evaluate the Guide and Plan for Improvements

After the product is released, involve users in evaluating the user guide. For example, you could include an email address to encourage users to provide feedback. Many products have some defect-tracking mechanism that allows users to report defects in the documentation as well as in the product. Plan improvements for the next release. You'll do it even better next time, and everyone will benefit. Conversely, if the team doesn't learn from its mistakes, it's doomed to repeat them.

HOW MUCH TIME IS REQUIRED FOR PRODUCING DOCUMENTATION?

Technical communicators have developed many schemes for estimating the time required to produce various forms of documentation. In this section, we summarize some estimating methods that work well as a starting point for a group that has never worked together before. If your department has data on past performance, use that data instead of these generic figures.

Time estimates depend on the number of pages or topics of material that must be written, edited, reviewed, and tested. From your audience and task analysis, you can estimate the number of conceptual, task, procedural, tutorial, or other topics that may be required. Plan to revise these estimates as the project develops.

JoAnn Hackos (*Managing Your Documentation Projects*) suggests an average total of four hours per topic to cover all activities: planning, researching, writing, editing, indexing, reviewing, and usability testing. This time estimate assumes that some topics will be quite short, while others will be longer and more complex. The average length of a "topic" is assumed to be between half a page and one printed page in a manual. For many years, the industry worked on an average of "a page a day," where a day is assumed to be eight hours. If your project consists mainly of long topics, you may need to increase the average time estimated.

Your situation may vary, so use these figures only as a starting point for developing your own estimates, and refine them as you gain experience with your clients and projects.

How Long Does Editing Take?

Technical editors know that the time required for editing depends very much on what level of editing is required. A detailed, heavy edit requiring major rewriting is obviously going to take a lot longer than a light copy edit. So the first thing you need to do is clarify some details:

- What are the document's main problems? Structure? Language? Inconsistent mechanical style? Other? Given limited time, what should be fixed first, from an editorial point of view?
- What does the project manager want to be done? This is often quite different from the editor's list of the document's main problems.

Be prepared to negotiate your "want list" when, as frequently happens, the project manager decides your initial estimate is too long or too expensive. When determining how much editing you can do for a project, consider

- the time available;
- what the manager wants;
- what the manager will accept;
- the ease or difficulty of fixing what needs to be done;
- what would best serve the needs of the document's readers;
- the importance of the document.

Accurate estimation is an important factor in running a successful business or publications department. You need to have a reasonably good idea of how long you or someone else will take to edit a document. Figure 22.1 breaks down editing time estimates by task, based on our personal experience.

The estimates are based on these assumptions:

1. A "standard page" contains 500 words. This is independent of the page size, or whether the draft is single- or double-spaced. You can adjust the page count by converting the number of physical pages into the number of "standard" pages.
2. If a document contains a lot of illustrations, it has fewer words per page; but you still have to give the illustrations some attention, so try to take that into account when people give you a page count. For example, if someone tells you the document has 350 pages, most of which contain one screenshot, you may decide that one document page equals two-thirds of a "standard" page and adjust your estimate accordingly. If the paper size is unusually small or large, a document page may be more or less than a standard page.

Notes on Figure 22.1:

1. The time required depends on whether you're adapting or updating an existing specification and style guide, or starting from scratch; whether a lot of negotiation with other stakeholders (and revision of the specifications) is needed; and to some extent whether the project uses one writer or several. You'll need to allow time to revise these materials after the first edit of sample material or the first review of a prototype book.
2. You may need more time to edit a very large project or a badly organized table of contents, especially if more than one writer is involved or if you need to make detailed suggestions on how to reorganize the contents.

Activity	Time required
Develop design specifications, style guide, templates	40–80 hours (see note 1)
Edit table of contents	1–2 hours (see note 2)
Edit substantively, including some rewriting	1–2 pages per hour
Copy edit	4–6 pages per hour
Light copy edit (skim document, correcting obvious errors in spelling, grammar, punctuation, consistency, and completeness)	8–10 pages per hour
Online documentation only: Check links against specifications (Do they go to the right place? Are they useful links?)	50–70 links per hour (see note 3)
Edit index (500 entries)	1 hour quick check; 4+ hours detailed check, no fixes; 5–20+ hours to fix problems
Quality or production edit	20 pages, or 60–100 help topics or more per hour (see note 4)

FIGURE 22.1
Time Required for Editing a User Guide (Our Estimates).

3. For electronic documents, depending on the complexity of the linking system used in your project, the time required could be considerably longer than suggested here.
4. Depending on the complexity of the project and the number of tables and graphics, the time required could be considerably longer than suggested here.

HOW MANY WRITERS CAN ONE EDITOR SUPPORT?

The time that editing takes is related to another question: how many writers can one editor support? JoAnn Hackos mentions that 10 to 15 percent of the total "writing time" is an estimate for editing time; we've seen similar figures in numerous other sources. Hackos also points out that the percentage of time spent on editing varies with the stages of a project, the type of editing required, and the skill of the writer. She recommends keeping records of how much time editors in your group actually spend, and on what tasks, so you can get a good idea of the dynamics of your particular situation. And last, having an on-call editor to handle extra work at peak traffic times, or when one of the staff editors is away sick or on vacation, can smooth out the workload without a company needing to hire another staff member.

Based on personal experience, Jean has found that one full-time editor can support around nine full-time writers, maximum (which fits into Hackos's 10–15 percent estimate: 1:9 = 11 percent)—but if all nine have a deadline at the same time, the editor is in deep trouble!

WHO DOES WHAT, WHEN?

You need to work with the software developers, testers, and other members of the team to make sure everyone understands the types of edits, reviews, and testing required for user documentation, and who is doing which type. Make these decisions during Step 1 of planning your project, and include the decisions and agreements in your plan. See "How Many Reviews Are Needed, and When?" for guidelines on the timing of edits and reviews.

STAGES OF DOCUMENTATION AND SOFTWARE DEVELOPMENT

The development stages for documentation run in parallel with the development stages of the software itself. Figure 22.2 compares the steps outlined in this chapter with the stages of software development and lists critical dependencies.

How Many Reviews Are Needed, and When?

All documentation should be reviewed for technical accuracy by software developers and SMEs, substantively edited, and copyedited. As we discuss in Chapter 3, reviewers and editors may work with soft copy or hard copy, but soft copy is becoming more common. On-line help, websites, and other materials should be edited and reviewed online, either in addition to a hard-copy review or in place of a hard-copy review, for the reasons we discuss in Chapter 19.

Do this documentation development step	*During this software development stage*	*Critical dependencies for each documentation development step*
1. Audience analysis, high-level task analysis, documentation plan 2a. High-level specs	Feasibility study Requirements definition Functional specifications High-level design specs	Requirements, audience, high-level task list
2b. Detailed specs	Detailed design specs	Functional specs, high-level design specs
3. Detailed task analysis 4a. High-level prototype	Software prototype	Functional specs, high-level design specs, detailed task list
4b. Detailed prototype 5. Outline 6. Write, index, edit	Coding, unit testing	Detailed software design specs
7. Technical reviews	Integration testing	Reviewers available
8. Test the guide	System testing Usability testing	Working software, testers available
9. Release the guide	Product release	Printer available
10. Evaluate	Evaluate	Feedback from users

FIGURE 22.2
Stages of Documentation Development and Software Development.

Review	Who	When
High-level specifications	Client, product developers, editor	Planning and product high-level design
Detailed specifications	Client, product developers	Planning and product detailed design
First prototype	Client, product developers, editor	Product prototype
Completed table of contents	Client, editor	Product development
Copyediting	Editor	Just before content review but after initial indexing
Content, accuracy, completeness, helpfulness	Product developers or SMEs	System test
Second copy edit	Editor	Just before second-draft review
Second-draft review (material 100 percent complete)	Product developers or SMEs	User acceptance testing, or its equivalent
Testing	Testers, writers, or editor	User acceptance testing

FIGURE 22.3
Number and Timing of Reviews.

For a new product or a major revision of an existing product, we suggest you perform the reviews and edits listed in Figure 22.3. Minor revisions normally need only copyediting and reviewing for accuracy.

What Types of Testing Are Required?

Technically correct user documentation is not necessarily helpful to the user. A usability expert, business analyst, or other suitably qualified person should review the documentation from the user's perspective. Does it answer questions that a user is likely to ask in a given situation? Can the user find required information using the table of contents and index? Are the cross-references useful or confusing? In general, is the documentation helpful?

In addition to other editing tasks, the technical editor may also be responsible for reviewing the user guide for helpfulness. Because many developers involve potential users in the product development cycle, the writing team should do everything possible to include a sample of users in the documentation reviews, too. Real users are the best people to judge the value of user documentation because they are usually not already familiar with the product and therefore don't "fill in the gaps" from other knowledge.

ORGANIZING THE WORKFLOW

Planning is a major step in the production style, but as a technical editor you will also need to know how to organize the writing process and the workflow, two major areas we highlight in the remainder of this chapter.

The best way to organize the division and flow of writing, editing, and reviewing depends on the individuals involved, the project, and the organization. Some questions to consider are

- How many people are involved in the project?
- What are their skill levels (as writers or editors, and as users of the tools)?
- Does everyone have access to the hardware and software required?
- What are the time constraints?
- How much other work does each person have to do?
- At what stage is a document reviewed by SMEs for accuracy of the content?
- How well do team members get along with each other? Do they respect and value each other's work (both methods and results)?
- What methods will be used for reviewing and editing?

TIP FOR TECHNICAL EDITORS

Be sure to get agreement on how the documentation will be reviewed, edited, and tested, and record the agreement in the documentation plan.

Some typical scenarios for technical editors are

- The editor types in changes, and the result goes directly to layout and production. This method is best when the writer is not available (for example, has left the company) or fast turnaround is required. It may be used in any situation where the editor has the final say. The writer may proofread a printout of the final layout.
- The editor types in changes and questions using a change tracking feature, and the file goes to the writer to accept or reject the changes or to request clarification if required. This method is best when the writers are experienced or will have their names on the resulting document, or when the writers are inexperienced and markup will assist in their education; it is equivalent to markup on paper.
- The editor types in questions and changes without using the change tracking feature, and the amended file or a printout is returned to the writer to check and answer questions. This method is best when the editor is dealing with writers who prefer to see only the revised version.
- The editor types in changes and questions and discusses the markup with the writer. This method is best when training inexperienced writers or those new to the company style or if the editor has many comments or questions that are best resolved through discussion.
- The editor provides comments in a separate file or as markup on a PDF and the writer or layout person inserts them into the document. This method is best when the editor doesn't have the appropriate software to edit the file directly and/or layout is most important (for example, in brochures). The writer usually proofreads a printout of the final layout.

The sequence of reviews and editing also varies with circumstances:

- Technical reviews may be conducted before editing. This method is best when a complex document is likely to be changed substantially by SMEs during their review, so that the copy is reasonably stable and accurate before it reaches the editor.

- First-pass editing may be done before technical reviews. This method is best when dealing with inexperienced writers or those unfamiliar with the company style, so that the review copy is free of grammatical, spelling, word use, and other errors that might distract the SMEs from focusing on the accuracy of the information.
- Time constraints often mean that the editor must edit pieces of a document separately (often interspersed with other work), rather than receiving the entire document to edit at one time. In this situation, the editor needs a detailed style sheet and must take extra precautions to spot inconsistencies. This method is not recommended but is quite common.

PRIORITIZING CHANGES

If writers have limited time available for rewriting or fixing other problems found during editing, divide the necessary changes into three categories—content, style, and minor grammatical issues—and suggest that the writers do them in this order of priority:

1. Correct factually incorrect *content,* wording that is unclear or ambiguous and may lead to serious misunderstanding, and changes that are required for legal reasons.
2. Improve *writing* or *presentation* where changes are not essential to understanding.
3. Correct minor instances of incorrect *grammar.*

MANAGING EDITORS

Many writers, editors, and other members of the production team work flexible hours, part-time, or off-site (at home or in their own offices). Experienced managers know that good management practices are much the same wherever people are located. Here we mention briefly a few points to keep in mind.

Managers and editors need to work together to

- Define the editor's role, both within the organization as a whole, and for each individual project, and ensure that all members of the group know what this role is.
- Define the required level of edit and a time frame for each project or document: what work is expected to be done each week?
- Set goals for editing, both for individual documents and overall within the department, by different time scales: weekly, monthly, six-monthly—whatever is appropriate for the job.
- Track progress against the agreed goals, using a spreadsheet.
- Determine who supplies whatever equipment is needed for off-site workers. Independents usually, but not always, provide their own equipment; employees usually, but not always, are provided with equipment by their employer.
- Schedule on-site meetings on specific days, as far in advance as possible, so all who need to attend can make arrangements to be there.
- Determine availability requirements for team members; for example, whether off-site workers are expected to be available for telephone or instant-message contact between certain hours, how flexible are everyone's work hours, and whether workers are expected to keep other team members informed of their availability.
- Specify any security requirements for materials kept on portable computers or at off-site locations, including nondisclosure issues regarding others (family, other members of the independent's company, and so on) at that location.

- If several editors are involved in a project, agree upon methods to ensure consistency in their work; for example, who is the lead editor—the person who is responsible for the contents of the style sheet, and has the authority to make and enforce decisions about consistency and style?

USING A FILE NAMING, TRACKING, AND ARCHIVING SYSTEM

Be sure you and your manager agree on when, where, and how often you save files on the company's server. Do you upload files only when you're finished editing them, or do you upload your latest (unfinished) version once a day? If you work off-site and can't upload files directly to the company's server, to whom do you send the files?

Does the company you're working for have a file naming, tracking, and archiving system? If so, use it. If not, encourage them to develop and enforce such a system. Many larger companies use a content management system (CMS), but many smaller companies do not. Here's a fairly simple manual tracking system that we use:

1. On your computer's hard disk or the company's network, create a folder for each project or document. Create several subfolders in each folder: *In, Pending, Out,* and *Archive* (or whatever terms you prefer).
2. If the project has multiple files (for example, separate graphic files), you might prefer to make subfolders to hold the graphics—we try to follow the convention used by the client.
3. When you receive a file, put it into the *In* folder for that document.
4. Copy the file into the *Pending* folder, rename it, and edit that copy. Don't touch the original.
5. When you've completed editing the file, move it into the *Out* folder. You know that anything in the *Out* folder is ready to send back to the client. We rename the file to indicate its changed status (for example, we might rename "Procedures" to "Procedures-edited").
6. After you send the edited file, move it to the *Archive* folder.
7. If you later edit the same files again (for example, after the writer has revised them), make a separate folder for the second edit.
8. At some point we'll remove all the files (originals and archived edited copies) from the hard disk; depending on the client, we may keep a backup copy or we may not. (If there's a security issue, we generally remove the files from our system and send all the backup copies to the client.)

You may also want to include tracking information in the file itself, for example in the header or footer of each page, or on a cover sheet that's included in the file but removed when the document is complete, approved, and ready for distribution.

You may also need to consider access controls and, particularly if you work in a group, an audit trail of editorial changes to documents if change tracking is not sufficient or some reviewers have marked up hard copy. A full discussion of the issues involved is well beyond the scope of this book.

Tracking Information You and Management Require

Does your employer have a system for tracking information required for accounting or other purposes? If so, use it. Otherwise, devise one for your own use. Even if the company has a system, the data required may not be what you need to know for your own use.

You need to keep track of your projected and actual editing schedule, including the projects you're working on, the length of time you estimate you'll need to do the work, when you expect to receive the files, when they need to be returned, what level of editing is required, how long you actually took to edit each file, and the accounting code to which you'll bill your time.

What software do you need to use? You probably don't need project management software, unless the client wants you to provide data in that format. A spreadsheet normally handles all of your needs, and if you set it up properly, you can extract a wealth of information about your editing projects, your clients, and your own work patterns. Even a table in your word processor is better than nothing!

Figures 22.4 and 22.5 show two simple tables of relevant information. You can set up a spreadsheet to calculate some of the details from data you enter and automatically insert data in several places at the same time. You need to consult the user guide or online help for your spreadsheet program to learn how to do this.

From this information, you can find out

- which people are habitually late getting work to you;
- how good your estimates are;
- how much project-related (billable) time you spend each day.

You can use some of this same information in other spreadsheets or your accounting program.

Client	Project	Estimate hours required	Planned start date	Actual start date	Days late starting	Actual hours required	Difference	Comments
AAA	AAA-1							
	AAA-2							
BBB	BBB-1							
	BBB-2							

FIGURE 22.4
Example Summary Table or Spreadsheet.

Date	Start time	End time	Hours spent	Running total (hours)	Hours left to finish work	Comments

FIGURE 22.5
Example Table or Spreadsheet for a Project (Repeat for Each Project).

Keeping Track of Multiple Text and Graphics Files

If multiple files are involved, you must be sure to bundle them all together when you send them to someone else. You'll need to know where your software stores files by default in places that are not useful to you as an editor (for example, text files might be saved in

"My Documents" but graphics saved in "My Pictures"), so you'll need to make sure the files are stored where you want them to go.

We generally use a system similar to the one we use for tracking and archiving files: we create a folder for each project. We may store all the project's files in that one folder, or we may create subfolders for, say, graphics and text.

If multiple files are involved, you must be sure to bundle them all together when you send them to someone else. Some software has utilities you can use to bundle multiple files into one for archiving or distribution purposes. Otherwise, if you know where everything is, you can use a program such as WinZip to compress all the files into one—being sure to keep the subfolder structure intact, if that is required by the publishing software.

SUMMARY

In this chapter, you learned and applied many practical skills as a technical editor. You sampled what it would be like to plan a software user guide project as well as estimate the time required to do a full edit of a document. As a technical editor, you need to be familiar with the more nuts-and-bolts aspects of the production cycle and some management issues. Finally, using a file naming, tracking, and archiving system is an excellent way to help you save time and stay organized during the production cycle.

CHAPTER 23

Usability Testing on a Budget

At the conclusion of this chapter, you will be able to

- apply a usability checklist to a project;
- plan goals and objectives for usability testing;
- appreciate the value of usability testing on a number of levels.

Software developers and their managers are generally aware of the need to test software and websites for usability, but they often don't understand the value of testing user guides and online help systems, or they don't understand the difference between usability testing and other forms of testing. They may think it takes too long and costs too much to fit into the project's schedule and budget. Reading this chapter[*] will help you learn how to find many problems quickly and inexpensively through simple usability testing.

In fact, user documentation (including online help) needs to be usability tested as much as the software does. Such testing can be done quickly and inexpensively, but the consequences of not conducting early usability tests can be quite expensive. According to experts such as Steve Krug, 80% or more of usability problems are found with only four or five subjects. If found early, these problems can be fixed at little cost and without delaying release of the product or the documentation. However, if found at the end of the development cycle, usability problems can be quite expensive and time-consuming to fix and often are not fixed at all.

Although usability testing is ideally planned and conducted by trained testers, in reality editors are often the only people available to do the job. Editors are also in a position to strongly recommend usability testing of documents if such testing is not already provided for in the documentation plan.

A full discussion of usability testing would take another book; in fact, several good ones have been published. Excellent information is also available on the Web. Particularly relevant are two chapters in Steve Krug's *Don't Make Me Think*, which describe how to conduct tests on websites, with a limited budget and using inexperienced people, if that's your only choice—and why that's a better choice than not doing

[*] This chapter is a slightly revised version of Chapter 11 in *Is the Help Helpful?* by Jean Hollis Weber and is used with the permission of the publisher.

any usability testing at all. The principles that Krug describes apply equally to printed documentation and online help.

WHAT IS USABILITY?

Usability is *the extent to which a product can be used by specified users to achieve specified goals with effectiveness, efficiency, and satisfaction in a specified context of use.* That's a somewhat ponderous way of saying a product is usable if people can use it quickly and easily to accomplish their goals.

For documentation, users' goals generally include finding required information, understanding the information, and applying the information to solving their immediate problems or answering their questions. Documentation can include printed or online user guides, tutorials, and help systems.

Documentation usability includes elements that are

- Easy to learn or understand. Can people use the documentation the first time they open it?
- Easy to remember. Can people use the documentation more easily the next time?
- Effective. Can people easily navigate through the documentation, understand the content, and put it to use to solve their problems?
- Efficient. Can people find what they need in the documentation and accomplish their goals in a reasonable amount of time?
- Satisfying. Do people have a good feeling about using the documentation? Do they feel it was worth their time to use it? Will they use it again?

Furthermore, usability is dependent on the users' perceptions and experiences. Documentation can pass objective tests for information included, links or cross-references within that information, links from the product to the information, and so on, but if the users' experiences are negative, then the documentation fails a usability test.

Usability testing, therefore, is a way to discover how users perceive and use documentation, by observing them actually using it and then analyzing the data to diagnose problems and recommend changes.

TIP FOR TECHNICAL EDITORS

Usability is not the same thing as accessibility, although the two concepts are related. A help system or website is considered accessible if people with disabilities can use it as effectively as people without disabilities. Thus a help system could be quite accessible but very unusable if everyone has difficulty understanding and using it.

PLANNING FOR USABILITY TESTING

Most people think of usability testing as something that's done at the end of the project during the system-testing phase of software development. Testing at that stage is certainly important and should be done, but you can also do a lot of valuable usability testing earlier in the product development cycle.

TIP FOR TECHNICAL EDITORS

Don't wait to test for usability until the end of the project, when it's too late to fix problems before the documentation is released for production. Test at several stages during documentation development.

In the early stages of documentation development, usability testing overlaps considerably with reviewing and developmental editing. In later stages of development, usability testing complements other forms of testing.

What you call this work (reviewing, editing, testing) doesn't matter, as long as you consider important usability issues:

- Terminology
- Icons
- Design of the table of contents
- Design of the internal navigation scheme (if online)
- Design of the index
- Page design
- Contents of topic types
- Writing style
- Use and presentation of lists, tables, and diagrams

TIP FOR TECHNICAL EDITORS

Don't skip usability testing just because you have no experience and can't afford to hire a usability expert to help you. Almost any testing is better than none, but you do need to plan a bit.

WHAT TO TEST WHEN

Divide your usability testing into four stages, as summarized in Figure 23.1 and described in the following sections.

Usability Testing During the Planning Stage

At the planning stage, you are either developing specifications and a style guide for the documentation of a new product or you are refining existing specifications for an upgrade from a previous version of the product. Editors should be fully involved in both aspects, whether developing or improving specifications.

You can include the testing described in this section during another stage of the project if you don't get a chance to do it during the planning stage.

Stage of documentation development	Items to be usability tested	Purpose of testing (questions you are trying to answer)
Planning	Task definition Terminology Icons	What should be in the specifications?
Prototyping	Overall design (fonts, colors, layout) Internal navigation Draft table of contents Draft index design	Are the specifications suitable, or do they need improvement?
Early draft	Contents of topic types Writing style Use and presentation of lists, tables, and diagrams	Are writers producing material that meets the specifications? Are specifications suitable, or do they need improvement?
Production	All of the above General helpfulness	Did we get it right?

FIGURE 23.1
What to Test at Each Stage of Documentation Development.

Documentation for an Upgraded Product

If you're working on documentation for an upgrade to an existing product, some informal usability testing has already been done for you. You may have already modified the specifications to take into account any known problems. Gather any available information about problems in the existing documentation—for example, calls to a help desk (e.g., an 800 number customer complaint or help line) or questions sent to a users' group (e.g., email to customer service or other users). You could also test the existing documentation in the same way as covered under "Usability Testing during the Production Editing Stage" later in this chapter.

Documentation for a New Product

If you're working on documentation for a new product, the planning stage is a good time to ask some representative users about terminology and icons you plan to use. If your draft table of contents is sufficiently developed, test it too at this stage; otherwise, do that during prototyping.

You may have already completed much of this work when you were conducting your audience and task analysis, if you had the chance to observe or talk with real users. If, however, you had to rely on discussions with marketing people or software developers, it is a very good idea to test your decisions with real users. Although writers, editors, and subject matter experts (SMEs) should be involved at this stage too, you may be surprised at the results you get from users.

Usability testing at this stage is typically done as a pen-and-paper exercise. One method uses index cards. Print one word, phrase, or icon on each card. Show the cards one at a time to users and ask what they think the words or icons mean, or what they think they would find if they clicked on the word or icon or chose that item from an index or table of contents in a book. Or place a row of cards on a table and ask users which word or icon they would pick to do a particular task or find some specific information.

Use the results from this test to help you write the detailed specifications and the style guide for the documentation.

Usability Testing During the Prototyping Stage

During prototyping, you may have several opportunities to ask users for their input regarding navigation and design issues. If terminology and icons weren't tested at an earlier stage, include them now.

Paper Prototype

To usability test a paper prototype of an online system, print sample screen diagrams or mockups on landscape pages, so they look something like a computer screen. When a user chooses a link, show the printout of the topic that the link would lead to. If testing a book, use printouts of sample pages.

Electronic Prototype

To usability test an electronic prototype of an online system, use a mockup or working prototype on a computer. If the links work, let test subjects click the links after telling you what they think they'll find.

Ask users what they would click on to learn more about the topic at a conceptual level (to answer questions such as, "What the heck is XX anyway and why would I want to use it? What are the consequences of choosing X or Y?"), or at a more detailed level ("What format does this date need to be in? Is this field case-sensitive?").

Be careful not to ask leading questions—that is, questions that help test subjects figure out the answers. Ask questions like, "What are you thinking? What information do you think you would want next? What do you think you should do now? What do you like about this page? What do you dislike about this page?"

Record users' reactions. Did they find what they expected after clicking a link? Did they understand what was going on? Did the terms and icons make sense to them?

After analyzing their responses, incorporate changes into your specifications. Changes might include modifications to topic content, not just design and navigation.

If for any reason you need to make major changes to your original documentation design, conduct a second prototype test.

Usability Testing During the Early Draft Stage

Your specifications look good, but do they work with users? You want to find out at the early draft stage, when only a few representative topics have been written. If you need to make major changes, now is the time to decide on them—not when the entire book or help system is written. At this time, you are testing

- contents of topic types
- writing style
- use and presentation of lists, tables, and diagrams

Show some representative topics to the test subjects and ask them questions. You might say, "You want to do X using the software, but you need help. When you look in the user guide, here are the instructions you will see. Please tell me what you think of this topic. Does it tell you what you would want to know? Is it too simple or too complicated? Easy or difficult to follow? Why? Are you not sure what some of the words mean? Which ones?"

At this point, you're looking for users' reactions, not asking them to complete a task. Take notes on what they say, and also mark their reactions on a scale of 1 to 5, "Easy" to "Difficult," or something similar (see "What Are You Going to Measure?" later in this chapter). Incorporate changes into your specifications and the writing of new topics, as well as changing the draft topics.

Usability Testing during the Production Editing Stage

Here is where your test subjects get to work with the full user guide. Did you get it right?

Usability testing of the documentation at this stage is often done as part of usability testing of the software itself. If that's the way it's done in your organization, work with your QA group to ensure that the documentation is well tested. Perhaps you can contribute the test questions. As a last resort, do some additional testing yourself.

Testing at this stage is often more formal than some of the tests you conducted at earlier stages. Ask the test subjects to complete specific tasks. Some tasks might be entirely within the documentation itself—for example, "Find out what part of the software you would use to do XX." Other tasks would involve doing something with the software that requires looking up some information in the documentation.

CONDUCTING USABILITY TESTS

If possible, be part of the overall product usability testing team, because many problems with the documentation turn out to be software bugs or problems with the user interface.

Here are the steps to follow when conducting a usability test:

1. Define the testing objectives and methods.
2. Write the test materials (scenarios for testers to follow).
3. Recruit the participants.
4. Set up the test environment.
5. Conduct the test.
6. Analyze and report on the results.

You can do all of this in three days: one day for Steps 1–4, one and a half days to conduct the test (with three or four subjects), and half a day to analyze the results and prepare a report. You might need more test subjects and more time for a major product, but you are mainly looking for categories of problems, not every specific instance.

TIP FOR TECHNICAL EDITORS

Sample test plans, forms, and test materials are available on the "Usability Toolkit" page of the Society for Technical Communication's Usability and User Experience group's website (http://www.stcsig.org/usability/resources/toolkit/toolkit.html).

Defining Test Objectives and Methods

Defining your objectives and methods is an important step that often gets overlooked by inexperienced or rushed testers.

- What are your goals for the test?
- Who is the audience?
- What are you going to measure?
- How will you collect the data?

What Are Your Goals for the Test?

Goals might vary at different stages of usability testing. Some typical goals include answers to the following questions:

- What are users' first impressions of the documentation?
- What do users want to find in the documentation?
- What are the trouble spots in the documentation?
- Can people find what they want to know? How quickly?
- Is the level of detail correct—not too much or too little?
- Is any important information (from the users' point of view) missing?
- Is the information clear, unambiguous, easy to understand, and easy to put to use?

TIP FOR TECHNICAL EDITORS

Usability testing is not specifically looking for accuracy of information or language problems, although incorrect information (if discovered) must be noted. Technical accuracy should be covered by technical reviews and language problems by copyediting.

Who Is the Audience?

Refer to the audience analysis that was done during early planning. Specify characteristics for test subjects to use when recruiting people.

Although it's a good idea to have your test subjects match the audience you're designing for, don't get too worried about this aspect if you can't find suitable people. You can learn valuable information even from people who don't fit your audience analysis.

What Are You Going to Measure?

Measurements can be qualitative (subjective) or quantitative (can be counted). Choose the measurements that are appropriate for the stage of testing, the audience, and your test goals.

Qualitative measurements are often on a scale of 1 to 5, "Easy" to "Difficult," or "Completely satisfied" to "Totally unsatisfied," or they may be subjective comments recorded by the person conducting the test. If you use a scale, choose one that suits your needs, using terms that fit the question. In addition, make notes on test subjects' reactions and comments.

Examples of qualitative measurements and reactions include

- Did the user find the task easy or difficult?
- Did the user think selected topics (and the documentation as a whole) included too much, not enough, or just the right amount of information?
- What are the user's concerns or problems?
- What are the user's expectations and desires?

Examples of quantitative measurements include

- How long did it take the user to do the task?
- Did the user successfully complete the task on the first try (yes/no), or on another try (which one)?
- If the user was unable to complete a task—for example, locating information in the documentation—was it because the information was missing, incorrect, or could not be found?

How Will You Collect the Data?

Decide how you are going to collect the data. For example, the tester usually marks the responses rather than having the test subject fill out a questionnaire.

If you plan to use a video camera or audiotape to record the session, don't rely on the tape as your only record of the session; take notes as well. Equipment can fail, and you might not have time to go over hours of tape afterwards. Use recordings to refresh your memory of what happened, if it's not clear from your notes.

During the test, ask subjects to tell you what they are thinking while they are doing a task. Take notes on their responses. After the test, ask a few planned questions and then let them talk freely about their experience, what they found easy or difficult, and any suggestions they might have for improvements.

You might have two people conducting the test—one taking subjective notes (e.g., recording his or her impressions), and the other recording time taken to complete tasks, number of errors per task, and so on. One person can do both jobs, but it's often easier for two people to share the work.

Writing Test Materials (Scenarios for Testers to Follow)

Writing good test scenarios is an art as well as a science. If possible, get an experienced usability tester to write them for you, but don't let a lack of experience stop you from conducting the test. Almost any testing is better than none, unless you deliberately try to skew the results to get the answers you want instead of those you need.

It's important to have a script to follow when conducting tests, so that you ask each subject the same questions and you don't lead them to the answers you want.

When testing at early stages, your questions may not relate to specific tasks, either in the software or the documentation. You're looking for general problems with terminology, icons, navigation, and organization. At later stages, questions should relate more closely to user tasks.

Base your test materials on task lists. You may be able to reuse the user scenarios that were developed when planning the documentation, or you may need to develop new ones.

Choose a few tasks that require subjects to find obscure information or make use of advanced features, so the subjects need to consult the documentation to complete a task. This approach helps you find usability problems with the documentation, not the software itself. If you usability test the documentation at the same time as the software, you may want to sometimes suggest that subjects look in the documentation (but don't say where or how) and other times leave subjects to decide themselves whether to use the documentation when struggling with a difficult task.

You could vary your task questions with the knowledge and experience level of the test subject, but if you do that, be sure to test several subjects for each group of task questions.

Review and edit the test materials to make sure they focus on significant tasks and are of a suitable level of difficulty.

Examples of Test Tasks

Figure 23.2 shows some sample tasks for a usability test of a software product. Two of the three tasks explicitly involve using the documentation.

---

1. Using the Contacts database, look up the California phone number of a business called Cozbiz, with offices in California, New York, France, and Japan. Write the California number here. _____

2. Locate all the contacts in Canada, and print out a list showing their names, addresses, and phone numbers, sorted by postal code. You may want to look in the documentation about finding information, sorting, and printing.

3. Find out what the command Match (on the Utilities menu) can do for you, using the Glossary and Index. Write the answer here. _____

FIGURE 23.2
Sample Tasks for a Usability Test.

Recruiting Test Subjects

A good sample of typical users is best. Five people should be enough. A "typical user" depends on your product. What personas did you develop? (As we discuss in Chapter 17, a persona is a fictional person who has the characteristics of a typical user in a specific user category.) Find people like the personas, either among your existing customer base or, if appropriate, among students or members of the public who may be willing to act as test subjects.

If you have the budget, you could get a recruiter or temporary agency to find the test subjects for you. Be sure the agency understands your needs; some agencies specialize in providing test subjects, but others may have no idea what you're talking about. Your company's HR department may be able to help, or you may need to find the test subjects yourself.

If you can't get representative users, at least find someone in another part of your company who is not familiar with the product but has some of the characteristics of your audience. Be sure that staff members understand they are not being "graded" on their performance— that the idea is to find problems with the documentation, not with the people using it.

Outside test subjects should be paid for their time and out-of-pocket expenses. Staff members should at least get a mug of tea or coffee and a snack.

Set up a schedule with plenty of time between tests for you to write up your notes, get the room and the software ready for the next person, and discuss initial impressions with other members of the team.

SETTING UP THE TEST ENVIRONMENT

You don't need special equipment, just a room with a table or desk, two chairs, a computer (if this isn't a pen-and-paper test), and an Internet connection if necessary. If you can use a video camera to record the session, do so. If possible, set the camera up with a cable to a monitor in a nearby room, so other people can observe the test without disturbing the subject.

Although usability experts say to conduct tests in a quiet, uninterrupted environment, consider doing some testing in users' real environments using the same equipment they use. For example, are typical users working in a noisy "cube farm" or factory floor, in bad lighting conditions, or being constantly interrupted? Do they have small or badly lit monitors such as those on some laptops or handheld machines? Do they have a well-lit place in which to open and read printed documents?

Conducting the Test

If at all possible, choose a calm, patient person to conduct the tests. This person's main job is to observe, take notes, put the test subjects at ease, and encourage them to talk, without leading them to the "right" answers.

The facilitator does not need to be part of your team. Perhaps someone from your company's training or HR departments could fill this role. Don't let lack of a suitable tester stop you from conducting the test.

More about the role of the tester (also known as a *facilitator*) is given in the books on usability testing. Steve Krug has a particularly good summary.

Analyzing and Reporting the Results

Immediately after conducting each test, write down any observations you didn't have time to record during the test (or expand on cryptic notes so you'll be able to understand what they mean later). When all of the test subjects are done, summarize your observations in a list or table.

Go over the results and write down your conclusions, which may be in general terms like "xxx needs more work"; then decide what needs to be done about each identified problem and assign a priority to each change. You don't need to write an elaborate document complete with statistical analysis—a simple table can be quite sufficient.

If you're working with a team of writers, organize a meeting to discuss your findings and decide what to do.

Prioritizing Problems and Making Changes

Classify the usability problems by severity level (see Figure 23.3).

Severity level	Description
1. Life-threatening	Problem may cause someone to be injured or killed.
2. Major	Problem may cause a loss of data or a mistake that results in lost work. User's workflow is severely disrupted. Legal requirements are not met.
3. Minor	Problem may cause a minor interruption of the user's workflow.
4. Annoyance	Problem may cause customer to perceive a lack of quality standards in your organization but causes no harm, lasting or temporary.

FIGURE 23.3
Severity Levels of Documentation Problems.

As with editorial changes, if you have limited time available for fixing problems found during usability testing, divide the problems into three categories and fix them in order of priority. Don't simply fix the easy ones first—they are usually the most trivial.

1. Changes that must be made because content is factually incorrect, unclear, or ambiguous and may lead to serious misunderstanding (severity levels 1 and 2).
2. Changes that improve the writing or presentation but are not essential to understanding (severity levels 3 and 4).
3. Changes that the majority of the audience won't care about or even notice, especially if test subjects' reactions or opinions are contradictory (severity level 4).

EXERCISE 23.1

1. Using the students in your class, run a sample usability test of your university's homepage.
2. Conduct a sample usability test of the web page for your major. For example, you could have potential students (high schoolers or transfer students) try to find information about department scholarships, requirements for the major, information about classes, etc.
3. Choose any government website, such as the Library of Congress (http://www.loc.gov/index.html) and give the page a "stress test" by asking the questions in the chart at http://instone.org/navstress. Remember that Krug noted that you need only four or five people to conduct a small usability test to find most usability problems. Use yourself and three or four classmates.

SUMMARY

Any usability testing is better than none. Do it early and often. A small number of test subjects will generally find most of the problems. If resources are limited, fix the problems in order of their severity, not in order of the ease of fixing them.

CHAPTER 24

Technical Editing Careers

At the conclusion of this chapter, you will be able to

- determine the compatibility of your personality traits to characteristics common to technical editing;
- understand the similarities and differences between editing and technical editing;
- recognize the meaning behind emotional intelligence;
- search for careers in technical editing;
- classify technical editors' earnings and editing speeds.

EDITING: IS IT RIGHT FOR YOU?

Look at the job descriptions in Figure 24.1; you will discover that these descriptions are fairly typical. Notice that many technical editors are required to write technical documents as well. More importantly, notice that beyond editing expertise, editors must be able to interact closely with other professionals and members of the production team.

Throughout this book, we have discussed the wide range of skills necessary to become a technical editor. In addition to these skills, good editors possess certain personality traits as well. What type of personality is best suited for the job? Do you believe your own personality is suited to begin (or continue) such a career? If you are interested in understanding more about your suitability for a technical editing career, we recommend beginning with http://www.online.onetcenter.org/ (O*Net OnLine, formerly the Occupational Network), and clicking on "Find Occupations." Type in the keywords "technical editor" or just "editor" and read the results under the various related fields. Editors, according to O*Net OnLine, should possess the personality traits of being attentive to details, dependable, stress-tolerant, ethical, responsible, self-starting, adaptable, persistent, independent, and in control of themselves. Does this sound like you?

Of all the desirable personality traits for a career in technical writing, be sure you can accept change and acclimate to different demands specific to your place of employment. As you gain editorial experience, you will discover that each organization adheres to its own culture. That is, the procedures and interpersonal dynamics in Organization X are likely to be different from those in Organization Y, especially as more organizations transition toward the production team paradigm. In addition, the role of editor may be remarkably different from one place to another.

427

Technical Editor
Job Responsibilities
Perform editorial reviews of:
1. Guidance to be added to the company's accounting manuals.
2. One or more technical publications relating to accounting developments. Publications are aimed at professionals, clients and other parties with interest in financial reporting matters. Responsible for content accuracy, tone and voice and for consistency of the guidance and publications. Coordinate with technical professionals as well as production personnel on publishing of guidance and publications.

Experience/Qualifications
Education—Associate's or Bachelor's Degree
2–3 years technical editing experience, preferably with accounting/business topics
Knowledge of Microsoft Word
Ability to work on multiple projects at a time
Ability to work beyond normal working hours when needed

Technical Writer/Editor
Looking for a contractor to write and edit promotional materials and technical documentation for external and internal audiences, employing a variety of communication styles and techniques. Researches, writes, formats and edits materials for proposals, sales and markup promotions, speeches, news articles, film and video productions, operation and administrative procedures, process documents, software documentation and reports. Confer with customers and technical experts to ensure technical accuracy and completeness of the content. Apply knowledge of grammar, language, and research processes in order to research, format, write, and produce materials. Perform business strategy, communication and markup documents, end-user training and information materials. Skills required: MS Office, Visio, Dreamweaver, Visual Basic, and MS Project.

FIGURE 24.1
Employment Ads for Technical Editors.

Whereas one organization may poorly define the responsibilities of their editors—who they work with, who they report to—and offer little managerial support, other organizations are quite specific about the levels, responsibilities, authority, and performance expectations of their editors.

Clients, managers, and coworkers often don't understand what technical editors can contribute to a project or how valuable that contribution can be. Even worse, they may have a strong idea of what editors do, but that idea is limited and out of date.

Personality Tests

No test can provide a definitive assessment of an individual's intelligence, personality, or ability. However, personality tests can inform our reasoned judgments and provide some insights into our character strengths and limitations.

Although we do not endorse any specific test, you might consider taking one to identify occupations that might make a good fit with your personality type. For example, the *Myers-Briggs Type Indicator* (MBTI) is a widely employed personality test based on eight psychological personality types developed by Carl Jung. Depending on responses to the multiple-choice questions, the MBTI assesses an individual's preference toward a particular personality type. Figure 24.2 shows the MBTI's categories.

Personality Type	Characteristics (People who . . .)
Extraversion	tend to focus on the outer world of people and things
Sensing	take in information primarily through their five senses
Thinking	base decisions on logical analysis
Judging	make planned and organized approaches to situations
Introversion	focus on the inner world of ideas
Intuition	see the "big picture" and future possibilities
Feeling	make decisions based on values and subjectivity
Perceiving	take a flexible and spontaneous approach to life

FIGURE 24.2
Myers-Briggs Type Indicator Categories.

After developing a personality profile, the MBTI lists occupations that are most compatible with a given profile, challenges that someone with this personality profile will encounter, and strategies on how to deal with those challenges.

Whether or not you choose to take the MBTI or another personality assessment, it's important for you to have some insights into your own personality strengths and limitations. Many schools, universities, and corporations offer personality test and career interest tests; some are free while others cost a considerable amount of money. If you are currently in school, consult with your career counselors about taking these tests for free or for little cost. If you are employed, your human resource office might administer these tests. One of the most common tests given is the Strong Interest Inventory test that you can take online. Other websites that you might find helpful include http://www.funeducation.com/Tests/CareerTest/TakeTest.aspx.

WHAT IS YOUR EMOTIONAL INTELLIGENCE?

After you have assessed your personality strengths and weaknesses, you will need to examine how these traits affect how well you work with others including writers, managers, and engineers. Before we explore the specific dynamics of the writer-editor relationship further, we would like to make some general observations regarding working with people in all areas. Regardless of your choice of professions, you will probably work with people who may be friendly, timid, helpful, aggressive, generous, obnoxious, empathetic, arrogant, willful, emotional, heartless, and any number of other personality combinations. Some of these individuals may be high-IQ types; others may be average in intellectual ability. How does an individual prepare to engage such a wide spectrum of people?

In his bestseller *Emotional Intelligence,* Daniel Goleman maintains that high achievers on IQ tests, SAT and ACT tests, and the myriad battery of other intelligence tests are no more successful than others in their chosen profession. He cites a 1995 study conducted by Karen Arnold of Boston University, who tracked eighty-one valedictorians and salutatorians from high school. These students generally performed well in college; however, her research indicates that ten years after high school graduation, only one in four was as successful in his or her profession as others of comparable age. Goleman concludes from this and other studies that "academic intelligence offers virtually no preparation for the turmoil—or opportunity—life's vicissitudes bring." Instead, one of the best predictors of success is a person's interpersonal skills, or as psychologist Howard Gardner (as quoted in *Emotional Intelligence*) puts it, "the capacities to discern and respond appropriately to the moods, temperaments, motivations, and desires of other people." That is, what makes people successful is their *EI*—their *emotional intelligence,* or simply their ability to interact with people.

Obviously, academic ability is still important. But, academic ability is only one small part of the equation. Possessing the emotional skills to understand others, their needs and motivations, and to act wisely in response to others is tantamount to success or failure. Psychologists Mayer and Salovey categorize these skills into five domains of emotional intelligence:

1. *Self-awareness:* knowing one's own emotions and being able to monitor them as they occur.
2. *Managing emotions:* being able to control and modify one's emotions through self-awareness.
3. *Motivating oneself:* redirecting and prompting initial, destructive emotions toward emotions that are beneficial toward completing a goal.
4. *Others' emotions:* recognizing how others feel and showing empathy.
5. *Relationships:* managing the emotions in others to create interpersonal effectiveness.

TIP FOR TECHNICAL EDITORS

Communicate with the author from the very beginning of the project and make certain you understand the author's expectations about the level of editing: basic copyediting or comprehensive editing?

EDITORS AND TECHNICAL EDITORS

The concept of an editor as someone who publishes a book or "prepares the literary work of another person, or number of persons for publication, by selecting, revising, and arranging the material" has been around for three centuries. Although technical editors are not usually editing literary works, they are, in essence, still editing texts—technical documentation, in this case—for publication. The list of duties at the beginning of Chapter 1 and the job descriptions given in this chapter show that technical editing requires all the language expertise of the typical editor—grammar, style, mechanics, and cohesion—as well as excellent people skills.

Furthermore, technical editors today must be familiar with technology. According to the Professional Communication Society (PCS)—a group of technical writers, editors, and engineers within the Institute of Electrical and Electronics Engineers (IEEE)—many technical editors "first became engineers, software developers, and scientists before getting interested in technical communication. After achieving preeminence in technical fields, they saw the need to develop their communication skills to share information with their peers, clients, or students." Many technical editors have humanities backgrounds, but they often actively take classes, read, volunteer, or work in technical fields. Many job advertisements for technical editors still call for the job candidate to have at least a bachelor's degree in English, journalism, or communications; other job ads prefer the applicants to have strong language skills but an educational background in the sciences or in engineering.

Although subject matter experts (SMEs) are expected to have the most technical expertise, it is crucial for technical editors to at least understand the technology so that they can be

Editing Ernest Hemingway was like wrestling with a god.
—*Tom Jenks*

sure the documentation accurately explains how the technology functions. Some believe that technical editors should not be responsible for technical accuracy of the product being documented; however, we have found that having technical skills and knowledge definitely helps. Editors should, of course, question any item that they suspect may be incorrect and ensure that the responsible person verifies its accuracy.

Technical editors who have language skills *and* know technology get better jobs and save companies millions of dollars each year. Think about it: if a technical editor changes aspects of a technical document to help the customer use the technology, the company does not have to spend as much money on technical support (800 numbers or website staffing and maintenance), on revising erroneous or unclear documentation, or on customer service (for customer complaints). If the technical documentation is clear, concise, and accurate, the product or procedure will work as planned, and the company will benefit from happy customers.

The Changing Editorial Role

A college education is almost essential for a technical editing job today. Recent surveys of organizations, businesses, and agencies that hire editors indicate that without any previous experience, an individual must have at least a certificate or an associate's degree, but a bachelor's degree is preferred. That degree, however, may vary widely, from a B.A. degree in liberal arts (such as English) to a B.S. degree in engineering or science.

If you are thinking about a career in technical editing, here are some aspects you will want to consider. Most technical writers are also SMEs and therefore have expertise in their particular area. Civil engineers, for example, more than likely are directly responsible for writing the technical documents related to their specific projects. After the engineer completes the document, a copy editor, who may or may not have the technical expertise, may be asked to review it with light editing.

Often, technical documents are generated by a production team. For example, several engineers, working with technical editors and graphic artists, collaborate on producing the desired documents. A senior editor, who probably has been a technical writer, may become responsible for directing production. Depending on the organization and its specific needs, a technical editor's role may range anywhere from copy editor to manager of editorial operations. To flourish professionally after acquiring firsthand experience, you should strive to be knowledgeable about different media, computers, applicable software programs, procedural testing, and so on, and of course possess excellent communication skills.

WHAT ELSE IS IMPORTANT?

It should be clear by now that an editor must possess a wide range of attributes: technical abilities, language skills, and people skills. We also believe that being able to see the "big picture" is as important as being able to analyze a document for grammatical errors. Editors need to have a good understanding of document design, not only for printed documents but for the computer screen as well. Document design calls for seeing the whole and how the various parts interact together to form that whole. No doubt you have visited many web pages that were confusing, unpleasing to the eye, and difficult to navigate. Perhaps you have read technical manuals that seemed to be nothing more than a series of unrelated sentences or topics.

College Reputation

How important is the reputation of the college you attend toward securing a technical editing job? Not as much as you might think, unless your school has an internship or work-study program that helps you network with local firms that hire editors. It may surprise you, but recent research indicates that a college's reputation has little influence on long-range job performance. Initially, employers may tip their hats toward a graduate from an Ivy League or prestigious school, but what employers want most are individuals with academic track records that reflect good communication skills as well as coursework that suitably fits the organization's mission. Organizations desire individuals who show initiative, can solve problems, learn new skills and tools quickly, are thorough and responsible, and will be able to work on a team. A good sense of humor and an ability to adjust to events as they happen are also definite assets.

Work Experience

Increasingly, experience—both volunteer and paid work in your field or related field—outweighs school performance in terms of value to employers. But how do you get experience without first having a job? (It's the proverbial Catch-22.) One solution for students seeking degrees is to take advantage of their school's *cooperative programs* (co-ops). Students who enroll in co-op programs may intern in their chosen field while attending school. The advantages of co-ops are many. Students either participate in a *parallel program* by attending school and working simultaneously for an organization, usually on a part-time basis, or they participate in an *alternating program* by going to school full-time one semester and then working full-time the next. Not only do co-ops provide important work experience, but they often provide additional income. Frequently, co-op students receive job offers from the organizations where they interned.

Students interested in computer documentation can also gain experience (and a portfolio of completed projects) through less formal programs such as volunteering to write or edit documentation for open-source software projects such as OpenOffice.org.

Co-ops

Participating in cooperative programs while in college allows students to get practical experience to determine if they are really suited for a particular field. More than likely, your local community has many co-op opportunities for editors and technical editors. Most schools have a career services center that sponsors these programs, helps students write their résumés, and places students in jobs after they graduate. Government agencies such as the Army Corps of Engineers and the military often participate in co-op programs. Businesses, publishers, newspapers, and manufacturers also hire co-op students. Even if you are not presently in school, many universities provide information on co-ops. Schools may refer to these programs as *professional internship* or, in some cases, *service-learning*. Regardless of the name, you will want to get some experience, whether paid or unpaid, before you graduate. The sooner you start a cooperative program in your career, the better. A volunteer opportunity that is five hours a week could quickly develop into a paid internship, which may eventually lead to a full-time job. Co-ops also teach you to *network,* which simply means working to build a network of individuals who know you well enough to recommend you should a job become available. This networking can occur in a more informal fashion, such as through joining and participating in organizations to which other technical editors belong.

TECHNICAL EDITING CAREERS

The skills that you learn in this book will help prepare you for a career in technical editing. Many of these skills overlap with those needed for more traditional editing jobs, such as working for a publishing house or for a newspaper. The Society for Technical Communication (STC) is an excellent resource for career information about the field of technical communication.

Most companies prefer to hire English majors for entry-level technical editing jobs, probably because such jobs tend to focus more on copyediting and text-based skills. Upper-level jobs, however, while requiring experience with technical editing, often also require a strong science or engineering background. This may be because upper-level jobs such as those found in Figures 24.3 and 24.4 entail management, supervisory, and training skills that require technical expertise and a knowledge of the technology with which all members of the company are working.

Some jobs in technical editing, whether entry-level or more advanced, are part-time. Although some technical editors work full-time for companies, many organizations still hire predominantly part-time or freelance technical editors to work LTE (limited-term

---

Supervisor, Technical Editing

Description: Supervisor, Technical Editing
Job Location: Collegeville, PA
Job Description: As a Supervisor, Technical Editing, you will follow the Technical Editor/Proofreading Checklist to create annotated, marked, and clean Word documents. You will perform electronic text compares of annotated, marked, and clean Word documents according to the checklist. These three final versions must be delivered in a timely fashion, be technically correct, and reflect the marked changes in the draft. You will work effectively with the Associate Director-Document Management to evaluate production workload, discuss timelines, and plan strategies for completing work on time. You will perform quality reviews of labeling documents created by editing team. You will also conduct the daily production meeting to review production issues with team members, identify issues, and seek resolutions and follow-up on assigned projects to assure a smooth process flow. You will gather metrics and prepare reports. You will work closely with the Technical Editors (TE) and Proofreaders (PR) so that instructions for Technical Editing are clear and accurate. You will perform quality review of all projects to ensure all steps have been completed, and that the final products meet required standards. You will work closely with the Associate Director-Document Management, the LM, the Global Labeling Operations, and the VP GLD to prioritize delivery of technically correct U.S. documents to the LM, WWRA, and/or the Global Labeling Operations team. You will also perform other duties assigned by your Supervisor. You will participate as a GLD representative on various task forces and working groups related to Regulatory or technology driven document initiatives. You will communicate updates/issues to your supervisor and seek guidance as appropriate for those initiatives.
This position requires a BA/BS degree with a minimum of 5 years of relevant experience, or an MS/MA with a minimum of 3 years relevant experience. Microsoft work certification is desired—Adobe Acrobat is considered a plus. Advanced word-processing i.e. Microsoft Word and Adobe Acrobat required. Must possess: excellent organizational skills; the ability to prioritize tasks; proofreading skills; active listening skills; and excellent grammar, verbal and written, as well as good spelling skills. Familiarity with the following is desirable, but not required: EDMS, Microsoft PowerPoint, and electronic regulatory submissions.
Compensation/Benefits: Competitive compensation and benefits programs including stock options, child-care subsidies, flex-time, business casual attire, educational assistance and professional development programs.

FIGURE 24.3
Advertisement for an Upper-Level Technical Editing Job.

Technical Editor
Photogrammetric Engineering and Remote Sensing

Duties and responsibilities: The **Technical Editor** serves as a member of the *PE&RS* journal staff, working closely with the Editor-in-Chief, Executive Editor, Manuscript Coordinator and printing company project manager. The Technical Editor is responsible for (1) conducting final edits on papers accepted for publication in *PE&RS,* (2) overseeing distribution of page proofs to authors, (3) checking corrected and final proofs prior to final production, (4) coordination and delivery of corrected proofs to the printer, (5) final decisions on quality and character of illustrations, and (6) scheduling papers for publication in monthly issues of *PE&RS,* (7) Delivery of final proofs to ASPRS for final production. The Technical Editor serves as a member of the ASPRS Journal Policy Committee.

Qualifications: Candidates should demonstrate broad familiarity with the spectrum of geospatial information technologies, and should have specialized academic training and experience in one or more of the following areas: photogrammetry, remote sensing, and/or geographic information systems (GIS). A background in mathematics sufficient to distinguish between variables, constants, and operators in equations, to identify vectors and matrices in equations and to recognize the validity of equations is critical. Candidates should be well-acquainted with word processing, including the ability to generate/edit equations using the Insert/Symbol capability of Microsoft Word and more complicated equations using Microsoft Equation 3 or Equation Editor. Preference will be given to candidates having previous experience in editing and publishing. Candidates who are demonstrably well-acquainted with *PE&RS* and who are ASPRS members are especially encouraged to apply.

Compensation and benefits: ASPRS provides limited compensation at the part-time level to the Technical Editor who is an independent contractor. ASPRS may also provide software upgrades, as needed and appropriate. The Technical Editor is expected to have and maintain his/her own computer system and email. If necessary, ASPRS may provide a travel allowance to cover attendance at ASPRS Journal Policy Committee meetings.

To Apply: Candidates for Technical Editor should provide a cover letter outlining the candidate's interest in the position and a resume including names and contact information of at least three professional references. All materials may be submitted via e-mail as attachments (applications received will be acknowledged) or by conventional surface mail.

FIGURE 24.4
Technical Editor Job Ad.

employment) on a project-by-project basis. Figures 24.5 and 24.6 are examples. Some technical editors like the freedom of freelance work, whereas others prefer the security of a full-time position with benefits. No matter what type of job they have, good technical editors need to know the importance of both language and technical expertise. For example, Nicole secured one job interview by submitting an edited version of the job ad itself (printed with errors in the local newspaper) with her job application letter and résumé. (Note that the job ads given in this chapter are real and may contain errors in the original.)

Notice how none of the example advertisements mentions pay. Most companies hire on a sliding scale based on expertise. Some organizations have a strict budget but many still allow some negotiations in terms of salary, and this is true for freelance or full-time work. It is important for you to know what you and your skills are worth in the field of technical editing, and salary is a frequent topic on editing listservs and websites. The mean salary for technical writers and editors is higher than for most jobs with only a bachelor's degree, and that includes freelance and part-time workers. As the need for superior technical documentation grows, particularly in the field of computers, technical editors and writers can expect to see that mean salary rise. For more information about other careers in editing and writing,

---

Contract Technical Proofreader/Editor

Description: **Technical Proofreader/Editor** for a five(5) to six(6) month contract assignment. This individual will proofread and edit new and updated user-centered documentation, including installation and homeowners guide manuals as well as roughing-in and specification sheets. For all technical literature, this person will ensure the technical accuracy, clarity, and conformance to acceptable style and usage guidelines. This individual will also visualize product installation and use considerations, and offer input to improve the effectiveness of all literature types. Superior language, editing, and proofreading skills are critical. Mechanical aptitude and diagnostic skills are important. Work experience in an engineering environment is helpful. Computer aptitude and direct experience with Arbortext, Interleaf, or Framemaker technical publishing software is desired. This person must at least be able to prove advanced ability with Pagemaker, Quark Express, Microsoft Word, or WordPerfect. Proficiency with at least one operating system, database, and spreadsheet is required. Competent keystroke skills are essential for productivity. A Bachelor's degree in English or a Technical Communications curriculum is required.

Job Title: Contract Technical Proofreader/Editor
Primary Skills: Edit/Proofing; Pagemaker; Quark Express; FrameMaker
Job Duration: 3–6 months
Degree Type: BA Degree
Area: English or Technical Communications
Experience Minimum: 1 Years

FIGURE 24.5
A Contract Job as a Technical Editor.

Technical Editor Wanted

Firm X is seeking candidates for a half-time paid position as Technical Editor. The responsibilities include:

- Technical editing of the Firm X Digest of Technical Papers
- Technical editing of the Firm X Advance Program
- Attendance at committee meetings in April, August, October, and February (travel and expenses paid)

This position requires nearly full-time effort between September and January. The ideal candidate would possess technical understanding of solid-state circuits, good organizational skills, good verbal and written language skills, and be a team player.

FIGURE 24.6
A Half-Time Paid Position as a Technical Editor.

visit the U.S. Department of Labor's Bureau of Labor Statistics web page on "Writers and Editors," http://www.bls.gov/oco/ocos089.htm.

Technical Editors' Earnings and Editing Speeds

For those of you who may choose freelance work, a per-page or hourly fee may be more relevant. In order to charge your clients by the hour or by how many pages you edit, though, you first need to establish how fast you edit. If you are a first-time technical editor, you may need some benchmarks to estimate how quickly you should be editing a document. However, please remember *that it is never the quantity but the quality of the editing that is important.* Too often, novice freelance technical editors underbid the project in an attempt to secure employment. Experienced technical editors know that a project always takes longer, usually

Estimator	Heavy (WPH)	Medium (WPH)	Light (WPH)
Jean Hollis Weber[a]	500–1,000	2,000–3,000	4,000–5,000
Amy Einsohn[b]	250–500 1,000–1,750	500–750	500–1,000
Jody Roes[c]	1,000	1,500	3,000
Gary Conroy[c]	500–1,000	2,000	4,000
W. Thomas Wolfe[d]	416	1,250	
Mary Jo David[c]	750–1,250		
Richard Ketron[e]	500–1,000		
Joanna Williams[c]			5,000

Notes:

[a] Original estimates given in terms of 500 words per page.

[b] Original estimates found in *The Copyeditor's Handbook*. Top row assumes difficult text; second row assumes standard text.

[c] Original estimates cited in pages per hour (250 words per page assumed).

[d] Original estimates given in pages per day (six hours per day assumed).

[e] Original estimate cited 15 to 30 minutes per page (250 words per page assumed).

FIGURE 24.7
David W. McClintock's (2002) Benchmark Editing Speeds, at Three Levels of Detail.

twice as long as anticipated, which is why they budget for more time. Therefore, until you become a more experienced technical editor, you may want to stick to a by-the-page or by-the-hour pay scale with your clients. Figure 24.7 shows a list of speeds for technical editors with years of experience. Nicole does only freelance editing and charges per typed, double-spaced page. Her fee is higher for substantive editing than for proofreading only. Substantive editing requires more thought, time, and revision cycles with the author.

SUMMARY

This chapter discussed factors to contemplate when considering a career in technical editing. These factors include skill level, pay scale, personality types, and emotional intelligence. It is important to understand the similarities and differences between editing and technical editing. If you do decide to pursue a technical editing career, it will be important to assess your personality strengths and weaknesses and how those traits will affect your performance as a technical editor.

Bibliography

Albers, Michael J. "The Technical Editor and Document Databases: What the Future May Hold." *Technical Communication Quarterly* 9, no. 2 (Spring 2000): 191–208.

Annett, Clarence H. "Improving Communication: Eleven Guidelines for the New Technical Editor." *Technical Writing and Communication* 15, no. 2 (1985): 175–179.

Bogdanovic, Gordana. "Publication Ethics: The Editor-Author Relationship." *Archive of Oncology* 11, no. 3 (2003): 213–215.

Boorstin, Daniel J. *The Discoverers: A History of Man's Search to Know His World and Himself.* New York: Vintage, 1983.

Browne, Glenda, and Jonathan Jermey. *Website Indexing: Enhancing Access to Information within Websites.* 2nd ed. Adelaide, South Australia: Auslib Press, 2004.

Buehler, Mary Fran. "Situational Editing: A Rhetorical Approach for the Technical Editor." *Technical Communication* 50, no. 4 (November 2003): 458–464.

Chapman, Tamara. "Survival Skills for the Technical Editor." *Corrigo: Newsletter of the STC's Technical Editing SIG* 5, no. 1 (March 2004): 1–8.

Clark, C. Scott. "Editing Statistical and Illustrative Graphics" (April 2005), http://members. peak.org/~cscottc/pdfs/graphics-primer.pdf.

Connatser, Bradford. "Coping with Wordslaughter and the 'Good Enough' Syndrome." *Intercom* (January 2004): 19–21.

Corbin, Michelle, Pat Moell, and Mike Boyd. "Technical Editing as Quality Assurance: Adding Value to Content." *Technical Communication* 49, no. 3 (August 2002): 286–300.

Dayton, David. "Technical Editing Online: The Quest for Transparent Technology." *Journal of Technical Writing and Communication* 28 (1998): 3–38.

Dayton, David. "Electronic Editing in Technical Communication: The Compelling Logics of Local Contexts." *Technical Communication* 51, no. 1 (2004): 86–101.

Dayton, David. "Electronic Editing in Technical Communication: A Model of User-Centered Technology Adoption." *Technical Communication* 51, no. 2 (2004): 207–223.

Doumont, Jean-Luc. "Gentle Feedback That Encourages Learning." *Intercom* (February 2002): 39–40.

Durham, Marsha, and J. H. Weber. "Beyond Copy-editing: The Editor-Writer Relationship." Seminar 91: Working Smarter Not Harder. Published in *Proceedings of the Technical Communication Seminar,* New South Wales Society for Technical Communication (October 1991): 47–50. Article available online at http://www.jeanweber.com/newsite/?page_id=26.

Eagleson, Robert. *Writing in Plain English.* Canberra: Australian Government Publishing Service, 1990.

Enquist, Anne. "Substantive Editing versus Technical Editing: How Law Review Editors Do Their Job." *Stetson Law Review* 30 (2000): 451–474.

Fishman, Stephen. *The Copyright Handbook: What Every Writer Needs to Know*. 10th ed. Berkeley, CA: Nolo, 2008.

Gatlin, Patricia L. "Visuals and Prose in Manuals: The Effective Combination." In *Proceedings of the 35th International Technical Communication Conference*. Arlington, VA: Society for Technical Communication, 1988.

Gerich, Carol. "How Technical Editors Enrich the Revision Process." *Technical Communication* (First Quarter 1994): 59–70.

Gerson, Sharon, and Steven M. Gerson. *Technical Writing: Process and Product*. 4th ed. Upper Saddle River, NJ: Prentice Hall, 2003.

Goleman, Daniel. *Emotional Intelligence: Why It Can Matter More Than IQ*. New York: Bantam, 2006.

Grady, Helen, et al. "How Editors Can Create Good Working Relationships with Authors." *Currents* (2004), http://www.stcatlanta.org/currents04/proceedings/grady.pdf.

Gregory, Judy. "Writing for the Web versus Writing for Print: Are They Really So Different?" *Technical Communication* 51, no. 2 (2004): 276–285.

Hackos, JoAnn. *Managing Your Documentation Projects*. New York: John Wiley & Sons, 1994.

Hackos, JoAnn, and Dawn M. Stevens. *Standards for Online Communication*. Hoboken, NJ: Wiley Computer Publishing, 1997.

Hart, Geoffrey J. S. "The Style Guide is Dead. Long Live the Dynamic Style Guide." *Intercom* (March 2000): 12–17.

Hart, Geoffrey J. S. "Editing Tests for Writers." *Intercom* (April 2003): 12–15.

Horton, William. "The Almost Universal Language: Graphics for International Documents." *Technical Communication* 40, no. 4 (1993): 682–693.

Hyland, Fiona, and Ken Hyland. "Sugaring the Pill: Praise and Criticism in Written Feedback." *Journal of Second Language Writing* 10 (2001): 185–212.

ITEDO Software. "Inside Technical Illustration," http://www.itedo.com/.

Jensen, Don, compiler, and Richard A. D'Angelo, ed. "An Early History of NASWA." *NASWA Journal* (July 1999), http://www.naswa.net/journal/1999/07/swc199907.

Kilian, Crawford. *Writing for the Web*. 4th ed. North Vancouver, BC: Self-Counsel Press, 2009.

Kohl, John. *The Global English Style Guide: Writing Clear, Translatable Documentation for a Global Market*. Cory, NC: SAS Institute Inc., 2008.

Krug, Steve. *Don't Make Me Think*. 2nd ed. Berkeley, CA: New Riders, 2006.

LabWrite Resources, North Carolina State University. "Designing Tables," http://www.ncsu.edu/labwrite/res/gh/gh-tables.html.

Lagon, Olin. "Culturally Correct Site Design." *Web Techniques* (September 2000): 49–51.

Lanier, Clinton R. "Electronic Editing and the Author." *Technical Communication* 51, no. 4 (2004): 526–536.

Lester, Paul. *Visual Communication: Images with Messages*. 3rd ed. Belmont, CA: Thomson Wadsworth, 2003.

Levie, W. H., and R. Lentz. "Effects of Text Illustrations: A Review of Research." *Educational Communication and Technology Journal* 30 (1982): 195–232.

Lunsford, Andrea. The New St. Martin's Handbook. With 2001 Apa Update. New York: St. Martin's Press, 2001.

Lynch, Patrick J., and Sarah Horton. Web Style Guide: Basic Design Principles for Creating Web Sites. 3rd ed. New Haven, CT: Yale University Press, 2009.

McClintock, David. "Benchmarks for Estimating Editing Speed." *Corrigo: Newsletter of the STC's Technical Editing SIG* (June 2002), http://www.wordsupply.com/portfolio/0206-estimate-edit-speed.html.

Mackiewicz, Jo, and Kathryn Riley. "The Technical Editor as Diplomat: Linguistic Strategies for Balancing Clarity and Politeness." *Technical Communication* 50, no. 1 (February 2003): 83–94.

Mackiewicz, Jo, and Rachel Moeller. "Why People Perceive Typefaces to Have Different Personalities." Professional Communication Conference, 2004. IPCC 2004. International. (29 September–1 October 2004): 304–13. *IEEE Transactions on Professional Communication.*

McMurrey, David. *Online Technical Writing,* http://www.io.com/~hcexres/textbook/.

Mancuso, Joseph C. *Technical Editing.* Upper Saddle River, NJ: Prentice Hall, 1992.

Mann, Gerald. "Planning and Editing Tables and Charts." *Writing and Editing.* STC Proceedings (1996), http://www.stc.org/ConfProceed/1996/PDFs/Pg554557.pdf.

Markel, Mike. *Ethics in Technical Communication: A Critique and Synthesis.* Vol. 14 of *Contemporary Studies in Technical Communication,* ed. M. Jimmie Killingsworth. Westport, CT: Ablex, 2001.

Meyerding, Henry W. "Facing Ethical Issues." Chapter 22 of *Getting Started in Consulting and Independent Contracting,* ed. T. Barker and K. Steele. Society for Technical Communication, August 1998.

Morrison, C., and W. Jimmerson. "Business Presentations for the 1990s." *Video Manager,* July 4, 1989.

Myers, Peter B., and Katharine D. Myers. "Myers-Briggs Type Indicator." *Myers-Briggs Type Indicator Career Report* (2004).

Nadziejka, David E. "Levels of Technical Editing." Council of Science Editors GuideLines No. 4. Reston, VA: Council of Science Editors, 1999.

Nice, Karim, and Julia Layton. "How Hybrid Cars Work," http://computer.howstuffworks.com/ hybrid-car.htm.

Nielsen, Jakob. *Designing Web Usability: The Practice of Simplicity.* Indianapolis: New Riders, 2000.

Nielsen, Jakob. *Eyetracking Research.* http://www.useit.com/eyetracking/.

Nielsen, Jakob, and Marie Tahir. *Homepage Usability: 50 Websites Deconstructed.* New Riders, 2002.

Nielsen, Jakob, and Hoa Loranger. *Prioritizing Web Usability.* Berkeley, CA: New Riders, 2006.

Nielsen, Jakob, and Kara Pernice. *Eyetracking Web Usability.* Berkeley, CA: New Riders, 2009.

O'Hara, Frederick M. Jr. "A Brief History of Technical Communication." *STC Proceedings* (2001): 48–52.

Olson, Corinne N. "Getting Edits Back on Time." *Intercom* (December 1999): 21–23.

Pfeiffer, William Sanborn. *Technical Writing: A Practical Approach.* 5th ed. Upper Saddle River, NJ: Prentice Hall, 2003.

Plotnik, Arthur. "Dealing with Attackese: How Do We Approach Tragedy-Related Language?" *Editorial Eye* 24, no. 11 (November 2001): 1–11.

Price, Jonathan, and Lisa Price. *Hot Text: Web Writing That Works.* Indianapolis: New Riders, 2002.

Redish, Janice C. *Letting Go of the Words: Writing Web Content That Works.* San Francisco: Elsevier, 2007.

Rew, Lois Johnson. *Editing for Writers.* Upper Saddle River, NJ: Prentice Hall, 1999.

Ribert, Roger M. "Proofing a Translated Document." *Intercom* (May 2005), http://www.stcsig.org/itc/articles/05-proof_transl.htm.

Robinson, Elizabeth, and Nancy Small. "Guide: Designing Graphics." *TAMU Tech Writing* (April 2002), http://www-english.tamu.edu/pubs/tamu_tech_writing/. (Go to "Guides" and then "Designing Graphics.")

Roper, Donna G. "How Much Technical Knowledge Do Editors Need? The Authors' Perspective." *STC Proceedings* (1993), http://www.stc.org/confproceed/1993.

Rude, Carolyn D. *Technical Editing.* 4th ed. New York: Longman, 2005.

Salovey, P., M. Brackett, and J. Mayer. *Emotional Intelligence.* New York: Dude Publishing, 2004.

Samson, Donald C., Jr. *Editing Technical Writing.* New York: Oxford University Press, 1993.

SAP America, SAP Design Guild. "Recommendations for Charts and Graphics," http://www.sapdesignguild.org/resources/diagram_guidelines/index.html.

Schriver, Karen. *Dynamics in Document Design: Creating Text for Readers.* Hoboken, NJ: John Wiley and Sons, 1996.

Schroeder, Will. "Testing Web Sites with Eyetracking." *User Interface Engineering* (1998), http://www.uie.com/articles/eye-tracking.

Schultz, Laura D., and Jan H. Spyridakis. "The Effect of Heading Frequency on Comprehension of Online Information: A Study of Two Populations." *Technical Communication* 51, no. 4 (2004): 504–516.

Society for Technical Communication, "Ethical Principles for Technical Communicators," http://www.stc.org/about/policy_ethicalPrinciples.asp.

Society for Technical Communication, AccessAbility Special Interest Group, page on Internet Accessibility, http://www.stcsig.org/sn/internet.shtml.

Society for Technical Communication, Usability Special Interest Group, Resources page, http://www.stcsig.org/usability/resources/index.html.

Sullivan, Danny. "How to Use HTML Meta Tags," http://searchenginewatch.com/2167931.

Sun Microsystems. *Read Me First! A Style Guide for the Computer Industry,* 3rd ed. Santa Clara, CA: Sun Microsystems, 2009.

Thatcher, Jim, Cynthia Waddell, Shawn Henry, et al. *Constructing Accessible Web Sites.* Birmingham, UK: Glasshaus Ltd., 2002.

Thatcher, Jim, Andrew Kirkpatrick, Mark Urban, et al. *Web Accessibility: Web Standards and Regulatory Compliance.* Berkeley, CA: Apress, 2006.

Tufte, Edward. *Visual Explanations.* Cheshire, CT: Graphics Press, 1997.

Tufte, Edward. *The Visual Display of Quantitative Information.* 2nd ed. Cheshire, CT: Graphics Press, 2001.

Turns, Jennifer, and Tracey S. Wagner. "Characterizing Audience for Informational Web Site Design." *Technical Communication* 51, no. 1 (2004): 68–85.

U.S. Department of Labor, Bureau of Labor Statistics. *Occupational Outlook Handbook, 2008–09 Edition.* "Writers and Editors," http://www.bls.gov/oco/ocos089.htm.

United States Department of Commerce, National Oceanic and Atmospheric Administration. "Federal Agencies Reach Consensus Ending Development of the Oregon Inlet Jerry Proposal," http://www.publicaffairs.noaa.gov/releases2003/ may03/noaa03r126.html.

Van Buren, Robert, and Mary Fran Buehler. *The Levels of Edit.* 2nd ed. Pasadena, CA: Jet Propulsion Laboratory, 1980.

Walsh, Barbara. *Communicating in Writing.* Canberra: Australian Government Publishing Service, 1989.

Web Accessibility Initiative, http://www.w3.org/WAI/.

Weber, Jean Hollis. *Is the Help Helpful? How to Create Online Help That Meets Your Users' Needs.* Whitefish Bay, WI: Hentzenwerke Publishing, 2004.

Weber, Jean Hollis. "Technical Editors' Responsibilities." *Technical Editors' Eyrie,* http://www.jeanweber.com/newsite/?page_id=24.

Weber, Jean Hollis. "The Role of the Editor in the Technical Writing Team." Seminar 90: Bringing Technology Closer. Published in *Proceedings of the Technical Communication*

Seminar, New South Wales Society for Technical Communication (October 1990): 67–69. Article available online at http://www.jeanweber.com/newsite/?page_id=25.

West, Charles, James Farmer, and Phillip M. Wolff. *Instructional Design: Implications from Cognitive Science.* Upper Saddle River, NJ: Prentice Hall, 1991.

Williams, Joseph M. *Style: Ten Lessons in Clarity and Grace.* 7th ed. New York: Longman, 2003.

Wong, Irene. "Parts of a Typical Statistical Table." *Technical Editors' Eyrie,* http://www.jeanweber.com/newsite/?page_id=49.

INDEX